Green Production Strategies for Sustainability

Sang-Binge Tsai
University of Electronic Science and Technology of China (Zhongshan Institute), China & Civil Aviation University of China, China

Bin Liu
Shanghai Maritime University, China

Yongian Li
Nankai University, China

A volume in the Advances in Environmental Engineering and Green Technologies (AEEGT) Book Series

IGI Global
DISSEMINATOR OF KNOWLEDGE

Published in the United States of America by
IGI Global
Engineering Science Reference (an imprint of IGI Global)
701 E. Chocolate Avenue
Hershey PA, USA 17033
Tel: 717-533-8845
Fax: 717-533-8661
E-mail: cust@igi-global.com
Web site: http://www.igi-global.com

Library of Congress Cataloging-in-Publication Data

Names: Tsai, Sang-Bing, 1973- editor. | Liu, Bin, 1971- editor. | Li, Yongian, 1973- editor.
Title: Green production strategies for sustainability / Sang-Bing Tsai, Bin Liu, and Yongian Li, editors.
Description: Hershey, PA : Engineering Science Reference, [2018] | Includes bibliographical references and index.
Identifiers: LCCN 2017020067| ISBN 9781522535379 (hardcover : alk. paper) | ISBN 9781522535386 (ebook)
Subjects: LCSH: Production management--Environmental aspects. | Manufacturing processes--Environmental aspects. | Green products. | Sustainable engineering.
Classification: LCC TS155.7 .G738 2018 | DDC 333.72--dc23 LC record available at https://lccn.loc.gov/2017020067

This book is published in the IGI Global book series Advances in Environmental Engineering and Green Technologies (AEEGT) (ISSN: 2326-9162; eISSN: 2326-9170)

British Cataloguing in Publication Data
A Cataloguing in Publication record for this book is available from the British Library.

All work contributed to this book is new, previously-unpublished material. The views expressed in this book are those of the authors, but not necessarily of the publisher.

For electronic access to this publication, please contact: eresources@igi-global.com.

Advances in Environmental Engineering and Green Technologies (AEEGT) Book Series

ISSN:2326-9162
EISSN:2326-9170

Mission

Growing awareness and an increased focus on environmental issues such as climate change, energy use, and loss of non-renewable resources have brought about a greater need for research that provides potential solutions to these problems. Research in environmental science and engineering continues to play a vital role in uncovering new opportunities for a "green" future.

The **Advances in Environmental Engineering and Green Technologies (AEEGT)** book series is a mouthpiece for research in all aspects of environmental science, earth science, and green initiatives. This series supports the ongoing research in this field through publishing books that discuss topics within environmental engineering or that deal with the interdisciplinary field of green technologies.

Coverage

- Policies Involving Green Technologies and Environmental Engineering
- Green Technology
- Contaminated Site Remediation
- Green Transportation
- Electric Vehicles
- Industrial Waste Management and Minimization
- Radioactive Waste Treatment
- Water Supply and Treatment
- Renewable energy
- Sustainable Communities

IGI Global is currently accepting manuscripts for publication within this series. To submit a proposal for a volume in this series, please contact our Acquisition Editors at Acquisitions@igi-global.com or visit: http://www.igi-global.com/publish/.

The Advances in Environmental Engineering and Green Technologies (AEEGT) Book Series (ISSN 2326-9162) is published by IGI Global, 701 E. Chocolate Avenue, Hershey, PA 17033-1240, USA, www.igi-global.com. This series is composed of titles available for purchase individually; each title is edited to be contextually exclusive from any other title within the series. For pricing and ordering information please visit http://www.igi-global.com/book-series/advances-environmental-engineering-green-technologies/73679. Postmaster: Send all address changes to above address.

Titles in this Series

For a list of additional titles in this series, please visit: www.igi-global.com/book-series

Hydrology and Best Practices for Managing Water Resources in Arid and Semi-Arid Lands
Christopher Misati Ondieki (Kenyatta University, Kenya) and Johnson U. Kitheka (South Eastern Kenya University, Kenya)
Engineering Science Reference • copyright 2018 • 266pp • H/C (ISBN: 9781522527190) • US $205.00 (our price)

Computational Techniques for Modeling Atmospheric Processes
Vitaliy Prusov (University of Kyiv, Ukraine) and Anatoliy Doroshenko (National Academy of Sciences, Ukraine)
Information Science Reference • copyright 2018 • 460pp • H/C (ISBN: 9781522526360) • US $205.00 (our price)

Advanced Nanomaterials for Water Engineering, Treatment, and Hydraulics
Tawfik A. Saleh (King Fahd University of Petroleum and Minerals, Saudi Arabia)
Engineering Science Reference • copyright 2017 • 384pp • H/C (ISBN: 9781522521365) • US $200.00 (our price)

Handbook of Research on Inventive Bioremediation Techniques
Jatindra Nath Bhakta (University of Kalyani, India)
Engineering Science Reference • copyright 2017 • 624pp • H/C (ISBN: 9781522523253) • US $260.00 (our price)

Handbook of Research on Entrepreneurial Development and Innovation Within Smart Cities
Luisa Cagica Carvalho (Universidade Aberta, Portugal & CEFAGE - Universidade de Évora, Portugal)
Information Science Reference • copyright 2017 • 661pp • H/C (ISBN: 9781522519782) • US $235.00 (our price)

Applied Environmental Materials Science for Sustainability
Takaomi Kobayashi (Nagaoka University of Technology, Japan)
Information Science Reference • copyright 2017 • 416pp • H/C (ISBN: 9781522519713) • US $205.00 (our price)

Environmental Issues Surrounding Human Overpopulation
Rajeev Pratap Singh (Banaras Hindu University, India) Anita Singh (University of Allahabad, India) and Vaibhav Srivastava (Banaras Hindu University, India)
Information Science Reference • copyright 2017 • 325pp • H/C (ISBN: 9781522516835) • US $200.00 (our price)

Reconsidering the Impact of Climate Change on Global Water Supply, Use, and Management
Prakash Rao (Symbiosis International University, India) and Yogesh Patil (Symbiosis International University, India)
Information Science Reference • copyright 2017 • 430pp • H/C (ISBN: 9781522510468) • US $215.00 (our price)

IGI Global
DISSEMINATOR OF KNOWLEDGE

701 East Chocolate Avenue, Hershey, PA 17033, USA
Tel: 717-533-8845 x100 • Fax: 717-533-8661
E-Mail: cust@igi-global.com • www.igi-global.com

Table of Contents

Section 3
Sustainability and Management

Detailed Table of Contents

In recent times, globalization policy eliminates the restriction of market boundaries, enabling all manufacturers to exploit the markets of developing countries, due to which, the environmental regulations in the developing countries are evolving to secure their nature with stringent regulation. In this view, this article is written to impart knowledge on sustainable manufacturing to all manufacturers. However, for successful implementation of sustainable manufacturing practices, formulation of strategies is found to be essential. In this view, critical insights have been presented by reviewing existing practices of sustainable manufacturing, how to deploy them and their correlation with the environmental benefits are mapped to support practicing managers. The practices discussed in the chapter include process optimization in view of the environment energy modeling, waste minimization options, better end-of-life policies, material efficiency and need to employee training.

This chapter on environmental management system (EMS) addresses the relationship between organizational factors and environmental goals. Because the increasing awareness about environment affects the related groups and forces the firms have an environmental approach. The global competition, regulations and other pressures can change the tools of the sustainability. So, the firms start to consider an environmental management system as an innovative instrument as well as an adaptation strategy. Because, environmental standards and policies give the firm an opportunity to improve the business performance. From this viewpoint, this chapter provides an overview of the context, organizational drivers and implementation of the EMS.

As the online purchasing is becoming prevalent, more and more corporates started to show their CSR information in the processes of online retail. For example: some CSR information could be showed in the "COMMODITY DETAILS". Could this kind of information affect consumer behavior? By recognizing three types of embedded CSRC information, according to the consumer behavior experiments, this paper will show CSR's influences on the consumer behavior such as evaluation of the company, purchasing

Preface

Since the United Nations Environmental Programme (UNEP) began actively promoting green production in 1989, GP and green environments have become global topics. Governments in every country are gradually focusing on directing industrial productions to GP, and have incorporated GP as a critical element in national development strategies. GP has gradually been developed into a full-range measure that incorporates process, product, and service dimensions for assessing complex topics regarding the environment, resources, ecology, economy, commercial trade, and sustainability.

Green production (GP) refers to a process through which management and technological approaches are used to minimize and control the amount of pollutants produced during production processes. The purpose of GP is to conserve energy, reduce energy consumption, and reduce pollution. Among manufacturing enterprises, pollution caused by the production of products, and their derived wastes in particular, are increasing. In response to the trend of corporate environmental management, manufacturing enterprises must execute green enterprise management on activities related to green production, green supply chain, green management, green marketing, and environmental friendliness. Thus, Green production is an important research issue.

This book aims to provide relevant theoretical frameworks and the latest empirical research findings in the area. The target audience of this book will be composed of professionals, students and researchers studying in the field of operation management and sustainability in various disciplines, e.g. management sciences, production management, supply chain management, environmental management, and sustainability.

This book contains 3 sections: Section 1, "Green Production and Application," with a total of five chapters. Chapter 1 discusses Green Production and Green Supply Chain Governance. Chapter 2 discusses the Assessment of Supply Chain Greenness. Chapter 3 discusses the Corporate Social Responsibility-based supplier selection process. Chapter 4 discusses Cleaner Production. Chapter 5 discusses Sensitization of Sustainable Manufacturing Strategies.

Section 2, "Green Environment and Management," has a total of five chapters. Chapter 6 discusses Green Environment and Management. Chapter 7 discusses the Influence on Consumer Behavior from the CSR Information Embedded. Chapter 8 discusses Pricing Decisions and governance mechanisms considering corporate social irresponsibility in supply chains. Chapter 9 discusses Opportunities and Challenges for Sustainable Business Strategic Planning in Small and Medium Enterprises. Chapter 10 discusses Green Innovation in the Absence of Social Trust.

Section 3, "Sustainability and Management," has a total of four chapters. Chapter 11 discusses Green Entrepreneurship in Transitional Economies. Chapter 12 discusses Improving Renewable Energy Sources in Automotive Applications. Chapter 13 discusses the Risk Implications of Heavy Metal Contamination in Agricultural Soil and Crop Productivity. Chapter 14 discusses Four Strategic Postures for Sustainability in the Project Based Organization.

There is a conclusion at the end of this book.

Sang-Binge Tsai
University of Electronic Science and Technology of China (Zhongshan Institute), China & Civil Aviation University of China, China

Bin Liu
Shanghai Maritime University, China

Yongian Li
Nankai University, China

Acknowledgment

We acknowledge IGI Global for their professional assistance in the publication of this book, as well as all the chapter authors for their contributions to this book. These editors were supported by University of Electronic Science and Technology of China Zhongshan Institute, China & Zhongshan Industrial Technology Research Institute and School of Innovation and Entrepreneurship, University of Electronic Science and Technology of China Zhongshan Institute, China & Civil Aviation University of China, China & Environment and Sustainable Development Center of China Civil Aviation University, China & Foshan University, China & Dalian University of Technology, China & Shanghai Maritime University, China & Nankai University, China & China Academy of Corporate Governance, Nankai University.

Section 1
Green Production and Application

Chapter 1
Green Production and Green Supply Chain Governance Under Extended Producer Responsibility

Xiukun Zhao
Tianjin University of Finance and Economics, China

Xuanming Bai
Nankai University, China

Xiaobing Lv
Nankai University, China

Yongjian Li
Nankai University, China

ABSTRACT

Extended Producer Responsibility (EPR) is a strategy designed to promote the integration of environmental costs associated with goods throughout their life cycles into the market price of the products. In order to explore the decision mechanism green production and the governance mechanism of green supply chain under EPR, this chapter makes a theoretical overview about the main issues of EPR and supply chain governance firstly. Generally, under EPR, the government uses financial incentives to encourage manufacturers to design environmentally friendly products by holding producers responsible for the costs of managing their products at end of life. Therefore, the green production framework under EPR is put up from the product life cycle perspective. Next, a supply chain governance framework is proposed to launch green supply chain when EPR policy is enforced. Finally, future directions of green production and green supply chain governance are developed.

DOI: 10.4018/978-1-5225-3537-9.ch001

INTRODUCTION

The United Nations points out that the green economy is an economic model aimed to improve human well-being and enable the broad masses of the people to fully enjoy the fruits of economic development while significantly reducing the adverse effects of human activities on the environment. Environmental economists believe that economic development must be affordable to the natural environment and the human self; it should neither lead to social division and ecological crisis because of blind pursuit nor can it stop the continual development of the economy because of the depletion of natural resources. The essence of green economy is sustainable economic development coordinated with ecological development. Green economy is a balanced economy based on the environment that promotes human survival and health and that reasonably protects resources and energy. In this economic model, environmental technologies, green production processes, and other environment-friendly technologies are transformed into productive forces, achieving sustainable economic growth and ultimately eliminating poverty. In the previous economic development model characterized by "the first pollution, post-governance," the national economic level and people's consumption level rapidly increased, but environmental problems became prominent. The pressure from asset scarcity gradually increased. In particular, the increase of waste products resulted in serious environment pollution.

All countries in the world consider waste disposal as a major problem that affects their sustainable development. Various policies and regulations have been enacted to solve this problem. Extended producer responsibility (EPR) is a waste-management system and principles that arise from the increasingly serious waste problem. The concept of EPR first proposed by the Swedish Lund University environmental economist Thomas Lindhquist in 1988 quickly became a global issue in management, environmental science, sociology, law, and other disciplines. As a harmonious concept of social development in the context of increasingly serous pollution and resource scarcity, EPR is applied to promote the basic principles of environmental policy in different countries and regions. In 1991, the German "Packaging Act" first extended the EPR system in the law, and many developed countries and regions established EPR-based waste recycling systems through legislation.

Examples include the European Union's Waste Battery Management Directive 91/157/EEC, Packaging and Packaging Waste Directive 94/62/EC, Waste Vehicle Management Directive 2000/53/EC, Waste Electrical and Electronic Equipment Directive (WEEE), Restriction of the Use of Hazardous Substances in Electrical and Electronic Equipment, Integrated Product Policy, and e-waste disposal. The Chinese government formulated a series of EPR policies to regulate the waste from electrical and electronic products, such as the Regulations on the Management of Recycling of Waste Electrical and Electronic Products, Measures for the Administration of Collection and Use of Waste Electrical and Electronic Products Processing Funds, and Implementation Plan for Producer Extension System. After nearly 30 years of academic research and policy practice, the scope of EPR expanded, and developing countries (such as India and Malaysia) also planned to introduce EPR in the field of product waste. Concrete measures to implement EPR are diversifying.

Issues on the amount of generated waste, resource depletion, and environmental pollution have become prominent issues affecting the national economy and sustainable social development. In this context, in-depth studies on green production and green supply chain management theory and practices that explore green production and green supply chain development strategies based on EPR under the guidance of enterprise green practice are necessary.

In this chapter, we discuss green production and green supply chain under EPR. The rest of this chapter is organized as follows. In Section 2, we propose a green product framework from the product life cycle perspective under EPR. In Section 3, we present a green supply chain governance framework when EPR policy is enforced. In Sections 4 and 5, we discuss related research on green production and green supply chain governance, respectively. In Section 6, we provide our concluding remarks and the future directions of green production and green supply chain.

EXTENDED PRODUCER RESPONSIBILITY

Since the mid-1980s, global recycling activities, with the influence of WEEE, have increasingly become an important issue for both management scholars and practitioners worldwide. EPR, as a policy principle, promotes the environmental improvement of products and product systems. Approaches for the implementation of EPR are key concerns in sustainable development. According to Prof. Lindhqvist (2000) in a report to the Swedish Ministry of the Environment in 1990, EPR is an environmental protection strategy to reach the environmental objective of decreasing the total environmental effect of a product by making product manufacturers responsible for the entire life-cycle of the product, particularly the take-back, recycling, and final disposal of the product. EPR is implemented through administrative, economic, and informative instruments. The composition of these instruments determines the precise form of EPR. The revised definition of EPR presents the concept as a policy principle (Lindhqvist, 2000); EPR is a policy principle that promotes the total life cycle and environmental improvements of product systems by extending the responsibilities of product manufacturers to various parts of the entire life cycle of the product, particularly the take-back, recycling, and final disposal of products. EPR is implemented through administrative, economic, and informative policy instruments.

The Organization for Economic Co-operation and Development (OECD) published several reports on EPR policy, namely, EPR: A Guidance Manual for Governments (OECD, 2001), Economic Aspects of EPR (OECD, 2004), and Analytical Framework for Evaluating Costs and Benefits of EPR Programs (OECD, 2004). OECD defines EPR as an environmental policy approach in which a producer's physical and/or financial responsibility for a product is extended to the post-consumer stage of a product's life cycle. The related features of EPR policy include 1) shifting responsibilities (physically and/or economically; fully or partially) upstream to producers and away from municipalities; and 2) providing incentives to producers for incorporating environmental considerations in the design of their products.

With knowledge on the practical background of China, Li et al. (2016) proposed EPR as a public institution that extends the environmental responsibility of producers to consumers. EPR focuses on the product waste stage and considers the design, production, distribution, retail, and other stages simultaneously. Finally, the environmental behavior of producers can be restrained and motivated in the entire product life cycle under EPR.

Based on the arguments of EPR, some studies have been conducted from the perspective of the full life cycle of products to provide additional insights into the EPR framework and to explore the mechanisms of green production and green supply chain governance. A commodity undergoes a number of stages from production to abandon, including the selection of raw materials, energy supply, production, transportation, sales, use, maintenance, and disposal. The adverse effects of a product on the environment may occur at any of the above stages, although the degree of hazard varies. For example, the fuel consumption of a vehicle during the use phase is higher than that in other stages, and plastic packaging is more harmful to the soil than other stages in the disposal stage, because of its difficulty to degrade.

EPR pursues the optimal environmental performance of a product throughout its life cycle (Lindhqvist, 2000). Thus, environmental products with different environmental characteristics must be formulated to match environmental strategies. EPR extends the responsibility of producers from the production stage of the life cycle to product use and disposal. This extended responsibility must internalize the external environmental costs and achieve the environmental protection requirements for industrial production. According to China, "Spend efforts where it is likely to do most good." The philosophy of dialectical materialism also requires us to seize the main and the main aspects' contradictions. Thus, identifying the relative importance of the environmental effect of a product life cycle is necessary. Life cycle assessment (LCA) is an environmental management tool that quantitatively analyzes the environmental effect of a product from production to disposal and identifies one or several stages that exhibited a prominent environmental effect of a product based on the analysis results. The LCA findings can be used for the various stages of the responsibility of the main body and its specific environmental responsibility characteristics, combined with economic and operational aspects, to implement the main responsibilities of the stage boundaries, which refer to an important theoretical tool for product environmental management.

EPR considers producers as the main body of responsibility and ensures that their responsibility extends to the stage that produces the highest environmental pollution. The responsibilities of the producer include material responsibility, economic responsibility, information liability, product liability, and title retention based on the LCA analysis from the life cycle perspective. In this analysis, the extent of environmental pollution at each extraction stage of the raw material stage, the manufacturing stage, the distribution and sales phases, the consumption phase, the recovery stage, the treatment stage, and the recycling phase is calculated. Relevant and responsible entities, including raw material suppliers, sellers, consumers, recyclers, processors, and government entities, should simultaneously bear some responsibility to ensure that EPR can be implemented smoothly to achieve an excellent life cycle of a product. The green production concept model based on the full life cycle perspective under EPR is shown in Figure 1.

Figure 1. Green production framework from the product life cycle perspective under EPR
Source: Li, 2016

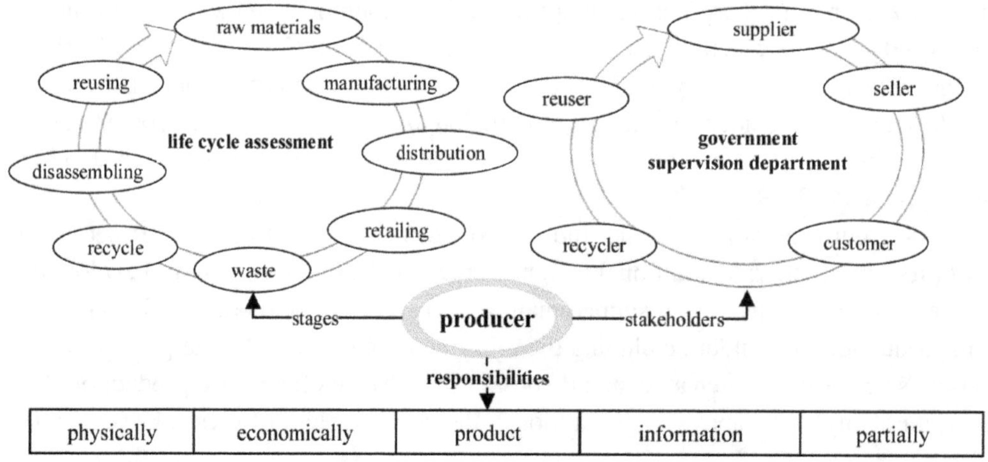

SUPPLY CHAIN GOVERNANCE

The formation of supply chain governance is a result of the dual effect of the characteristics of the supply chain and the rationality of decision makers. The concept of supply chain was first referred by Peter Drucker as "economic chain", which Michael Porter called "value chain" and eventually termed "supply chain". As a whole, supply chain is the functional network model of suppliers, manufacturers, distributors, retailers, and end users. However, from an individual point of view, the supply chain is a loose and open organization, and its members are independent of each other, and decision-making is decentralized. First, each enterprise is an independent legal entity, and member companies have independent decision-making power and control. Enterprises depend on individual rationality to pursue their own interests to the maximum, resulting in conflicting behavior objectives in the supply chain. Second, the mutual independence between members produces information asymmetry. To gain an advantage in negotiations, members usually retain private information, such as raw materials or product costs, product quality, and production capacity. Hence, "private information" unknown to other members exists in the supply chain. This private information is often either not verifiable or the cost of obtaining and verifying this information is high. Thus, signing an effective contract for supply chain members as part of the supervision and control is impossible. As a result, participants can become victims of opportunistic behavior, damage, and plunder.

Participants may exhibit conflicting and opportunistic behaviors because of the loosening of the supply chain and the limited rationality of decision makers, which will infringe the benefits of the participants in the supply chain and expose them to trading risks. However, a supply chain is an open organization with less constraint, which affects supply chain stability. Therefore, controlling the supply chain is necessary. The vulnerability of a supply chain and the limited rationality of decision-makers motivate the implementation of supply chain governance.

Although governance is discussed in different research areas (such as political science, sociology, economics, and management), the understanding of scholars on the connotation and attributes of "governance" is roughly the same. In terms of induction, "governance" can be seen as an institutional arrangement that allows conflicting interests to be reconciled and stakeholders to take joint action in maintaining a sustained and stable development of the coordination situation. Cooperation, coordination, and interconnection are the core attributes of governance. The following attributes can be observed from the concept of governance: (1) the basis of governance is not control but coordination; (2) governance involves interrelated stakeholders, including the private and public sectors, as well as economic organizations; (3) governance emphasizes the balance of interests; and (4) governance is a continuous interaction that aims to maintain the continuity of relations.

Based on the characteristics of the supply chain, (1) supply chain is a new model of cooperative partnership. This new model lacks strict organizational constraints and guarantees without sufficient and effective market rules, and it is typically characterized by the principal–agent problem. Administrative affiliation among members does not exist. Thus, opportunism will more likely occur in the supply chain than in companies. (2) Supply chain members are independent and decentralized. Although they are integrated in key business activities, members pursue the maximization of their own profits based on individual rationality, resulting in the conflicts between the supply chain operation and the overall target. (3) Aside from a seller–buyer relationship, communication and cooperation among supply chain members occur through information, technology, finance, and human resource. On the premise of the resource and capacity complementarity of the members, the supply chain creates synergistic effects to gain more profits compared with pure market transactions. However, concerns about the information asymmetry and "lock-in" will affect the long-term investments and cooperation among members than it will influence the overall profits of the supply chain.

In addition, external environment changes strengthen the vulnerability factors of supply chains, which mainly include the following: (1) Economic globalization leads to the supply chain's physical elongation and enlarges the uncertainty in supply chains; (2) the rapid development of science and technology shortens the product development cycle, shortens product replacement speed, and exacerbates the contradictions on the cooperation among members in the production capacity. (3) The diversification and personalization of consumer demands lead to the difficulties of accurately capturing consumer demand, thereby expanding the adverse effects of the "bullwhip effect" in supply chains.

Therefore, from the perspective of the supply chain development, measures must be taken to repress opportunism under information asymmetry, coordinate the target conflicts among members, facilitate long-term investment and integration, and ensure the sustainable development of a supply chain. Governance precisely realizes cooperation, coordination, and continuous interaction effect. Supply chain governance is defined as a continuous interaction process in the study. This process is based on the institutional arrangement of stakeholders consisting of the bind of economic contracts and the insertion of social relations; under the influence of environment governance, a series of governance mechanisms are designed to achieve the relationship of supply chain members in order to coordinate the target conflicts among the members and to ensure the sustainable and stable development of the supply chain.

The scope of the organizations involving the supply chain represents its boundary. With a focus on the core enterprise, the functional network structure that consists of suppliers, manufacturers, distributers, and retailer constitutes the core category of the supply chain governance, where the core enterprise plays a leading and organizational role and parties are connected through transactional relationships. With the expansion of the supply chain under the EPR, the increase in externality and social force poses a strong external shock to supply chain governance, which may involve parties (stakeholders), such as government, community, industry associations, and other groups (such as NGOs). These parties form a relationship with supply chain members through various bonds of interest. These stakeholders must include a complete endogenous variable into the supply chain governance but may be through an external mechanism and a supply chain oppression to adapt to the external pressure of the governance system arrangements, thereby forming the internal and the outer boundary of the EPR implementation, as shown in Figure 2.

Under the EPR system, the government, NGO organizations, and consumers become the stakeholders of an enterprise that pursues the balance between economic effects and environmental benefits. The new institutional pressure led to the evolution of the supply chain governance structure to achieve the internal efficiency of the system and the legalization of the system outside by reorganizing governance structure. Thus, the green production and green supply chain management concept model under EPR is developed, as shown in Figure 3.

GREEN PRODUCTION DECISION MECHANISM UNDER EPR

Green Production and Remanufacturing

The development of the remanufacturing industry has become significantly important in transforming the mode of economic development and relieving the increasing resources shortage problems. Remanu-

Figure 2. Boundary of the EPR implementation
Source: Li, 2016

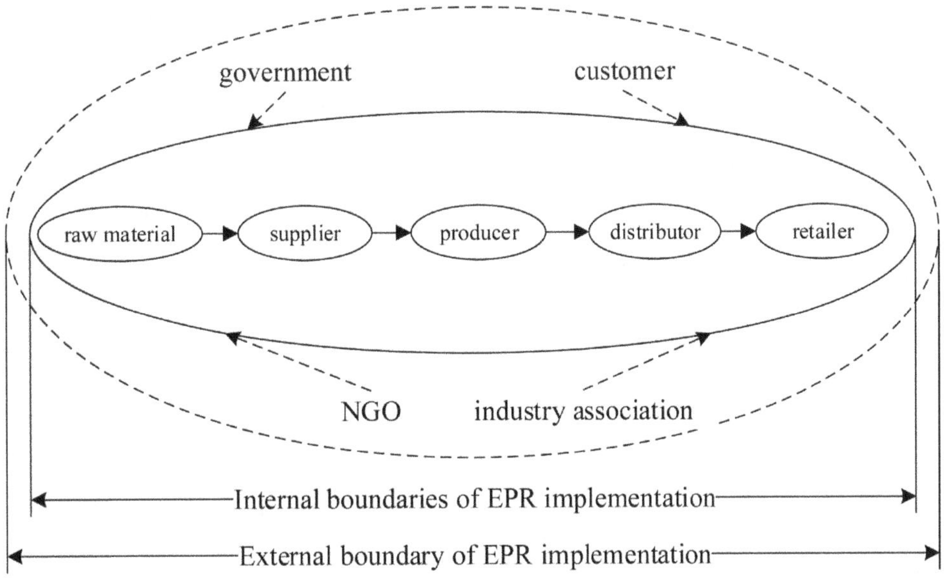

Figure 3. Green supply chain governance framework under EPR
Source: Li, 2016

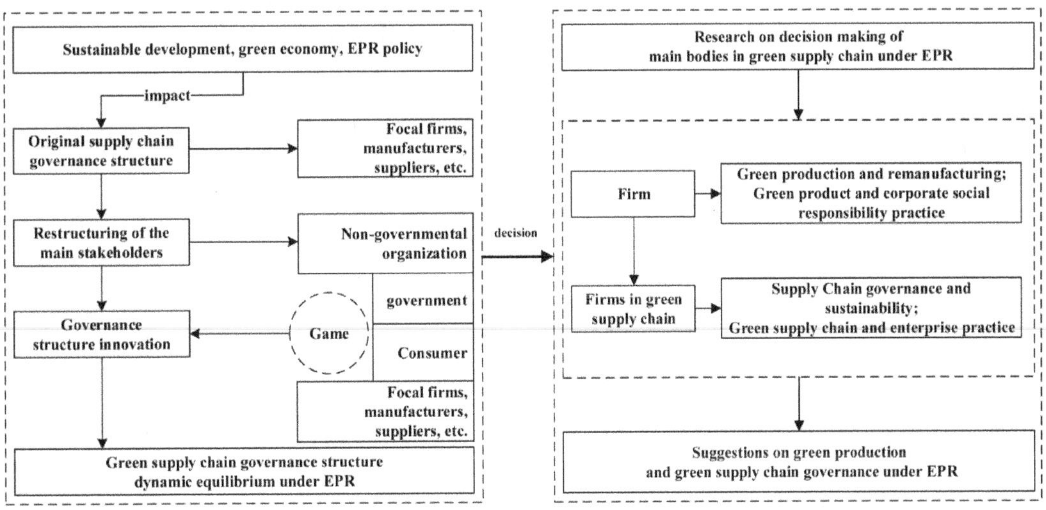

facturing issues are the focus of green production. Many topics related to green production and remanufacturing have been investigated from different perspectives. Based on input–output theory and previous studies, Feng et al. (2014) combined input–output/decision-making trials and analyzed the laboratory method for evaluating the effect of the remanufacturing sector development. This study establishes an input–output model with multiple remanufacturing sectors and develops a combined input-output and decision making trial and evaluation laboratory method. The input–output table of China in 2007 (Table 1) and the combined method are used to examine the role of the remanufacturing sector in the

Table 1. Sector classification in the 2007 input–output table of China

Sector	Code	Sector	Code
Agriculture, Forestry, Animal Husbandry, and Fisher	01	Scrap and Waste	22
Mining and Washing of Coal	02	Production and Supply of Electric Power and Heat Power	23
Extraction of Petroleum and Natural Gas	03	Production and Distribution of Gas	24
Mining of Metal Ores	04	Production and Distribution of Water	25
Mining and Processing of Nonmetal Ores and Other Ores	05	Construction	26
Manufacture of Foods and Tobacco	06	Traffic, Transport, and Storage	27
Manufacture of Textile	07	Post	28
Manufacture of Textile Wearing Apparel, Footwear, Caps, Leather, Fur, Feather(Down) and Its products	08	Information Transmission, Computer Services, and Software	29
Processing of Timbers and Manufacture of Furniture	09	Wholesale and Retail Trades	30
Papermaking, Printing and Manufacture of Articles	10	Hotels and Catering Services	31
Processing of Petroleum, Coking, Processing of Nuclear Fuel	11	Financial Intermediation	32
Chemical Industry	12	Real Estate	33
Manufacture of Nonmetallic Mineral Products	13	Leasing and Business Services	34
Smelting and Rolling of Metals	14	Research and Experimental Development	35
Manufacture of Metal Products	15	Comprehensive Technical Services	36
Manufacture of General Purpose and Special Purpose Machinery	16	Management of Water Conservancy, Environment And Public Facilities	37
Manufacture of Transport Equipment	17	Services to Households and Other Services	38
Manufacture of Electrical Machinery and Equipment	18	Education	39
Manufacture of Communication Equipment, Computer and Other Electronic Equipment	19	Health, Social Security, and Social Welfare	40
Manufacture of Measuring Instrument and Machinery for Cultural Activity & Office Work	20	Culture, Sports and Entertainment	41
Manufacture of Artwork, Other Manufacture	21	Public Management and Social Organization	41

Source: (Feng et al. 2014)

national economy, as well as its symbiotic effects on other sectors to save energy and reduce emissions. The results show that the remanufacturing development in China is far below the average development level of other sectors. With the rising market share of remanufacturing, the demands of energy-intensive and pollution-intensive sectors, including smelting and rolling of metals, and production and supply of electric power and heat power, can be significantly decreased. However, traffic, transport, and storage demands increase.

According to input–output analysis and relevant studies on remanufacturing, Zhu et al. (2014) conducted economic effect analysis on the remanufacturing industry in China using input–output method. This work establishes an economic effect model of the remanufacturing industry. The model is used to quantitatively analyze and evaluate the effect of the remanufacturing industry on China's macro economy and other related industries by applying the model on the latest Chinese input–output table in 2007. This

Figure 4. Effect of the price of remanufactured products on total economic input
Source: Zhu et al, 2014

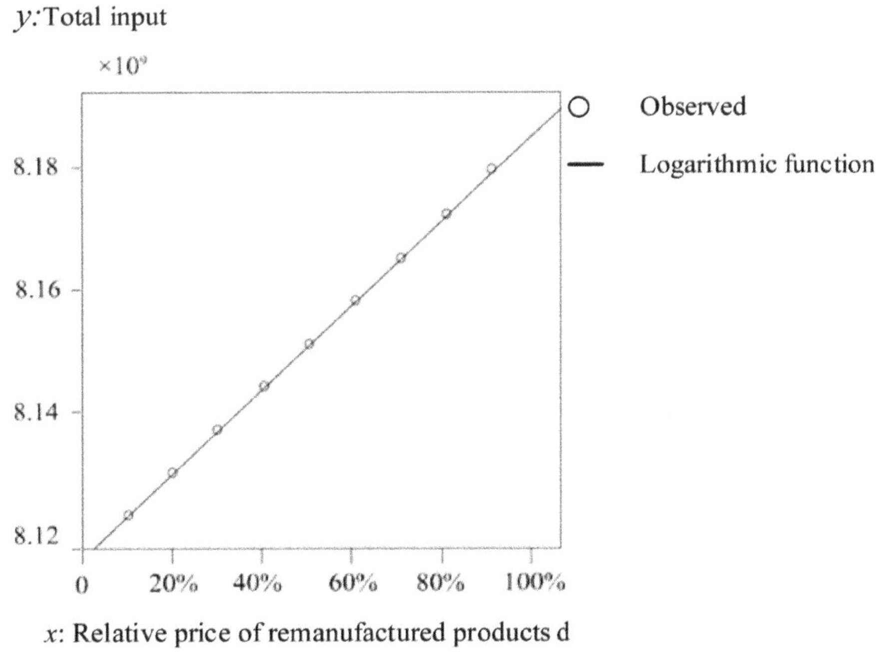

study reveals that the remanufacturing industry development will decrease the total input and the inputs of several related industries, such as smelting and rolling of metals, and manufacture of metal products. Moreover, the resource-saving effect will be further enhanced with the rising shares of remanufacturing products or the increasing price of remanufacturing products (Figure 4).

In China, remanufacturing automotive parts is still in its nascent stage, but the country is striving to implement a strategy in various industries to create clean production. Remanufacturing yields high financial, societal, and environmental returns, and it provides advantages, such as reduced costs (compared to virgin products) and increased jobs, as well as the conservation of scare resources. Initial studies examined Chinese remanufacturing issues but only focused on modeling concepts and failed to consider significant barriers. To bridge this gap, Xia et al. (2015) focused on analyzing the internal barriers for automotive parts remanufacturers in China using the Grey Decision Making Trial and Evaluation Laboratory (DEMATEL) approach. This study analyzes the internal barriers met by automotive parts remanufacturers and evaluates causal barriers using the proposed model framework (Figure 5). This objective is illustrated by employing the Grey DEMATEL approach. The findings indicate that remanufacturers can eradicate major internal barriers and can increase productive remanufacturing opportunities in the Chinese automotive environment. This study extends the scope of this problem through varied applications of different techniques such as a fuzzy approach, matrix laboratory, and operations research tools by diversifying and including additional barriers.

To explore the optimal strategies of two complementary products in a supply chain, Wei et al. (2014) discussed the price and warranty period decisions for the complementary products with the cooperation/noncooperation strategies. Five decentralized models are established by considering the cooperation/noncooperation strategies of two manufacturers and the different bargain powers of firms. The

Figure 5. Proposed framework for identifying causal internal barriers for automotive parts remanufacturers
Source: Xia et al., 2015

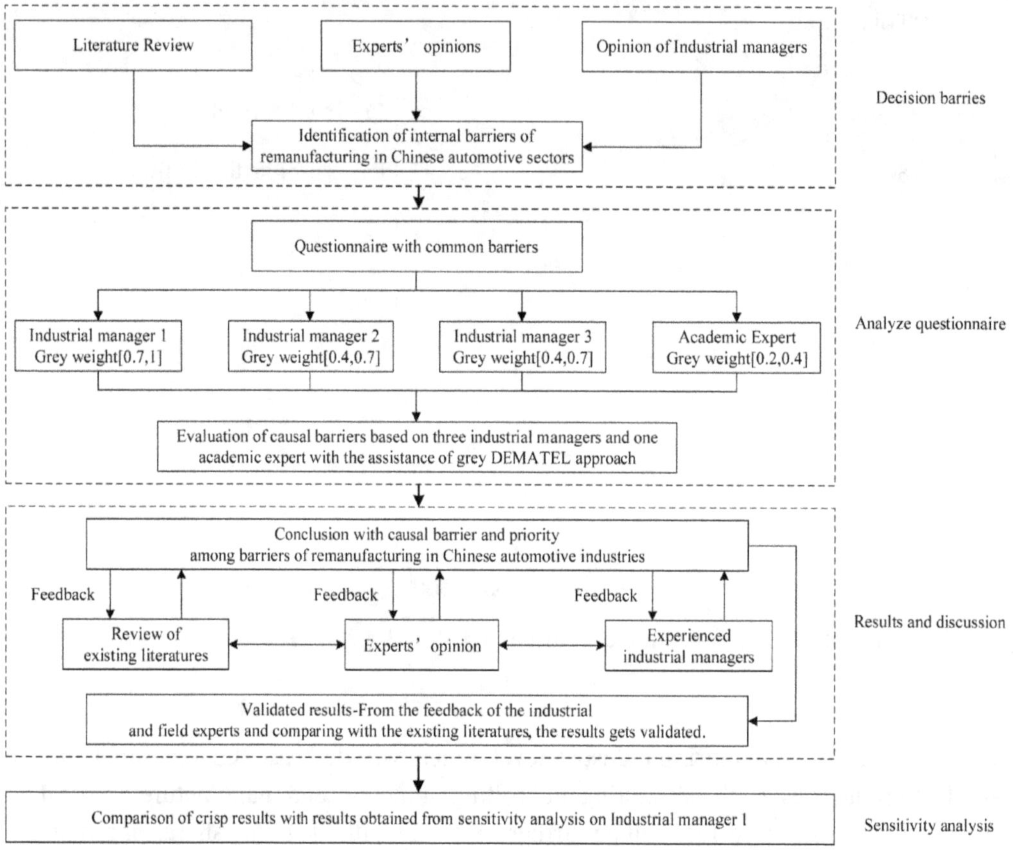

corresponding closed-form expressions for equilibrium decisions are obtained. A systematic analysis and comparison indicated some interesting and valuable managerial insights. A retailer always has the advantage to obtain a high profit when he or she is a leader; however, this assumption does not apply to two manufacturers. For example, the profits of two manufacturers when they are leaders and when their adoption of the noncooperation strategy in the supply chain is lower than when two manufacturers and the retailer have identical market power. All channel members and consumers benefit from low retail prices, long warranty periods, and large profits when the two manufacturers cooperate.

Green Production and Corporate Social Responsibility Practice

Corporate social responsibility (CSR) has gained increasing attention from both enterprises and the academia. In recent years, researchers have begun to examine if CSR practices can promote the financial performance of organizations (Lee et al., 2013; Wu and Shen, 2013).

Based on stakeholder theory, Zhu et al. (2016) developed a "7 + 2" dimensions to measure CSR practices among Chinese national state-owned enterprises (CNSOEs) and to examine if their CSR practices improve financial and social performances. By analyzing the 2011 CSR reports that cover 100 CNSOEs, CSR practices related to the environment, labor practices, political responsibility, and human rights are highlighted in their reports. However, three CSR practices in relation to managing supply chains, fair operating practices, and community development require attention. The social performance of CNSOEs is evaluated by using the two performance ranking systems developed by the Chinese Academy of Social Sciences. The regression results obtained through the structural equation model using the partial least square method indicate that CSR practices pertinent to organizational governance, human rights, and the environment benefit social performance (Figure 6). With secondary data inputs from the annual operating performance evaluation conducted by the State-owned Assets Supervision and Administration, four CSR practices that can enhance financial performance are identified. However, two of the four refer to communities, and the supply chains are managed at the lowest level among the others. "7 + 2" CSR practice dimensions, together with their underlying measurement items, extend the use of stakeholder theory for studying CSR practices, which can also serve as a checklist for evaluating how well CSR practices are implemented in CNSOEs.

CSR has gained popularity worldwide, but studies have mainly focused on developed countries. Therefore, Zhu and Zhang (2015) developed items on CSR practices mostly based on the internationally accepted ISO26000 framework while integrating characteristics of Chinese SOCs. The statistical results of the 146 usable questionnaires show that Chinese SOCs implement four types of CSR practices under the ISO26000 framework. Examples include practices related to employee rights, consumer issues, fair operation, and labor practices. Charity is a specific CSR dimension in China, but it is only implemented at the low level. Chinese SOCs have high internal CSR values, which promote most CSRs except those related to charity and employee rights. The external institutional drivers of CSR practices are relatively low. Normative drivers motivate most CSR practices, whereas competitive drivers only motivate con-

Figure 6. Path model of CSR practices and performance
Source: Zhu et al, 2016

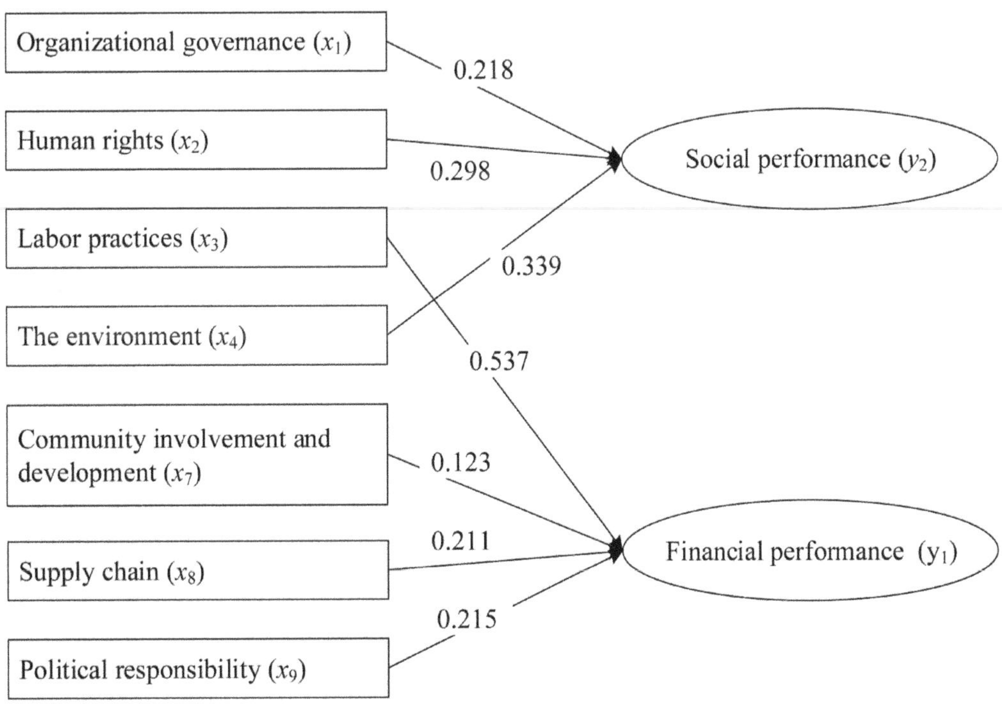

sumer issues-related CSR practices, as shown in Figure 7. This study extends previous studies mainly for developed countries to develop a CSR concept for Chinese SOCs by using both the international ISO26000 framework and by considering specific contexts. The statistical results indicate how CSR practices can be effectively promoted among Chinese SOCs, which can provide implications for the Chinese government, as well as multi-national companies that do or plan to conduct businesses in China.

To reduce the waste of resource and environment pollution caused by waste electrical and electronic equipment, Yang et al. (2016) used the LCA method to assess 1,000 refrigerators and compared the energy consumption and environment effects at various stages. The results show that the influence on energy consumption is largest in the use phase and second largest in the raw material extraction phase, and smallest in the waste processing phase. From the aspect of main environmental effect, when the safety rate is less than 93.73%, use phase is the largest, followed by manufacturing, raw materials extraction phase, waste processing stage, and transport phase, which is the smallest. Finally, this paper indicates that (1) the key to reduce the effect of the refrigerator industry on the environment and improve the

Figure 7. Relationships between antecedents and CSR practices among Chinese state-owned companies
*Note: Significance levels for regression coefficients are *p < 0.05, **p < 0.01, and ***p < 0.001*
Source: Zhu and Zhang, 2015

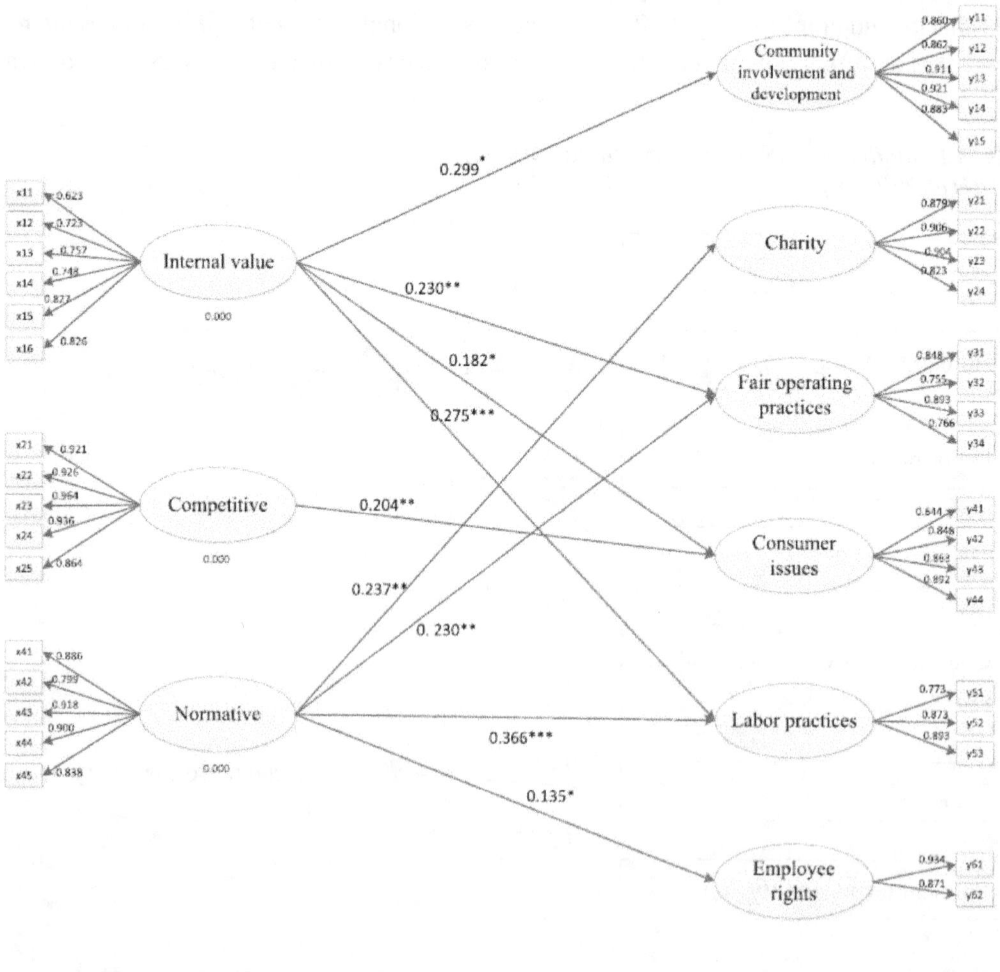

environment quality is reducing the power consumption of the use phase; (2) to decrease the energy consumption requirements of refrigerator production enterprises to improve the energy efficiency design of their products, enhance the level of refrigerator energy efficiency, and promote consumer consciousness of energy conservation.

Aside from focusing on the free rider problem arising from the conspiracy of producers and auditors, Ma et al. (2016) utilized a dynamic game model to study the conspiracy activities and the control measures of government authorities. Producers will increase their economic burden to fulfill their EPR responsibilities. With the incentive to evade responsibility, they may take risks in exchange of certain interest for an auditor's false report. Government departments must cover the waste product recycling costs that result in financial losses. Government departments can take a special supervision strategy to punish the behavior of enterprises including evading responsibility and their conspiracy with auditors, which will cost additional human and material resources. Three parties cannot guarantee an accurate judgment of each other's strategy choice, and employing a hybrid strategy is common. In this strategy, all parties determine their own action program with a certain probability distribution to achieve a mixed strategy equilibrium. Under the equilibrium condition, changing a party's own strategy to bring additional benefits is impossible. The study pointed out that only the monitoring strategy and supervision costs that government departments choose are closely related. In addition, the possibility of conspiracy can be reduced by increasing the reward and punishment within a certain range. The research also articulated that the proportion of the initial selection of the respective game strategies can be changed by the long-term propaganda of the EPR policy. Thus, equilibrium results can be changed. By developing green technology, producers can benefit and EPR effect can be improved.

GREEN SUPPLY CHAIN GOVERNACE MECHANISM UNDER EPR

Supply Chain Governance and Sustainability

With the rapid development of information technology and global economic integration, the competition between enterprises became a competition between supply chains. A broad system perspective is introduced into the supply chains study by applying the lens of governance. As a new governance mode, supply chain governance is the structural reflection of environment evolution and organization changes, which differs from traditional corporate governance. A theoretical framework on the concept and intention of governance in supply chains is developed. Li et al. (2016) focused on the theoretical research on supply chain governance: concept, intention, and normative analysis framework. First, the basic connotations of supply chain governance are defined based on the analysis of the governance necessity of the supply chain. By reviewing the related research on economics, management, sociology, and other disciplines, the theoretical basis is identified for the supply chain governance analysis. Based on the theories of TCE, resource dependence, social networks, and supply chain management, Li et al. (2016) extracted the analysis attributes of a supply chain and developed a standardized logical framework for the analysis of supply chain governance (Figure 8). Core concepts, such as governance boundary, governance mechanism, governance target, and governance structure, are integrated into this framework.

Governance structure refers to the configuration of node enterprises in a supply chain, and governance mechanism is a collection of rules/institutions/constraints that influence the exchange process between various organizations. The core issue of supply chain governance is the selection of governance structures

Figure 8. Framework of supply chain governance research
Source: Li et al., 2016

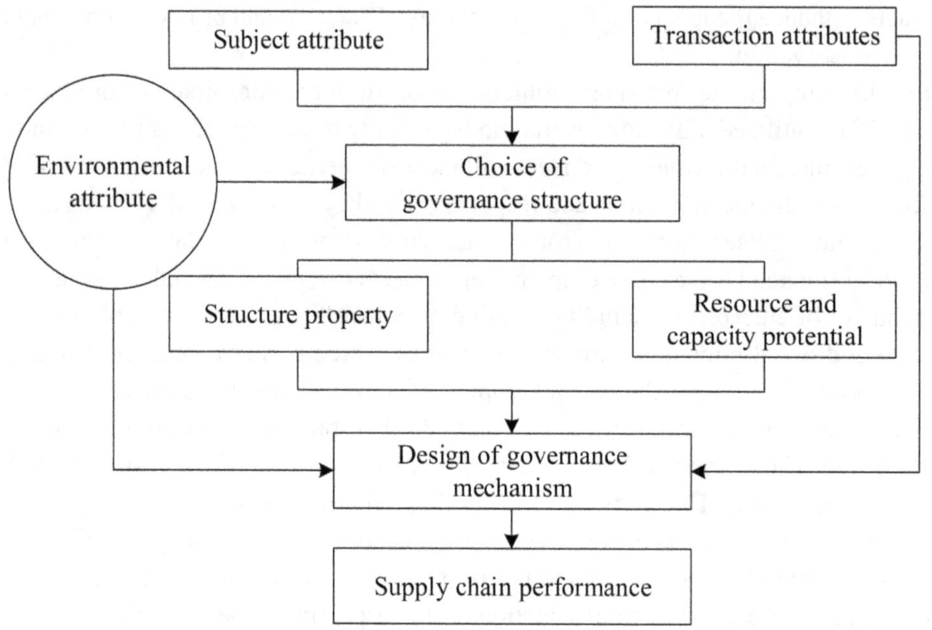

(Figure 9) and the design of governance mechanism (Figure 10) that coordinate conflicting objectives of participating organizations to maximize the output.

With the current trend of economic globalization, environmental sustainability, and social responsibility, enterprises have become aware of the importance of sustainable development in the economy, environment, and society. Li et al. (2015) examined the influence of corporate social responsibility behavior on the sustainability performance of focal companies and their partners in fast fashion supply chains. The attributes of sustainability (Figure 11) and the mechanism of sustainability governance in a fast fashion supply chain are also discussed.

From the perspective of strategic corporate social responsibility, the motives for adopting sustainability governance in fast fashion supply chains are first analyzed (Figure 12). Seven competitive sustainable attributes of a fast fashion product are identified based on sustainable development theory.

Figure 9. Governance structure of a supply chain under link
Source: Li et al., 2016

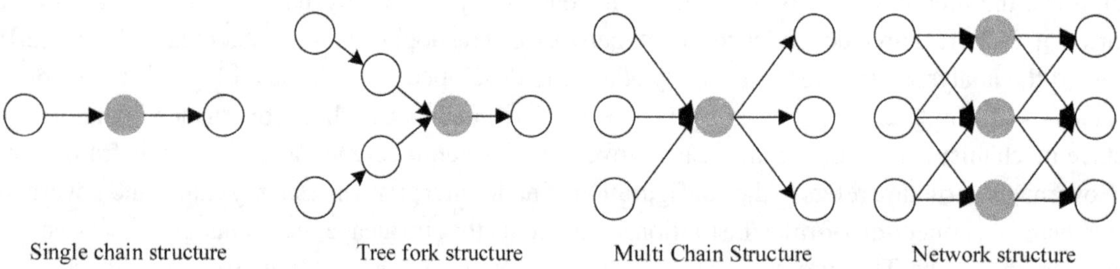

Figure 10. Supply chain governance mechanism and its mechanism
Source: Li et al., 2016

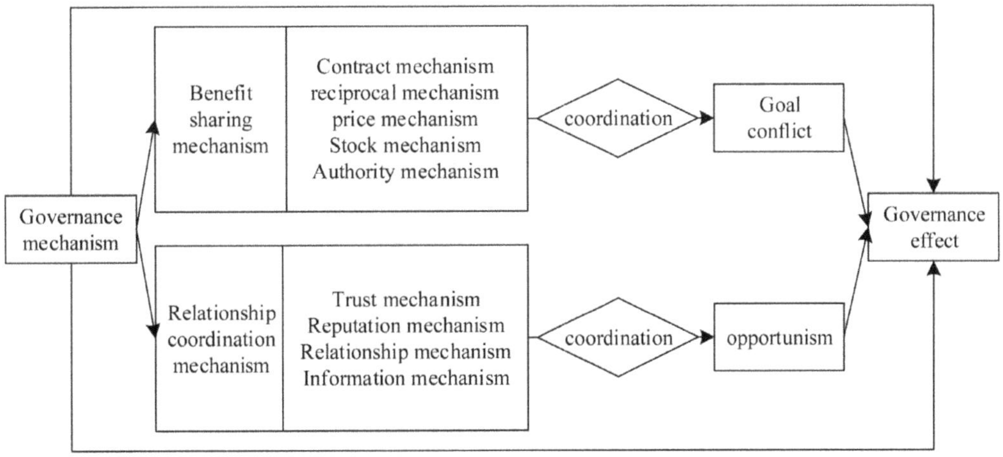

Figure 11. Sustainable attributes of a fast fashion product
Source: Li et al., 2015

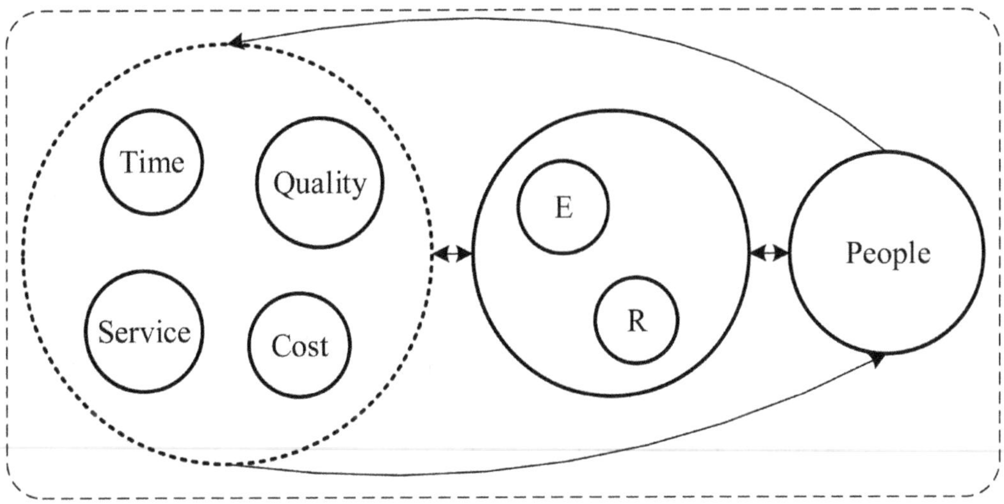

By establishing a sustainability governance framework (Figure 13), seven factors that affect the sustainability governance decision-making are identified and the efficiency and legitimacy mechanisms of sustainability governance from internal and external perspectives are evaluated.

The application of the governance mechanisms through a case study is explored based on the seven sustainability commitments of H&M. The findings suggest that the core influence and centrality of a corporation should be strengthened from the perspective of internal governance and that stakeholders should collaborate to achieve sustainability governance throughout an entire fast fashion supply chain from the perspective of external governance (Figure 14).

Figure 12. Motivations of sustainable supply chain governance
Source: Li et al., 2015

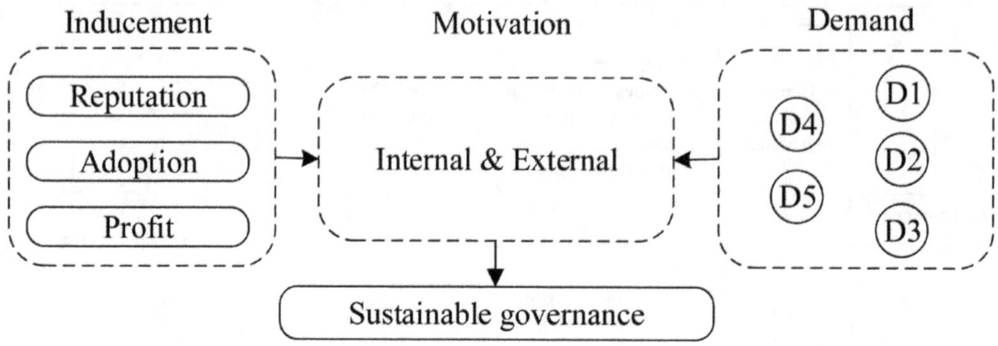

Figure 13. SSCG framework
Source: Li et al, 2015

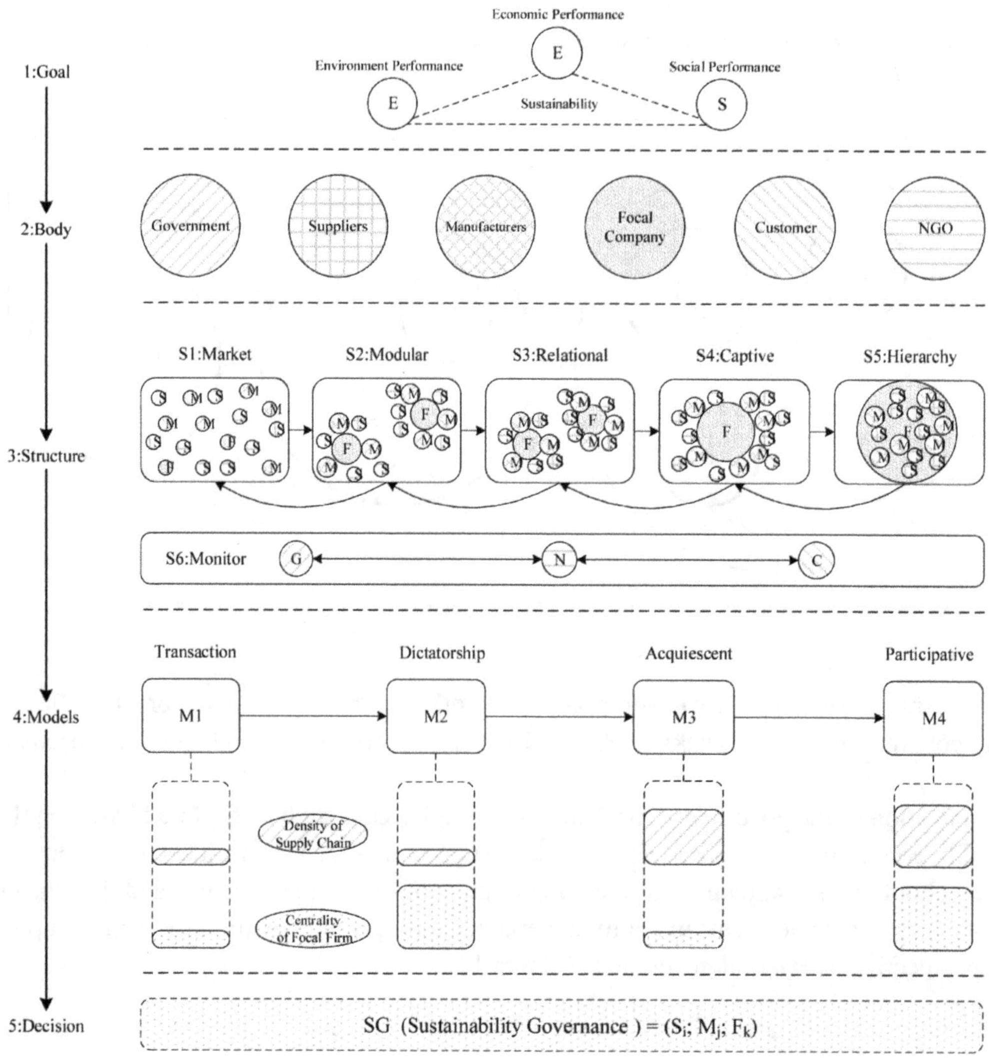

Figure 14. Mechanisms, influence factors, and sustainable commitments of the sustainability governance in H&M
Source: Li et al., 2015

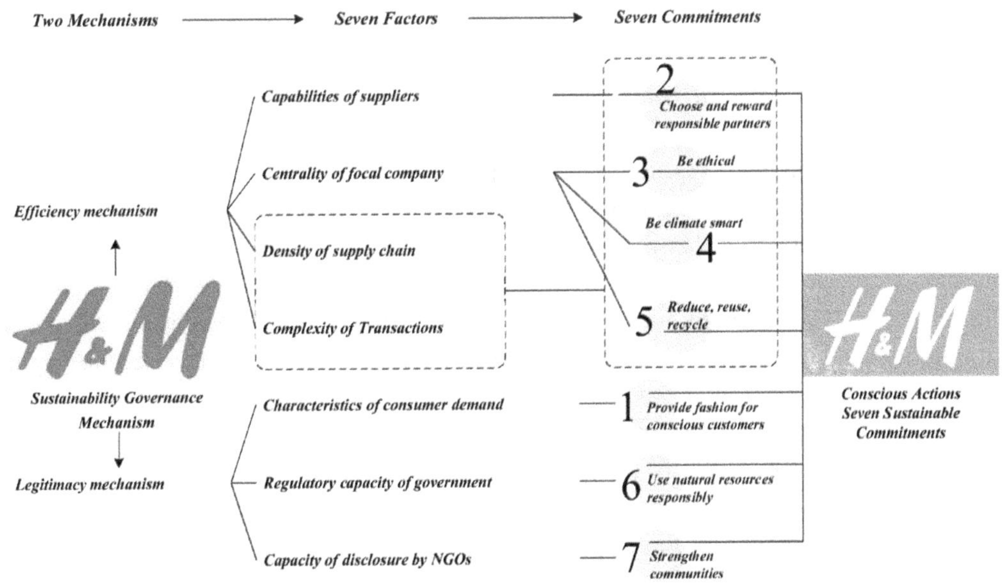

The implementation of extended producer responsibility is the simultaneous focus of enterprises, governments and scholars, particularly the responsibility of the main body of the entire supply chain system. The effective implementation of EPR will help enterprises and the entire supply chain reduce environmental costs, improve supply chain revenue, and enhance overall competitiveness from a long-term perspective. Zhao et al. (2014) constructed a framework of sustainable governance in the EPR implementation process based on the supply chain governance as a theoretical perspective to discuss its mechanism problem from three dimensions. These dimensions include individual producer responsibility, collective producer responsibility, and government regulation. The study pointed out some significant management insights. First, the effective implementation of the individual responsibility of an enterprise positively influences EPR implementation. From the perspective of individual responsibility, enterprises should pay attention to effective cost control mechanism. Second, the effective implementation of producer collective responsibility positively affects EPR implementation. From the perspective of organizational responsibility, fair cooperation mechanism plays an important role in sharing responsibilities and interests. Third, the effective establishment of a government regulation system positively influences an effective EPR implementation. At the national level, the government should focus on the safety of financial capital to effectively encourage enterprises to implement EPR policies. Therefore, the main body of the EPR implementation process, such as enterprises, governments, and scholars, should explore the mechanism of EPR implementation from the three dimensions of government regulation, corporate social responsibility, and supply chain governance.

Based on the above discussion, Zhao et al. (2015) established the individual producer responsibility (IPR) and collective producer responsibility (IPR) game modes and obtained optimal decisions of enterprise pollution governance technology in both modes. The results show that the optimal decision depends largely on the technological gap and degree of reliance among enterprises. The large technological gap and strong degree of reliance render the CPR mode more applicable. However, the IPR mode is more likely to be adopted among comparable companies with a weak degree of reliance. Finally, this study summarizes the efficiency and legitimacy mechanisms of dynamic equilibrium governance in the supply chain system from the internal and external perspectives under the EPR policy.

Green Supply Chain and Enterprise Practice

Increasing global environmental pressures from regulators, markets, and communities have caused focal companies in supply chains to recognize the significance of environmentally conscious management. Green supply chains are part of this recognition. Environmental supplier development is a valuable and viable strategy for greening supply chains. However, low carbon management is rarely explored in supplier development. Furthermore, formal tools and models for focal companies to evaluate environmental supplier development programs (ESDPs) are limited considering the low carbon management and their effect on supplier performance improvement. To help address these gaps in the literature, Dou et al. (2015) proposed a portfolio evaluation model for ESDPs that considers three types of supplier performance: traditional operational factors (Table 2), traditional environmental factors (Table 3), and low-carbon management factors (Table 4). This model applies the fuzzy scoring method to measure the effect of ESDPs on supplier performance and uses fuzzy DEMATEL to examine the cause–effect interrelationships among the ESDPs.

Green supply chain management (GSCM) is now a proactive approach for organizations to enhance their environmental performance and to achieve competitive advantages. Govindan et al. (2015) first used the decision-making trial and evaluation laboratory (DEMATEL) method with intuitionistic fuzzy sets to handle important and causal relationships between GSCM practices and performances (Figure 15). DEMATEL evaluates GSCM practices to find the main practices to improve both environmental and economic performances. This study uses intuitionistic fuzzy set theory to handle the linguistic imprecision

Table 2. Traditional operational factors for supplier performance evaluation

Factors	Description	Sub-Factors
Cost (CT)	Overall cost of a product that a supplier provides	Product price, logistics cost, ordering cost
Quality (QY)	A supplier can provide good quality products to make a good impression on the customer	Quality assessment system, rejection rate of products, remedy of quality problems
Time (TE)	A supplier delivers products to the customer on time	Lead time, on time delivery
Flexibility (FY)	A supplier can change according to the requirements of customers	Product volume changes, short set-up time, conflict resolution
Relationship (RP)	A supplier has a good inter-relationship with a customer	Long-term relationship, relationship closeness, reputation for integrity
Technology (TY)	A supplier can provide appropriate technological support according to the requirements of customers	Capability of R&D, capability of design, technological compatibility
Culture (CE)	A supplier has a cultural compatibility with a customer	Feeling of trust, strategic fit, top management compatibility, compatibility among levels and functions

Source:(Dou et al., 2015)

Table 3. Environmental factors for supplier performance evaluation

Factors	Sector
Pollution reduction (PR)	Waste water reduction, solid waste reduction, waste gas reduction
Resource consumption reduction (RCR)	Fresh water consumption reduction, energy consumption reduction, land use reduction, material consumption reduction
Green image improvement (GII)	Green product image improvement, green brand image improvement, green marketing image improvement
Capability of Design for the Environment (CDfE)	Capability of design for low consumption of raw material, capability of design for low toxicity, capability of design for low consumption during use stage, capability of design for environmental package
Implementation level of environmental management system (ILEMS)	Establishment of environmental commitment and policy Identification of environmental aspects Planning of environmental objectives Assignment of environmental responsibility Checking and evaluation of environmental activities

Source: (Dou et al., 2015)

Table 4. Low carbon management factors for supplier performance evaluation

Categories	Factors	Sub-Factors
Carbon emission reduction performance	Direct carbon emission (DCE)	Direct emissions from stationary combustion, direct emission from mobile combustion, direct emissions from process sources, direct emissions from fugitive sources
	Indirect carbon emission (ICE)	Indirect emissions from purchased/acquired components, indirect emissions from purchased/acquired electricity, indirect emissions from purchased/acquired steam, indirect emissions from purchased/acquired heating, indirect emissions from purchased/acquired cooling, indirect emission from purchased products, indirect emission from sold products
Carbon reduction practices	Carbon reduction planning (CRP)	Setting targets, determining a strategy, evaluating alternatives, justifying decisions, project planning
	Carbon reduction communication (CRC)	Communication from top to down in an organization, cross department communication within an organization, communication with upstream suppliers, communication with downstream customers, communication with other external stakeholders
	Carbon reduction information management (CRIM)	Information system construction, identifying the boundary of carbon emissions, information gathering, information analysis, information verification, information disclosure and report
	Carbon emission control (CEC)	Carbon reduction investment, carbon reduction training, low carbon supplier collaboration, low carbon customer collaboration

Source:(Dou et al., 2015)

and the ambiguity of human judgment. A case study from the automotive industry is presented to evaluate the efficiency of the proposed method. The results reveal ''internal management support'', ''green purchasing'', and ''ISO 14001 certification'' as the most significant GSCM practices. The practical results of this study offer useful insights for managers to become more environmentally responsible while improving their economic and environmental performance goals. Furthermore, a sensitivity analysis of results, managerial implications, conclusions, limitations and future research opportunities are provided.

Industrial and business activity problems that influence the environment have been closely scrutinized by the governments of developing countries in recent years (Mathiyazhagan et al., 2014; Govindan et al., 2014). Tian et al. (2014) discussed a system dynamics model based on evolutionary game theory for green supply chain management diffusion among Chinese manufacturers (Figure 16). In this study, system dynamics models are developed to guide the subsidy policies in order to promote the diffusion of green supply chain management (GSCM) in China. The relationships of stakeholders such as the government, enterprises, and consumers are analyzed through evolutionary game theory. Finally, the GSCM diffusion process is simulated by the model with a case study on the Chinese automotive manufacturing industry. The results show that the subsidies for manufacturers are better than that for consumers to promote GSCM diffusion and that environmental awareness is another influential key factor.

Figure 15. Steps of the intuitionistic fuzzy DEMATEL method
Source: Govindan et al, 2014

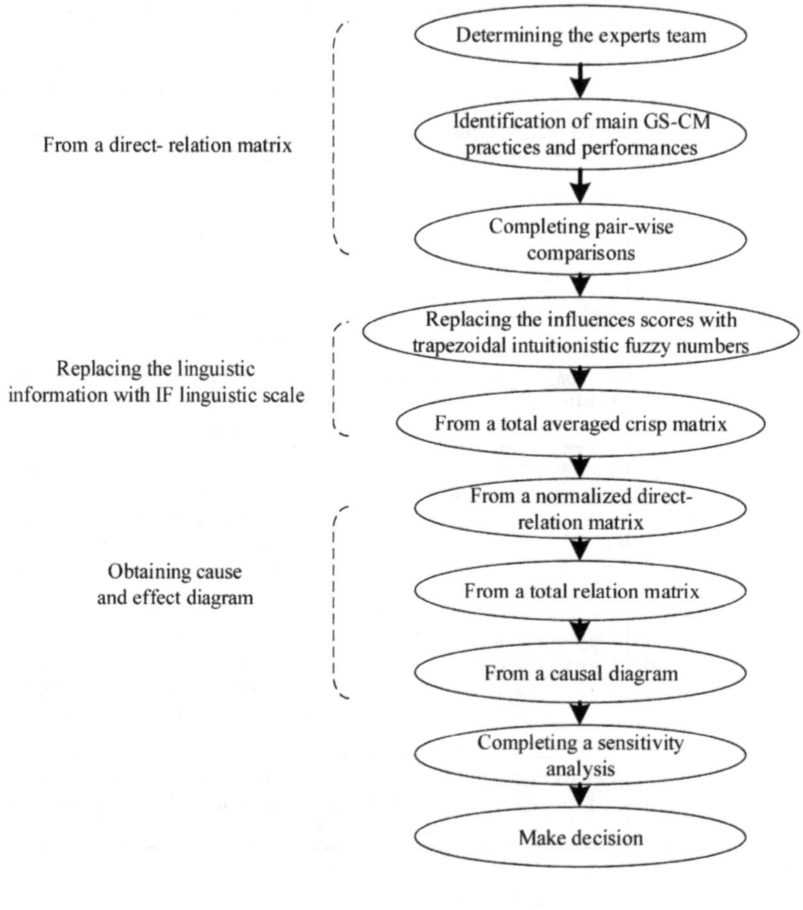

Figure 16. System analysis framework of GSCM diffusion among manufacturers
Source: Tian et al, 2014

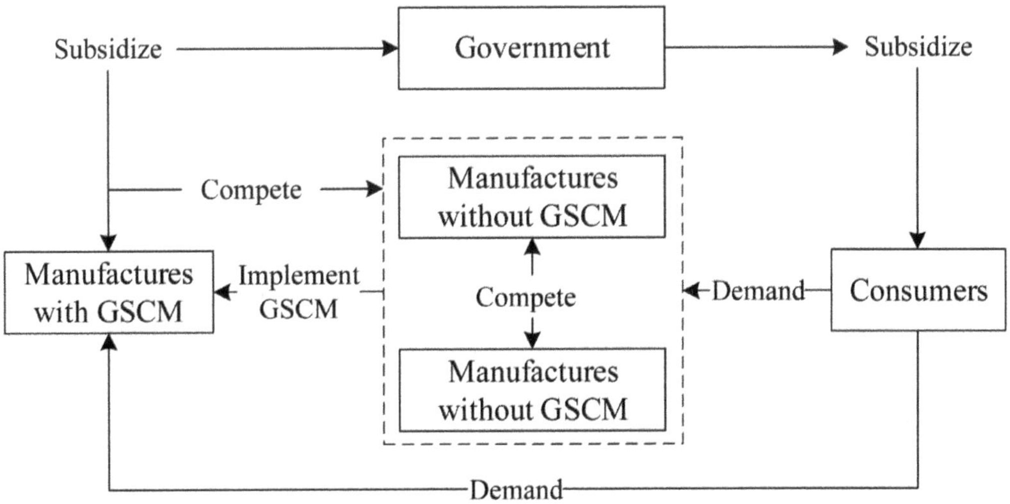

Increasing consumer demand further drives activities across all sectors of products and services. Governmental regulations are enforced, and countries worldwide are adopting means of promoting environmental policies to mitigate pollution by improving environmental performance and increasing efficiency. However, the implementation of green supply chain management is hardly an easy task for most industries. Wang et al. (2015) used a decision-making trial and evaluation laboratory approach to analyze the barriers to green supply chain management adoption in a food packaging company (Figure 17).

The study results aim to assist the packaging industry to better understand major barriers, as this is the first crucial step in overcoming these barriers. The results demonstrate that lack of adequate training and progress monitoring, as well as poor customer awareness and lack of pressure for widespread adoption of GSCM, are the most influential barriers.

Moreover, the adoption of environmental management practices is essential in each industry to fulfill the requirements of environmental regulations and customer demands. Indian industries have initiated actions to adopt GSCM practices in their traditional activities due to increased pollution from automobiles. Responding to all the pressures simultaneously is difficult, making decision-making difficult for managers. To help managers identify the primary pressure among the available pressures for GSCM adoption, Mathiyazhagan et al. (2015) studied the application of analytical hierarchy process in evaluating pressures to implement green supply chain management. Indian mining and mineral industries, especially small and medium enterprises (SMEs), focused on developing their environmental performance by integrating sustainability practices into traditional supply chain management. However, difficulties emerge in their attempts to analyze and identify influential sustainable supply chain management (SSCM) practices to enhance sustainable performance. Hence, Jia et al. (2015) analyzed and identified dominant SSCM practices in the mining and mineral industries utilizing interpretive structural modeling (Figure 18).

The study uses three different research phases: identification of SSCM practices from the literature, interviews with various department managers of the Indian mining and mineral industries, and a survey conducted within the mining and mineral industries. In this paper, 25 SSCM practices were considered in less than six categories. Finally, the approach was applied to 15 mining and mineral industries in India. The

Figure 17. Research flowchart to analyze and identify the influential barriers to GSCM adoption
Source: Wang et al, 2015

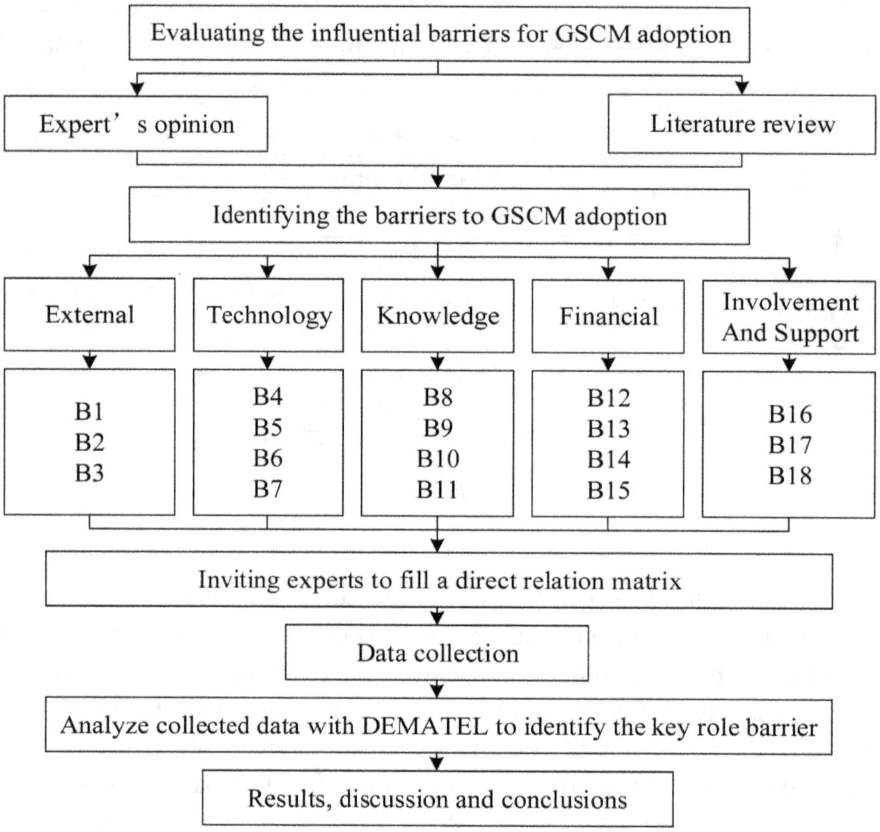

results of this study show that suppliers' ISO14000 certification practice acts as an influential role over the recommended 25 SSCM practices, which may be used as environmental management certification.

CONCLUSION

Practical Implications

Green production and green supply chain for enterprises is not a compulsory implementation of the environmental strategy but a sustainable development strategy consistent with the economic interests of enterprises. Furthermore, it is a significant social behavior, as well as an effective means of achieving significant economic benefits. The current business focuses on both its competitiveness and social images. Important customers proposed that a supplier should set effective environmental management measures.

Green production and green supply chain management can avoid wasting resources, enhance corporate social responsibility, provide a good reputation and green product brand image to enterprises, and expand the product market. The savings in the production of raw materials reduce the life cycle cost of the final product. Thus, consumers have access to safe and environmental-friendly products at a

Figure 18. Flowchart of the preparation of an ISM model for SSCM practices
Source: Jia et al., 2015

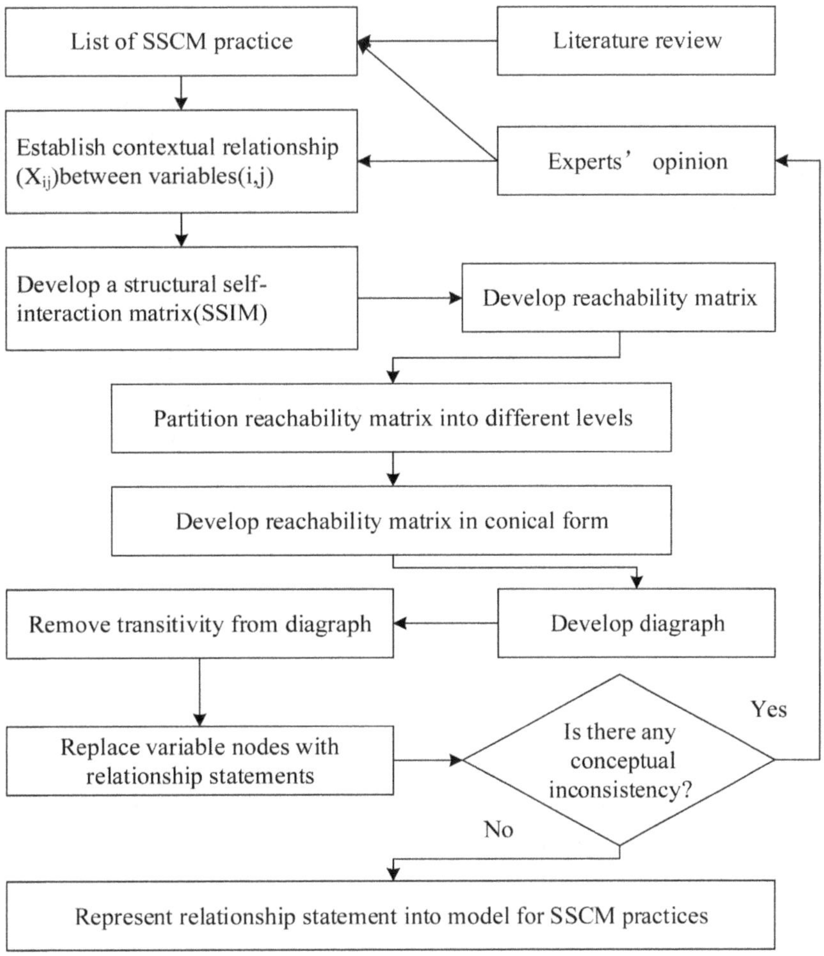

low price. Green production and green supply chain governance will improve and enhance the working environment of employees, improve the health conditions of employees, promote the safety of work, and reduce unnecessary expenses. The staff can be given opportunities to become comfortable. The subjective initiative and work efficiency of the staff can be improved to increase profits, further creating a better social image for enterprises to add intangible assets for enterprises.

In addition, the behavior of enterprises is based on the dialogue among all supply chain nodes, which is the optimal decision of a supply chain system. The green production achievements of the supply chain upstream can be expanded in the follow-up process. Companies with green production and green supply chain management not only easily meet their own environmental standards, they can also promote the environmental performance of the other enterprises in the supply chain. Green supply chain establishes long-term cooperation relationships among members to achieve both commercial profits and environmental protection purposes.

Future Directions

Several potential areas can be pursued to extend this line of research. First, studies on EPR and supply chain governance are still on the starting point relative to green production and green supply chain. Some problems were determined for future studies regarding this subject. Our research only explores the application of influence factors. However, the evaluation of the effectiveness of the supply chain governance framework in real-world situations still requires further research. Therefore, the effectiveness of operating a green supply chain using supply chain governance principles under EPR can be further surveyed.

Second, we used secondary data from CSR reports to evaluate their green production and CSR practices. However, a gap may exist between CSR disclosure and real CSR practices (Font et al., 2012). The manner in which CSR practices are evaluated is an interesting topic for further studies. Future research may focus on this CSR practices using quantitative evaluation to justify the validity of the study. Some studies have explored only one case study in green supply chain. Hence, conclusions may not generally suit various sectors. Future works should also investigate the green products and GSCM practices in different industrial sectors.

Finally, the influencing factors in green products and the GSCM model are relatively simple. Additional stakeholders who can potentially affect the GSCM diffusion system, such as NGOs, can also raise the environmental awareness of both manufacturers and consumers. In the future, a complicated model becomes necessary to comprehensively explore the GSCM system.

ACKNOWLEDGMENT

We gratefully acknowledge the support of grants from The Major Program of the National Social Science Fund of China (Grant No. 13&ZD147).

REFERENCES

Agamuthu, P., & Victor, D. (2011). Policy trends of extended producer responsibility in Malaysia. *Waste Management & Research*, *29*(9), 945–953. doi:10.1177/0734242X11413332 PMID:21771873

Agrawal, V. V., Ferguson, M., Toktay, L. B., & Thomas, V. M. (2012). Is leasing greener than selling? *Management Science*, *58*(3), 523–533. doi:10.1287/mnsc.1110.1428

Agrawal, V. V., & Ülkü, S. (2012). The role of modular upgradability as a green design strategy. *Manufacturing & Service Operations Management*, *15*(4), 640–648. doi:10.1287/msom.1120.0396

Atasu, A., Özdemir, Ö., & Van Wassenhove, L. N. (2013). Stakeholder Perspectives on E-Waste Take-Back Legislation. *Production and Operations Management*, *22*(2), 382–396. doi:10.1111/j.1937-5956.2012.01364.x

Atasu, A., Sarvary, M., & Van Wassenhove, L. N. (2008). Remanufacturing as a marketing strategy. *Management Science*, *54*(10), 1731–1746. doi:10.1287/mnsc.1080.0893

Atasu, A., & Souza, G. C. (2013). How does product recovery affect quality choice? *Production and Operations Management, 22*(4), 991–1010. doi:10.1111/j.1937-5956.2011.01290.x

Atasu, A., & Subramanian, R. (2012). Extended Producer Responsibility for E-Waste: Individual or Collective Producer Responsibility? *Production and Operations Management, 21*(6), 1042–1059. doi:10.1111/j.1937-5956.2012.01327.x

Atasu, A., Van Wassenhove, L. N., & Sarvary, M. (2009). Efficient Take-Back Legislation. *Production and Operations Management, 18*(3), 243–258. doi:10.1111/j.1937-5956.2009.01004.x

Branco, M. C., & Rodrigues, L. L. (2006). Corporate social responsibility and resource-based perspectives. *Journal of Business Ethics, 69*(2), 111–132. doi:10.1007/s10551-006-9071-z

Cao, M., & Zhang, Q. (2011). Supply chain collaboration: Impact on collaborative advantage and firm performance. *Journal of Operations Management, 29*(3), 163–180. doi:10.1016/j.jom.2010.12.008

Das, D., & Dutta, P. (2015). Design and analysis of a closed-loop supply chain in presence of promotional offer. *International Journal of Production Research, 53*(1), 141–165. doi:10.1080/00207543.2014.942007

De Giovanni, P. (2014). Environmental collaboration in a closed-loop supply chain with a reverse revenue sharing contract. *Annals of Operations Research, 220*(1), 135–157. doi:10.1007/s10479-011-0912-5

Dou, Y., Zhu, Q., & Sarkis, J. (2015). Integrating Strategic Carbon Management into Formal Evaluation of Environmental Supplier Development Programs. *Business Strategy and the Environment, 24*(8), 873–891. doi:10.1002/bse.1851

Feng, Y., Tian, Y., & Zhu, Q. (2014). A combined input–output/decision making trial and evaluation laboratory method for evaluating effect of the remanufacturing sector development. *Journal of Cleaner Production, 114*, 103–113. doi:10.1016/j.jclepro.2014.10.048

Govindan, K., Khodaverdi, R., & Vafadarnikjoo, A. (2015). Intuitionistic fuzzy based DEMATEL method for developing green practices and performances in a green supply chain. *Expert Systems with Applications, 42*(20), 7207–7220. doi:10.1016/j.eswa.2015.04.030

Jia, P., Diabat, A., & Mathiyazhagan, K. (2015). Analyzing the SSCM practices in the mining and mineral industry by ISM approach. *Resources Policy, 46*, 76–85. doi:10.1016/j.resourpol.2014.04.004

Lee, S. K., Seo, K. L., & Sharma, A. (2013). Corporate social responsibility and firm performance in the airline industry: The moderating role of oil prices. *Tourism Management, 38*(13), 20–30. doi:10.1016/j.tourman.2013.02.002

Li, W. A., Li, Y. J., & Shi, D. (2016). Theoretical research on supply chain governance: Concept, intention, and normative analysis framework. [in Chinese]. *Nankai Business Review, 19*(1), 4–15.

Li, Y., Zhao, X., Shi, D., & Li, X. (2014). Governance of sustainable supply chains in the fast fashion industry. *European Management Journal, 32*(5), 823–836. doi:10.1016/j.emj.2014.03.001

Ma, L., Wang, Y. H., & Ma, M. S. (2016). The Tripartite Evolutionary Game Analysis of Collusive behavior Regulatory in EPR System. [in Chinese]. *Science and Management, 36*(3), 24–29.

Mathiyazhagan, K., Diabat, A., Al-Refaie, A., & Xu, L. (2015). Application of analytical hierarchy process to evaluate pressures toimplement green supply chain management. *Journal of Cleaner Production, 107*, 229–236. doi:10.1016/j.jclepro.2015.04.110

Seuring, S. (2013). A review of modeling approaches for sustainable supply chain management. *Decision Support Systems, 54*(4), 1513–1520. doi:10.1016/j.dss.2012.05.053

Tian, Y., Govindan, K., & Zhu, Q. (2014). A system dynamics model based on evolutionary game theory for green supply chain management diffusion among Chinese manufacturers. *Journal of Cleaner Production, 80*(7), 96–105. doi:10.1016/j.jclepro.2014.05.076

Wang, Z., Mathiyazhagan, K., Xu, L., & Diabat, A. (2015). A decision making trial and evaluation laboratory approach to analyze the barriers to Green Supply Chain Management adoption in a food packaging company. *Journal of Cleaner Production, 117*, 19–28. doi:10.1016/j.jclepro.2015.09.142

Wei, J., Zhao, J., & Li, Y. (2014). Price and warranty period decisions for complementary products with horizontal firms cooperation/noncooperation strategies. *Journal of Cleaner Production, 105*, 86–102. doi:10.1016/j.jclepro.2014.09.059

Wu, M. W., & Shen, C. H. (2013). Corporate social responsibility in the banking industry: Motives and financial performance. *Journal of Banking & Finance, 37*(9), 3529–3547. doi:10.1016/j.jbankfin.2013.04.023

Xia, X., Govindan, K., & Zhu, Q. (2015). Analyzing internal barriers for automotive parts remanufacturers in China using grey-DEMATEL approach. *Journal of Cleaner Production, 87*(1), 811–825. doi:10.1016/j.jclepro.2014.09.044

Yang, H., Cui, L., & Zhu, Q. H. (2016). Full life cycle impact assessment of refrigerators considering their security processing proportion. [in Chinese]. *Systems Engineering*, (5): 136–143.

Zhao, X. K., Li, Y. J., & Shi, D. (2015). Game analysis on supply chain governance mechanism based on extended producer responsibility. [in Chinese]. *Journal of Systems Engineering, 30*(2), 231–239.

Zhao, X. K., Li, Y. J., Wang, W., & Abdullah, A. T. M. (2014). Framework for sustainability implementation of EPR: Supply chain governance perspective. *International Conference on Service Systems and Service Management. IEEE*,1-5. doi:10.1109/ICSSSM.2014.6943415

Zhu, Q., Liu, J., & Lai, K. H. (2016). Corporate social responsibility practices and performance improvement among Chinese national state-owned enterprises. *International Journal of Production Economics, 171*, 417–426. doi:10.1016/j.ijpe.2015.08.005

Zhu, Q., & Zhang, Q. (2015). Evaluating practices and drivers of corporate social responsibility: The Chinese context. *Journal of Cleaner Production, 100*, 315–324. doi:10.1016/j.jclepro.2015.03.053

Zhu, Q.H., Feng, Y.T., &Tian, Y.H. (2014). Economic Impact Analysis of remanufacturing Industry in China Based on Input-output Method. *China Soft Science,* (6), 34-43. (in Chinese)

Chapter 2
Assessment of Supply Chain Greenness:
A Literature Review

Mete Han Topgul
Marmara University, Turkey

Huseyin Selcuk Kilic
Marmara University, Turkey

Gülfem Tuzkaya
Marmara University, Turkey

ABSTRACT

Having become conscious about the importance of the environment much more than past nowadays, the environmentally benign applications also called as green applications have increased tremendously in the recent years. Although there are specific and individual studies in various fields from design of product to final delivery of it, there is a need to handle the subject as a whole and construct green supply chains. Before constructing a green supply chain, the first step is to assess the existing structure from the viewpoint of greenness. Hence, it is important to develop a systematic assessment approach and this study aims to analyze the existing literature by mainly clarifying the criteria and methods used in such assessment systems and provide directions for future studies. Moreover, this study also aims to provide a base for the academics and practitioners that are willing to develop a greenness assessment methodology for supply chains. For this aim, numerous studies have been analyzed under a proposed structure including four sub-sections which consist of a green supply chain.

INTRODUCTION

Greenness which mainly reflects "the respect to the environment" is one of the most popular topics in the last decade. It requires the consideration of the environmental aspects from various points for all the systems including manufacturing and service sectors. Although there are individual applications to

DOI: 10.4018/978-1-5225-3537-9.ch002

provide greenness, the topic is also handled as a whole under supply chain logic which starts from the supply of raw materials to final delivery of it and as van Hock and Erasmus (2000) state, in all phase of the life cycle of the finished goods, it is directly related to the environment. Depending on this relationship with the supply chain management, this concept is called as Green Supply Chain Management (GSCM). GSC have become an important model for the companies to stay in the market and make profit while respecting to environmental impacts. As a result, a large body of literature for GSCM have been published and the aim of this study is to review the related literature and to give an insight for the researchers and practitioners.

The study of Wilkerson (2005) is regarded as one of the first studies including "Green Supply Chain (GSC)" concept, but, actually before him, many researchers studied the topic under various parts of GSC. One of those studies was performed by Barnes (1982) and Pohlen (1992). They focused on the reverse logistics which is also one of the main parts of GSC. Afterwards, Murphy et al. (1994) and Szymankiewicz (1993) addressed firstly the logistics related Green Supply Chain. Then, Drumwright (1994) published an article about purchasing part of GSC. After these preliminary studies related to the different parts of GSC topic, Sarkis (1995) put the effort on the concept and systematical integration of them including purchasing, operations, marketing, logistics and reverse logistics with environmental aspects. As can be inferred from the studies, there are various parts of GSC that need to be considered for a good design. However, in this study, the first stage of GSC design which is the assessment of greenness is the main focus.

Performance assessment of a system for different indicators is one of the main and well known principles of management. Hence, to provide a green supply chain, an evaluation system is a requirement and as Theyel (2006) mentioned "environmental performance is a concern of managers due to reasons ranging from regulatory and contractual compliance, to public perception and competitive advantage" in "Greening the Supply Chain". With this way of considering the green supply chain, measuring some of the processes by key performance indicators (KPI) is critical for the companies for different purposes. After checking the defined KPIs continuously, measuring the process or situation is another important factor for the companies if some of the environmental labels or certificates, like ISO 14000 series, are aimed to be gained and kept. These labels or certificates are considered as good indicators by the customers and the governments.

Many studies on performance related measurements have been published for the classic supply chains, nevertheless these key performance indicators are not capable of measuring the green supply chain. For this reason, different type of comprehensive KPIs are required to be developed for green supply chain performance measurement. Consequently, in this important area, many researchers cooperated with the business to measure the Green Supply Chain activities. First green supply chain performance measurement related investigation was performed by Sarkis et al. (2001) and it was related to the waste management which is a subdivision of Green Supply Chain. However, first study on the performance evaluation for the whole green supply chain was prepared by Lakhal et al. (2006) for lumber industry and by Zhu et al. (2004) in Chinese Manufacturing Enterprises in the same year. Similarly, Zhu et al. (2005) described GSCM drivers and evaluated the performance of a number of Chinese manufacturing organizations. Hervani et al. (2005) prepared an overview of the various issues related to environmental (green) supply chain management performance measurement. Tsai et al. (2009) and Büyüközkan et al. (2012) used different techniques to evaluate supplier selection and purchasing process for the green supply chain.

In addition to the greenness assessment studies, a limited number of literature reviews that focus on the assessment of green supply chains have been performed. Up to the knowledge of the authors, there are seven review studies performed between the years 2012 and 2015. The detailed analysis of these studies are given in Section 2.

Regarding the existing literature review studies on the assessment of supply chain greenness, this study differs from them with respect to various points. The main difference is the structure that the review is based on. GSC is firstly divided into four logistical subsections as supported by Sarkis (2002), namely inbound, in-plant, outbound and reverse logistics. Hence, this literature review is performed on this structure. Firstly, an overall analysis is performed among the sub-sections regarding the Sarkis (2002)'s model, inbound (IL), in-plant (IPL), outbound (OL) and reverse logistics (RL). Afterwards, for each sub-section, a detailed analysis is made with respect to "Variety of *Key Performance Indicators (KPI)*", "*Decision Level*", "*Methods for the Assessment*" and "*Country and Sector.* All reviewed papers and the mentioned classifications are shown in Appendix 1, Table 17. Moreover, future trends and directions are provided in the conclusion section.

The remainder of the paper is organized as follows: The detailed literature review about the assessment of supply chain greenness with sub-sections is provided in Section 2. There are the conclusions, trends and advances in Section 3 and the Appendix 1 is provided finally.

PROPOSED METHODOLOGY

A systematic procedure was followed within the literature review as shown in Figure 1. Determining the main and sub-parts of the topic (GSCM), determining the key words, determining the proper databases and realizing the search, reviewing the literature and analysis/synthesis, and discussing the gaps and future directions are the main steps that are performed.

Key words, "*Green Supply Chain Assessment*", "*Green Supply Chain Evaluation*", "*Green Supply Chain Measurement*", "*Supply Chain Environment Assessment*", "*Supply Chain Environment Evaluation*", and "*Supply Chain Environment Measurement*" were selected for the literature search. Afterwards, the selected key words were searched in "*Google Academic*", "*Taylor & Francis Online*", "*Science Direct*", "*Elsevier*", and "*Google*" in the web. The outputs were checked, non-related scientific papers are eliminated and 46 scientific papers are selected for this literature review. After discussion and determination of the sub-parts, classification of the papers is realized. In the last stage, the final results and future research directions are provided.

First analysis is made based on the type of the reference. Five different types, which are "*Conference Paper*", "*Journal Paper*", "*Book Chapter*", "*Master Thesis*", and "*Web Page*" are considered and shown in Table 1.

Second analysis is made based on the publication type, which are "*Literature Review*", "*Modelling*", "*Modelling & Survey*" etc. as shown in Table 2. Most of the papers are based on "*Modelling & Case Study*" with an amount of 13.

As mentioned above, seven literature reviews were published by the researchers between 2012 and 2015. Hassini et al. (2012) reviewed 707 studies published between 2000 and 2010. Hassini et al. (2012) reviewed the highest number of papers according to other literature reviews. Then Fahimnia et al. (2015) follow them with 247 papers published between 1996 and 2013. These three studies reviewed GSC in a general perspective. On the other hand, other four literature reviews focused on the specific areas in

Figure 1. Literature Review Writing Stages

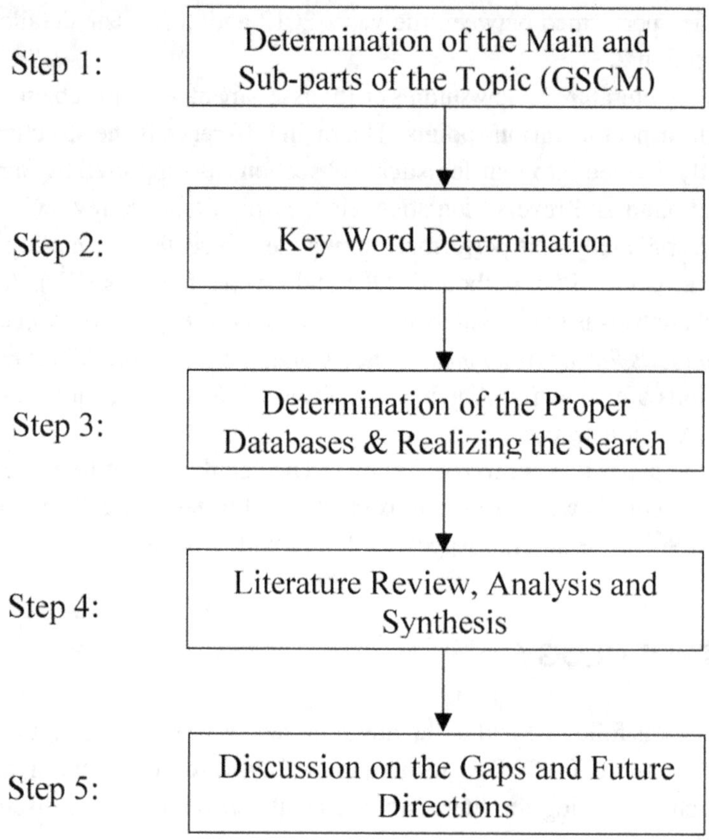

Table 1. Classification of the Reviewed Papers with respect to "Reference Type"

Reference Type	# of Publications
Journal Paper	40
Conference Paper	3
Book Chapter	1
Master Thesis	1
Web Page	1
Total	**46**

GSC. Govindan et al. (2015) reviewed the least number of papers, 33, directly related with the decision making approaches for green supplier evaluation and selection. Ntabe et al. (2015) studied the supply chain operations reference (SCOR) model with 45 studies, Brooks et al. (2012) focused on the unpacking Green IS with 53 papers and Miemczyk et al. (2012) concentrated on the sustainable purchasing and supply management via 73 papers. Table 4 shows the literature reviews with the number of studies analyzed in them and their review period.

Table 2. Classification of the reviewed papers with respect to "Publication Type"

Publication Type	# of Publications
Literature Review	7
Modelling	8
Modelling & Case Study	13
Modelling & Case Study & Survey	7
Modelling & Survey	8
Survey	1
White Paper	2
Total	**46**

Above mentioned literature reviews were published in the journals shown in Table 3. As can be seen from the table, most of the literature reviews were published in International Journal of Production Economics.

In this literature review study, logistics stages were also considered for the analyses. Many models have been developed by the researchers and implemented by the business in Supply Chain Management area. Cox et al. (1995) defined SCM as "Functions within and outside a company that enable the value chain to make products and provide services to the customer" According to The Supply Chain Council (1997), SCM is "A term increasingly used by logistics professionals - encompasses every effort involved in producing and delivering a final product, from the supplier's supplier to the customer's customer". A short period later, environmental consciously improved concept, called as GSCM, mentioned in previous section, became an important strategy for the companies to become competitive in the market. Companies implement GSCM applications due to regulatory reasons, customer requirements or missional reasons. Considering its importance, various studies are prepared for GSCM and proposed methodologies were implemented in real world by many companies.

Aim of the proposed GSCM methodologies are different. While some of them are theoretical and conceptual models, the others are practical solutions implemented to enhance the existing models. Sometimes, existing models were enriched with the cross-disciplinary methods to solve the barriers, such as environmental, technological, organizational and economical etc., faced in the real world. Besides that, some of GSCM models discussed the design issues of some operational areas like as logistics, production etc. Also in some models, the scope of GSCM ranges from reactive monitoring to more proactive

Table 3. Journal classification of the literature reviews

Journal Name	# of Publications
International Journal of Production Economics	3
Supply Chain Management: An International Journal	1
Journal of Cleaner Production	1
Green Business Process Management	1
Journal of Advances in Management Research	1
Total	**7**

Table 4. Analysis of the literature reviews

Author	Paper Name	# of Publications	Review Period
Brooks et al. (2012)	Unpacking Green IS: A Review of the Existing Literature and Directions for the Future	53	2007-2010
Hassini et al. (2012)	A literature review and a case study of sustainable supply chains with a focus on metrics	707	2000-2010
Miemczyk et al. (2012)	Sustainable purchasing and supply management: a structured literature review of definitions and measures at the dyad, chain and network levels	73	1982-2010
Luthra et al. (2014)	Green supply chain management Implementation and performance – a literature review and some issues	189	1995-2012
Fahimnia et al. (2015)	Green supply chain management: A review and bibliometric analysis	247	1996-2013
Govindan et al. (2015)	Multi criteria decision making approaches for green supplier evaluation and selection: a literature review	33	1997-2011
Ntabe et al. (2015)	A systematic literature review of the supply chain operations reference (SCOR) model application with special attention to environmental issues	45	2001-2012

practices. One of the practices was implemented with various Rs (Reduce, Re-use, Rework, Refurbish, Reclaim, Recycle, Remanufacture, Reverse logistics, etc.). These practices become not only a major part of GSCM, but also business management's main principles. Moreover, a combination of some of the models are developed by the researchers. One of these models published by Sarkis (2002) is roughly shown in Figure 2. This model demonstrates that operational activities consist of four stages in the logistics point of view, inbound, in-plant, outbound and reverse logistics. The reviewed academic papers are classified according to these stages in this study. Detailed analyses are made under this classification in the following sub-sections.

The related logistics stages are analyzed under four dimensions as stated below and shown in Figure 3.

- Variety of KPIs (Studies are classified with respect to the number of fields considered within the evaluation process: various KPIs, specific KPIs)

Figure 2. The four main logistics stages

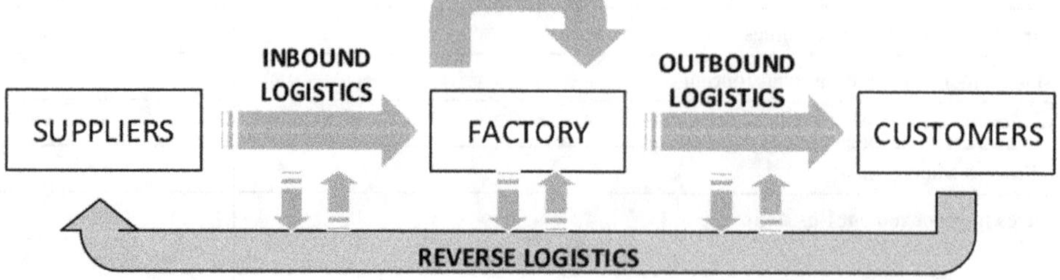

Figure 3. Four Dimensions for the Analysis of Logistics Stages

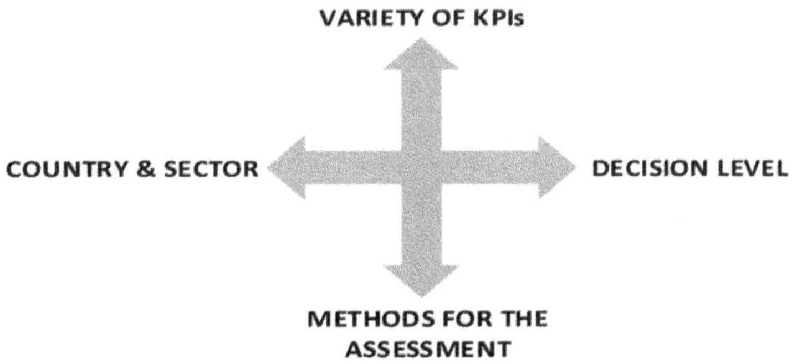

- Decision Level (Decision level in the studies are classified into three classes considering their planning horizons: strategic, tactical and operational decisions)
- Methods for the Assessment (Studies are classified considering the methods used like as ANP, AHP, statistical, etc.)
- Country and Sector (Studies are classified considering the country and sector where the applications are performed)

LITERATURE REVIEW: ASSESSMENT OF GREEN SUPPLY CHAINS

The main structure of the literature review is based on the classification of four logistics stages proposed by Sarkis (2002). Initially, 46 papers were classified according to the determined logistics stages and sub-stages. Most of the papers, 28 of 46, are related to "*All*" of the logistics stages. Afterwards "*Inbound*" logistics stage follows with 7 papers, then "*In-plant*" logistics stage comes with 4 papers. The percentage share of the logistics stages is shown in Figure 4. In addition to the logistics stages, these papers are also classified according to the sub-stages. "*Supplier/Purchasing*" sub-stage under "*Inbound*" logistics stage and "*Remanufacturing*" sub-stage under "*In-plant & Outbound & Reverse*" are some sub-stages existing in the Table 5 (for details please check Appendix 1, Table 17).

Inbound Logistics (IL) Literature

The first logistics stage is the inbound logistics. As shown in Table 5, there are 37 publications related to the Inbound Logistics (IL). 28 of them are also related to the whole stages and 7 of them are just related to the IL. 5 of 7 papers are directly related to the Supplier/Purchasing sub-stages.

The Variety of KPIs

Key Performance Indicators (KPIs) play an important role while determining the greenness level. When the studies are analyzed, it is seen that 26 of them have various KPIs, whereas 6 of them have specific KPIs as shown in Figure 5. Lakhal et al. (2004) mentioned the assessment of the "*five zeros*", zero emissions, zero waste of resources, zero waste in activities, zero use of toxics, zero waste in product life-cycle,

Figure 4. Logistics Stages Percentage Shares

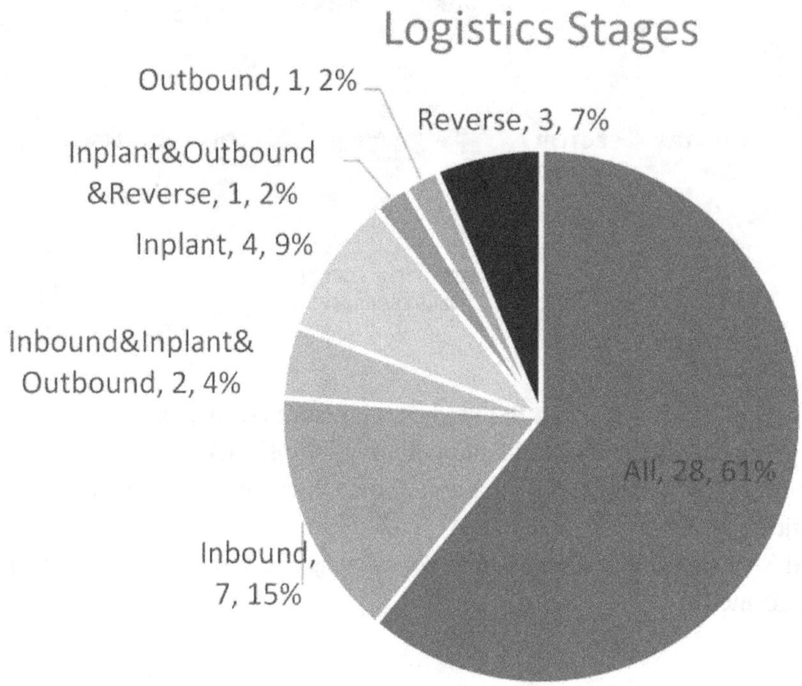

Table 5. Logistics Stages & sub-stages classification in the review

Logistics Stage	Logistics Sub-stage	T# of Publications
All	All	27
All	Transportation	1
Inbound	All	2
Inbound	Supplier/Purchasing	5
Inbound & In-plant & Outbound	All	2
In-plant	All	2
In-plant	N/A	1
In-plant	Waste	1
In-plant & Outbound & Reverse	Remanufacturing	1
Outbound	Packaging	1
Reverse	All	2
Reverse	Packaging	1
Total		**46**

and zero waste. Afterwards Tsai et al. (2009) measured the suppliers with value-chain activities in IL according the greenness of their production, recycling/disposal rates, logistics approach, and distribution channel structure. In 2012, Hassani et al. (2012), Buyukozkan et al. (2012) and Saridogan, M. (2012) proposed methodologies for the assessment of KPIs. Hassani et al. (2012) prepared a literature review study directly related to the supplier and purchase order KPIs. Other study prepared by Buyukozkan et al. (2012) analyzed the supplier selection criteria with an integrated fuzzy MCDM methodology which combines the fuzzy Decision Making Trial and Evaluation Laboratory Model (DEMATEL), the Analytic Network Process (ANP), and Technique for Order Performance by Similarity to Ideal Solution (TOPSIS). Saridogan, M. (2012) considered "*Transportation Cost*" assessment for evaluating the GSC. Also Tuzkaya (2013) proposed a supplier evaluation methodology based on intuitionistic fuzzy Choquet integral operator which considers cost, capacity, delivery reliability, flexibility, and responsiveness.

Decision Level

Another classification is prepared based on the level of decisions considered in the studies such as strategic, tactical and operational. As it is shown in Figure 6, 81% of the publications included all business decision categories. Whereas, strategic and tactical level studies are at 8% and lastly operational level studies are at 3%.

Methods for the Assessment

IL related papers are also analyzed with respect to the methods used. Most of the researchers used statistics in their studies. Analytic Network Process (ANP) approach is also used very often. In Sarkis (1998), ANP is used as an individual technique, on the other hand, Buyukozkan et al. (2012), Tuzkaya et al. (2009), and Sarkis (1999) used combinations of ANP with other techniques. Sarkis (1998) used

Figure 5. Variety of KPIs in IL

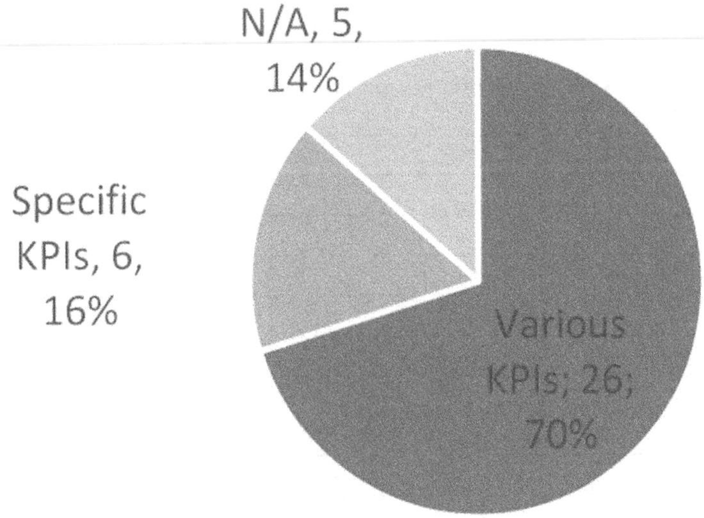

Figure 6. Business decision level in IL

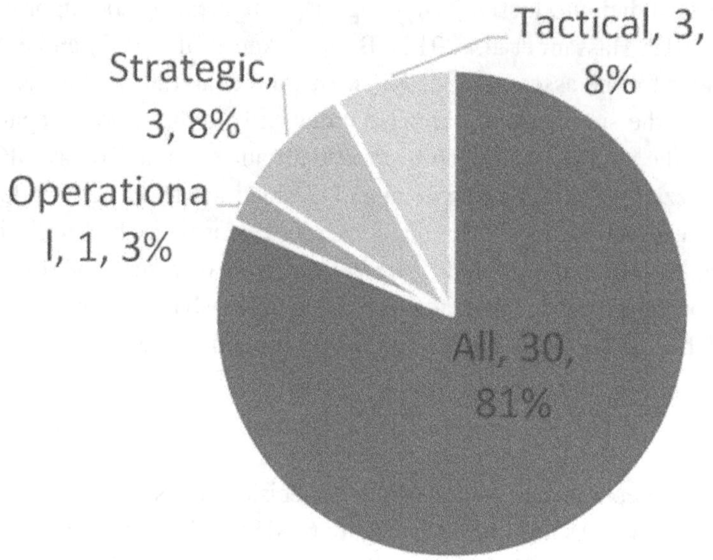

ANP for the integration of the elements and their attributes for a decision support tool and for detailed assessments. Analytic Hierarchy Process (AHP) is also used as an individual technique and integrated technique with different techniques by three researchers: Olasunkanmi et al. (2014), Odeyale (2014) and Tramarico et al. (2016). The methods used in IL by the researchers are shown in Table 6.

Country and Sector

The last classification is prepared based on the country and sector of the applications. The analysis of 15 publications are provided in Table 7. China is the mostly selected country by the researches such as Zhu et al. (2005), Zhu et al. (2007a), Zhu et al. (2007b), Li, Y. (2011) and Zhang et al. (2016). Additionally,

Table 6. Methods for the assessment used by researchers in IL

Method	# of Publications
Statistics	9
ANP/Fuzzy ANP	3
AHP/Fuzzy AHP	3
DEMATEL	2
Survey	1
Balance Score Card (BSC)	1
Goal Programming	1
Other	4
Total	**24**

automotive sector is the mostly studied sector by the researchers such as Zhu et al. (2007a), Olugu et al. (2011), Buyukozkan et al. (2012), and Luthra et al. (2016) in IL.

In-Plant Logistics (IPL) Literature

Second stage mentioned by Sarkis (2002) is In-Plant Logistics (IPL). As shown in Table 5, there are 35 publications related to the IPL. Among them, 28 papers consider the whole stages and 4 papers are only related to the IPL and the rest deals with a combination of various sub-stages.

The Variety of KPIs

IPL related 35 publications were also reviewed and categorized with respect to the variety of KPIs. 26 of them have various KPIs, whereas 5 of them have specific KPIs as shown in Figure 7. Among the publications which only deal with IPL, Sarkis (1999) studied on the cost, quality, recyclability, waste reduction, waste of package, and compliance KPIs and Sarkis et al. (2001) aimed to minimize the cost of waste.

In his study, Sarkis (1999) proposed an integrated ANP and DEA methodology which aims to help managers and organizations for measuring the cost, quality, recyclability, waste reduction, waste of package, and compliance. Besides, Sarkis et al. (2001) while studying the implementation of environmental friendly waste management technologies, aimed to minimize the cost of waste with a methodology using data envelopment analysis (DEA) in their case study for energy sector in USA.

Decision Level

IPL related 35 publications are analyzed with respect to decision levels. As shown in Figure 8, 30 of the publications include whole business decision categories. Other researches assessed the KPIs for specific business decision categories, 2 of them are related to the tactical, 2 of them are related to the strategic and one of them is related to the operational decisions as shown in Table 8.

Methods for the Assessment

Similar to the IL literature, statistics is the mostly used method in the researches. Other preferred methods are "*Analytic Hierarchy Process (AHP)*", "*Analytic Network Process (ANP)*" and "*Data Envelopment Analysis (DEA)*". They are used either individually or integratedly with other techniques. AHP is used in three studies: Olasunkanmi et al. (2014), Odeyale (2014) and Tramarico et al. (2016). ANP is used in Sarkis (1998) and Sarkis (1999). Last mentioned method DEA is used by Sarkis (1999) and Sarkis et al. (2001). Sarkis et al. (2001) studied the implementation of environmentally friendly waste management technologies, aimed to minimize the cost of waste with a data envelopment analysis (DEA) methodology in their case study for energy sector in USA. The methods used in IPL are shown in Table 9.

Country and Sector

In 14 publications, different sectors and countries are analyzed by the academicians as shown in Table 10. China is the mostly analyzed country, and the studies on China are prepared by Zhu et al. (2005), Zhu et al. (2007a), Zhu et al. (2007b), Li (2011) and Zhang et al. (2016). Additionally, automotive sector

Table 7. Country and sector analyzed by the researchers in IL

Sector	Country							
	China	**Unknown**	**Canada**	**Turkey**	**India**	**Japan, China**	**Italy**	**Total**
Automotive	Zhu et al. (2007a)	Olugu et al. (2011)		Buyukozkan et al. (2012)	Luthra et al. (2016)			4
Many	Zhu et al. (2005), Zhu et al. (2007b), Li, Y. (2011)					Zhu et al. (2010)		4
Unknown	Zhang et al. (2016)						Parisi et al. (2010)	2
White Goods				Tuzkaya et al. (2009)				1
Electronic		Tsai et al. (2009)						1
Lumber			Lakhal et al. (2004)					1
Electricity			Hassini et al. (2012)					1
Energy		Ahmad et al. (2016)						1
Total	**5**	**3**	**2**	**2**	**1**	**1**	**1**	**15**

Figure 7. KPIs in IPL

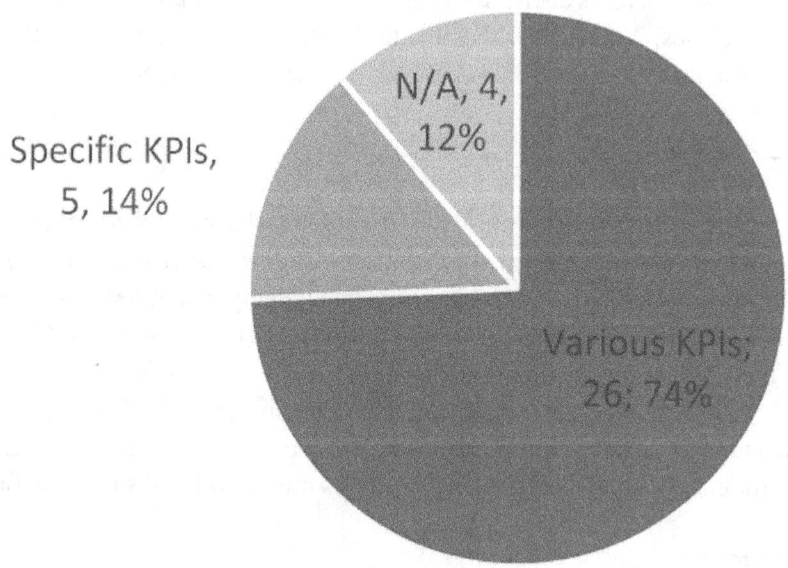

Figure 8. Business decision level in IPL

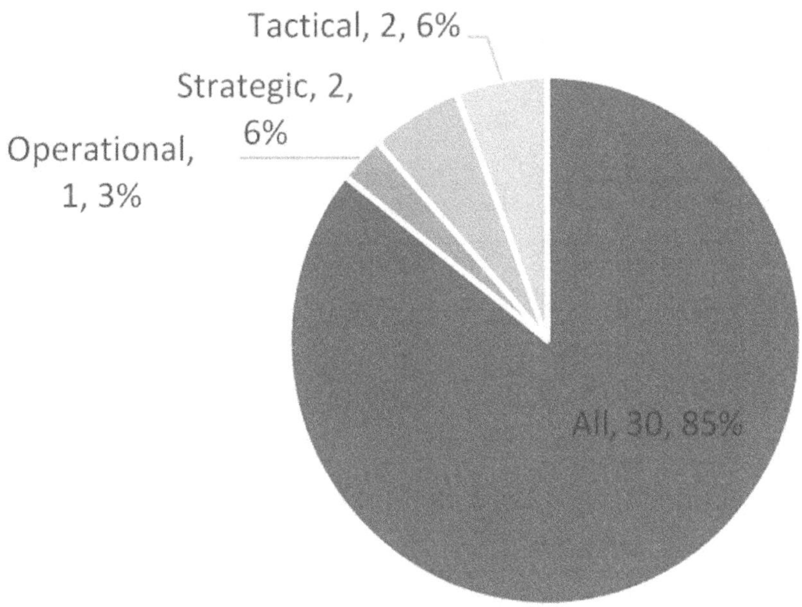

Table 8. Business decision level in IPL

Business Decision Category	Publication(s)
Strategic	Sarkis, J. (1998), Brooks et al. (2012)
Tactical	Sarkis et al. (2001), Lakhal et al. (2004)
Operational	Saridogan, M. (2012)

Table 9. Methods for assessment used by researchers in IPL

Method	# of Publications
Statistics	9
AHP/Fuzzy AHP	3
ANP/DEA/ANP & DEA	3
Survey	1
Balance Score Card (BSC)	1
DEMATEL	1
Other	5
Total	**22**

Table 10. Country and sector analyzed by the researchers in IPL

Sector	Country							
	China	India	Canada	Unknown	USA	Japan, China	Italy	Total
Many	Zhu et al. (2005), Zhu et al. (2007b), Li, Y. (2011)					Zhu et al. (2010)		4
Automotive	Zhu et al. (2007a)	Luthra et al. (2016)		Olugu et al. (2011)				3
Unknown	Zhang et al. (2016)						Parisi et al. (2010)	2
Energy				Ahmad et al. (2016)	Sarkis et al. (2001)			2
Mnfg.		Digalwar et al. (2013)						1
Electricity			Hassini et al. (2012)					1
Lumber			Lakhal et al. (2004)					1
Total	5	2	2	2	1	1	1	14

is the mostly studied sector, and papers on this sector are prepared by Zhu et al. (2007a), Olugu et al. (2011), Buyukozkan et al. (2012), and Luthra et al. (2016).

Outbound Logistics (OL) Literature

Third stage mentioned by Sarkis (2002) is Outbound Logistics (OL). As shown in Table 5, 32 of the reviewed publications are related to the OL. 28 of them are related to the whole stages and one of them (Wang et al., 2016) is just related to the OL. In Wang et al. (2016), "*Packaging*", sub-stage of the OL, is comprehensively analyzed by using Decision Making Trial and Evaluation Laboratory (DEMATEL) methodology. Besides, Odeyale (2014) proposed a model for assessing the green supply chain with many KPIs, Guide et al. (1997) studied on the repairable inventories, and Ahmad et al. (2016) worked on the factors relevant to the sustainability strategies and supply chain functions in energy sector.

The Variety of KPIs

As in the previous logistics stages, OL related 32 publications were also reviewed and categorized according to the KPIs. 26 of them have many KPIs, whereas 3 of them have specific KPIs as shown in Figure 9. Researches, which have specific KPIs, published by Lakhal et al. (2004), Hassani et al. (2012), and Saridogan (2012) are mentioned in IL section.

Figure 9. KPIs in OL

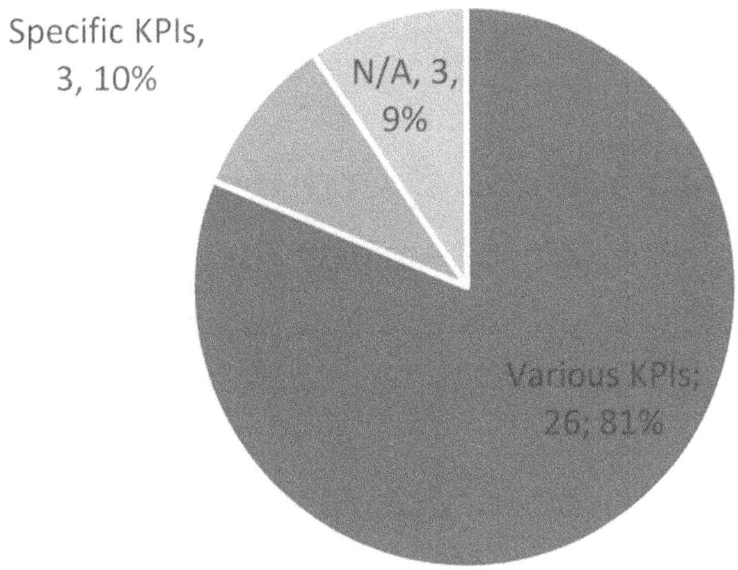

Decision Level

OL related 32 publications are also classified based on business decision categories. As it is shown in Figure 10, 28 of the publications studied whole business decision categories. Whereas others have specific business decision categories, 2 of them are related to the tactical, one of them is related to the strategic and one of them is related to the operational decisions as shown in Table 11.

Methods for the Assessment

As it is for the previous categories, "*statistics*" is the mostly used tool for this area. "*Analytic Hierarchy Process (AHP)*", and "*Decision Making Trial and Evaluation Laboratory (DEMATEL)*" are also used very often. They are used individually or integratedly with other methodologies. AHP is used by three researchers: Olasunkanmi et al. (2014), Odeyale (2014) and Tramarico et al. (2016). DEMATEL is used by two researchers: Wu et al. (2012) and Wang et al. (2016). Wu et al. (2012) used fuzzy DEMATEL method to find influential factors in selecting GSC criteria. All of the methods used in OL literature are shown in Table 12.

Country and Sector

OL related 13 publications from different sectors and countries are shown in Table 13. Similar to IPL, China is the mostly analyzed country by the researchers: Zhu et al. (2005), Zhu et al. (2007a), Zhu et al. (2007b), Li (2011) and Zhang et al. (2016). Besides, similar to the previous sections, automotive sector is the mostly studied sector in OL literature, Zhu et al. (2007a), Olugu et al. (2011), and Luthra et al. (2016).

Figure 10. Business decision level in OL

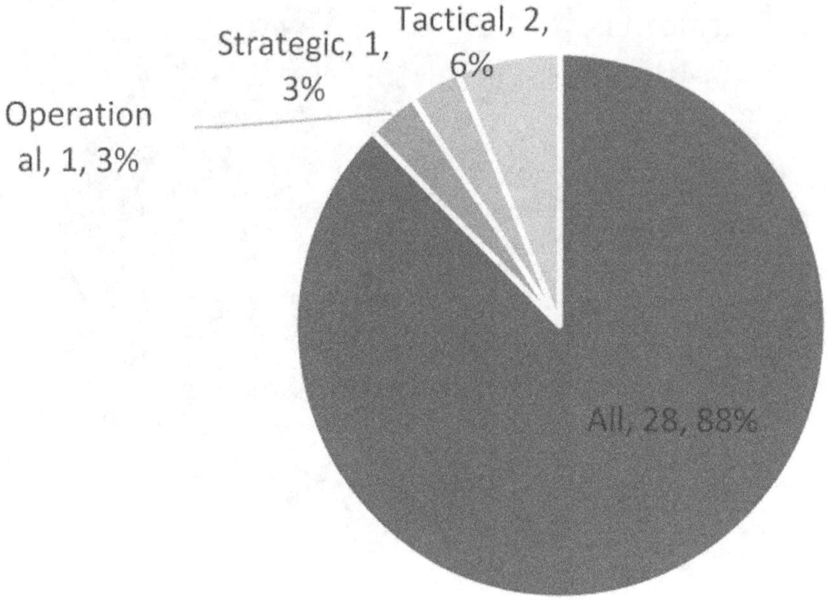

Table 11. Business decision level in OL

Business Decision Category	Publication(s)
Strategic	Sarkis, J. (1998)
Tactical	Wang et al. (2016), Lakhal et al. (2004)
Operational	Saridogan, M. (2012)

Table 12. Methods for assessment used by researchers in OL

Method	# of Publications
Statistics	8
AHP/Fuzzy AHP	3
DEMATEL/Fuzzy DEMATEL	2
ANP/Fuzzy ANP	1
Survey	1
Balance Score Card (BSC)	1
Exploratory Factor Analysis (EFA), Confirmatory Factor Analysis (CFA)	1
Other	3
Total	**20**

Table 13. Country and sector analyzed in OL literature

Sector	Country						
	China	**Canada**	**Unknown**	**India**	**Japan, China**	**Italy**	**Total**
Many	Zhu et al. (2005), Zhu et al. (2007b), Li, Y. (2011)				Zhu et al. (2010)		**4**
Automotive	Zhu et al. (2007a)		Olugu et al. (2011)	Luthra et al. (2016)			**3**
Unknown	Zhang et al. (2016)					Parisi et al. (2010)	**2**
Electricity			Ahmad et al. (2016)				**1**
Packaging				Wang et al. (2016)			**1**
Energy		Hassini et al. (2012)					**1**
Lumber		Lakhal et al. (2004)					**1**
Total	**5**	**2**	**2**	**2**	**1**	**1**	**13**

Reverse Logistics (RL) Literature

The fourth and last stage mentioned by Sarkis (2002) is Reverse Logistics (RL). As shown in Table 5, 32 of the publications are related to the RL. 28 of them are related to the whole stages and just 3 of them are related to the RL stage. Casillejo Tinedo et al. (2011) analyzed the returnable plastic packaging flow in automotive industry. Kongar (2005) applied a balance scorecard approach for a case study in Turkish automotive industry to make assessment for reverse logistics stage. Knemeyer et al. (2002) examined and proposed a qualitative method for computers' reverse logistics process.

The Variety of KPIs

As shown in Figure 11, RL related 32 publications are also classified according to the KPIs. 25 of them have various KPIs, whereas 4 of them have specific KPIs as shown in Figure 11. Knemeyer et al. (2002), Lakhal et al. (2004), Hassani et al. (2012), and Saridogan, M. (2012) are the researches which have specific KPIs. Specifically, Knemeyer et al. (2002) studied the evaluation of the strategic cost, overall quality, customer service, environmental and legislative concerns of refurbishment and recycled computers. Other three publications' details are mentioned in IL section.

Decision Level

As shown in Figure 12, for the business decision categories classification, 32 publications are reviewed in the RL research area. 28 of the publications include whole business decision categories. Whereas others

Figure 11. KPIs breakdown in RL

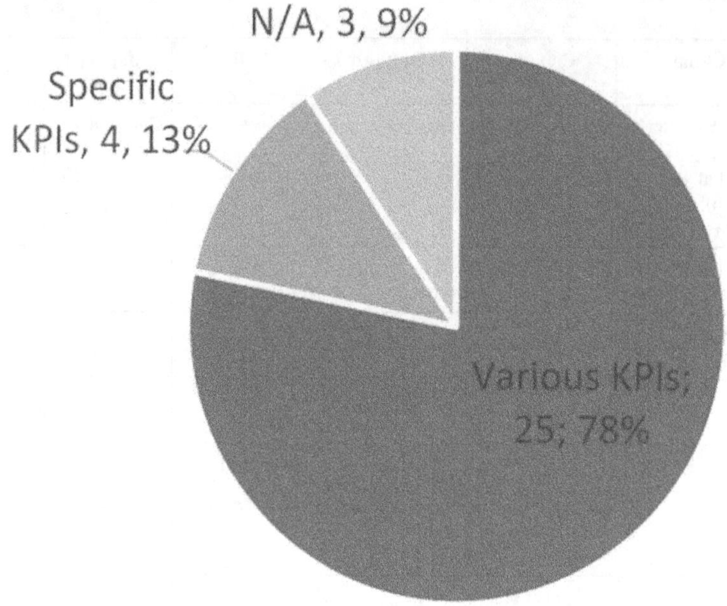

have specific business decision categories: 2 of them are related to the tactical, one of them is related to the strategic and one of them is related to the operational decisions as shown in Table 14. Knemeyer et al. (2002) evaluate environmental and legislative concerns of refurbishment and recycled computers and at the same time gives the tactical business decision perspective for managers. Also Lakhal et al. (2004) studied on KPIs for tactical business decisions. In addition to the tactical business decisions, Saridogan (2012) studied on business decisions with transportation cost evaluation. A strategic business decision related research is prepared by Sarkis (1998).

Methods for Assessment

Table 15. shows the methods used in RL literature. Like as the previous sections, "*Statistics*" is the mostly used method in the studies: Zhu et al. (2004), Kainuma et al. (2006), Zhu et al. (2007a), Zhu et al. (2010), Li (2011), Saridogan (2012), and Luthra et al. (2016). "*Analytic Hierarchy Process (AHP)*" is used in its crisp or fuzzy version. AHP is used by Olasunkanmi et al. (2014) and Tramarico et al. (2016). All the methods implemented in RL literature are shown in Table 15.

Country and Sector

Table 16. shows a country and sector based classification in RL literature in which 13 publications are analyzed. Similar to other logistics stages, China is the mostly analyzed country in the RL stage with the following researches: Zhu et al. (2005), Zhu et al. (2007a), Zhu et al. (2007b), Li (2011) and Zhang et al. (2016). Similar to the previous classifications, automotive sector is the mostly studied sector in RL stage with the studies of Kongar (2005), Zhu et al. (2007a), Olugu et al. (2011), and Luthra et al. (2016).

Figure 12. Business decision level in RL

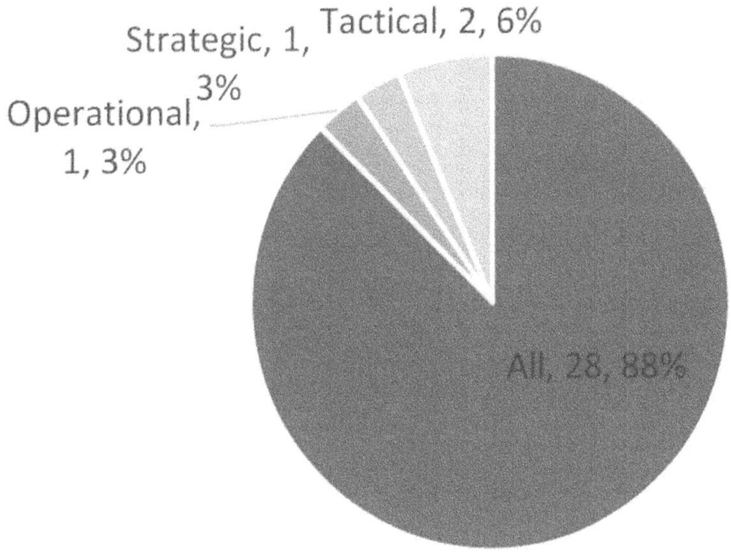

Table 14. Business decision level in RL

Business Decision Category	Publication(s)
Strategic	Sarkis, J. (1998)
Tactical	Knemeyer et al. (2002), Lakhal et al. (2004)
Operational	Saridogan, M. (2012)

Table 15. Methods for assessment used by researchers in RL

Method	# of Publications
Statistics	7
AHP/Fuzzy AHP	2
ANP	1
Survey	1
Balance Score Card (BSC)	1
DEMATEL	1
Exploratory Factor Analysis (EFA), Confirmatory Factor Analysis (CFA)	1
Other	5
Total	**19**

Table 16. Country and sector analyzed by the researchers in RL

Sector	Country								
	China	**Canada**	**USA**	**India**	**Italy**	**Japan, China**	**Unknw**	**Turkey**	**Total**
Many	Zhu et al. (2005), Zhu et al. (2007b), Li, Y. (2011)					Zhu et al. (2010)			**4**
Auto.	Zhu et al. (2007a)			Luthra et al. (2016)			Olugu et al. (2011)	Kongar, E. (2005)	**4**
Unknw	Zhang et al. (2016)				Parisi et al. (2010)				**2**
Comp.			Knemeyer et al. (2002)						**1**
Lumber		Lakhal et al. (2004)							**1**
Electricity		Hassini et al. (2012)							**1**
Total	**5**	**2**	**1**	**1**	**1**	**1**	**1**	**1**	**13**

CONCLUSION

Environmentally conscious supply chains or green supply chains have become a popular research area in recent years. The reason behind this interest may be related to the governmental regulations, customer awareness or economical, etc. Additionally, assessment of the greenness in GSC, which includes determining and measuring the KPIs, has become another important aspect for the companies. In the literature, a significant progress is observed for the definition and monitoring models of the KPIs. Moreover, either depending on regulatory reasons or standardization, states and organizations focus on green applications. In this point of view, International Organization for Standardization (ISO) has published ISO 14000 series of standards, which also includes documents that provide necessities, responsibilities, specifications, procedures and characteristics, for environmental responsibilities of the businesses.

In the literature, it is observed that researchers mostly either focus on restricted stages or consider specific KPIs in detail. As Luthra et al. (2014) mention, in the RL stage, the number of studies which focus on closed loop supply chain (CLSC) is relatively low. Additionally, more KPIs would be defined and measured in the GSC including the design phase for the businesses. Overall review and assessment of greenness would be helpful in the business environment and be an interesting future research direction.

In the sector based classification of the literature, it is concluded that, 20 studies are prepared for different applications. Those studies would generate an awareness on the states, organizations, businesses, and customers. Moreover, considering the KPIs, it is seen that KPI set coverage for business evaluation may be enlarged for the future researches. Moreover, a Decision Support System (DSS) should rather be proposed for facilitating tactical and strategic decisions on environmental applications.

Implementation of the Internet of Things (IoT) and Machine to Machine (M2M) in the GSC area would be another study area for the academicians. Besides, influencing the output received from the IoT and M2M into the GSC evaluation, KPIs would be another new research area. The output received from the physical devices and connected devices would be directly imported in an evaluation system and proactively, the DSS providing a solution and implementing an action to increase the GSC value into a higher level would be an evolution for the business in the era of the Industry 4.0. Hence, this literature review, which classifies the publications in the logistics perspective proposed by Sarkis (2002), would be a reference for the states, organizations, academicians, practitioners, and especially business, who are eager to study in this area theoretically or practically.

REFERENCES

Ahmad, W. N. K. W., Rezaei, J., Tavasszy, L. A., & de Brito, M. P. (2016). Commitment to and preparedness for sustainable supply chain management in the oil and gas industry. *Journal of Environmental Management, 180*, 202–213. doi:10.1016/j.jenvman.2016.04.056 PMID:27233046

Beamon, B. M. (1999). Designing the green supply chain. *Logistics Information Management, 12*(4), 332-342.

Brooks, S., Wang, X., & Sarker, S. (2012). Unpacking green IS: a review of the existing literature and directions for the future. In *Green Business Process Management* (pp. 15–37). Springer Berlin Heidelberg. doi:10.1007/978-3-642-27488-6_2

Buyukozkan, G., & Çifçi, G. (2012). A novel hybrid MCDM approach based on fuzzy DEMATEL, fuzzy ANP and fuzzy TOPSIS to evaluate green suppliers. *Expert Systems with Applications, 39*(3), 3000–3011. doi:10.1016/j.eswa.2011.08.162

Casillejo Tinedo, A., & Stensson, F. (2011). *Returnable plastic packaging flow in the automotive industry-an evaluation of the washing from a green logistics' perspective.* Academic Press.

Cox, J. F., Blackstone, J. H., & Spencer, M. S. (Eds.). (1995). *American Production and Inventory Control Society.* Falls Church, VA: APICS Dictionary.

Digalwar, K., A., R. Tagalpallewar, A., & K. Sunnapwar, V. (. (. (2013). Green manufacturing performance measures: An empirical investigation from Indian manufacturing industries. *Measuring Business Excellence, 17*(4), 59–75. doi:10.1108/MBE-09-2012-0046

Duarte, S., Cabrita, R., & Machado, V. C. (2011, January). Exploring lean and green supply chain performance using balanced scorecard perspective. *Proceedings of the 2011 international conference on industrial engineering and operations management*, 520-525.

Fahimnia, B., Sarkis, J., & Davarzani, H. (2015). Green supply chain management: A review and bibliometric analysis. *International Journal of Production Economics, 162*, 101–114. doi:10.1016/j.ijpe.2015.01.003

Govindan, K., Rajendran, S., Sarkis, J., & Murugesan, P. (2015). Multi criteria decision making approaches for green supplier evaluation and selection: A literature review. *Journal of Cleaner Production, 98*, 66–83. doi:10.1016/j.jclepro.2013.06.046

Guide, V. D. R. Jr, & Srivastava, R. (1997). Repairable inventory theory: Models and applications. *European Journal of Operational Research, 102*(1), 1–20. doi:10.1016/S0377-2217(97)00155-0

Hassini, E., Surti, C., & Searcy, C. (2012). A literature review and a case study of sustainable supply chains with a focus on metrics. *International Journal of Production Economics, 140*(1), 69–82. doi:10.1016/j.ijpe.2012.01.042

Hervani, A. A., Helms, M. M., & Sarkis, J. (2005). Performance measurement for green supply chain management. *Benchmarking: An International Journal, 12*(4), 330–353. doi:10.1108/14635770510609015

Kainuma, Y., & Tawara, N. (2006). A multiple attribute utility theory approach to lean and green supply chain management. *International Journal of Production Economics, 101*(1), 99–108. doi:10.1016/j.ijpe.2005.05.010

Knemeyer, A. M., Ponzurick, T. G., & Logar, C. M. (2002). A qualitative examination of factors affecting reverse logistics systems for end-of-life computers. *International Journal of Physical Distribution & Logistics Management, 32*(6), 455–479. doi:10.1108/09600030210437979

Kongar, E. (2005). Evaluation criteria for environmentally benign forward and reverse supply chain management. *International Journal of Operations and Quantitative Management, 11*(2), 147.

Lakhal, S. Y., H'Mida, S., Béland, M., & Kassogue, C. O. (2004). *Green Supply Chain Parameters for Companies in the Lumber Industry.* Supply Chain Discussion Paper.

Li, Y. (2011). Research on the performance measurement of green supply chain management in China. *Journal of Sustainable Development, 4*(3), 101. doi:10.5539/jsd.v4n3p101

Luthra, S., Garg, D., & Haleem, A. (2014). Green supply chain management: Implementation and performance–a literature review and some issues. *Journal of Advances in Management Research, 11*(1), 20–46. doi:10.1108/JAMR-07-2012-0027

Luthra, S., Garg, D., & Haleem, A. (2016). The impacts of critical success factors for implementing green supply chain management towards sustainability: An empirical investigation of Indian automobile industry. *Journal of Cleaner Production, 121*, 142–158. doi:10.1016/j.jclepro.2016.01.095

Miemczyk, J., Johnsen, T. E., & Macquet, M. (2012). Sustainable purchasing and supply management: A structured literature review of definitions and measures at the dyad, chain and network levels. *Supply Chain Management: An International Journal, 17*(5), 478–496. doi:10.1108/13598541211258564

Ntabe, E. N., LeBel, L., Munson, A. D., & Santa-Eulalia, L. A. (2015). A systematic literature review of the supply chain operations reference (SCOR) model application with special attention to environmental issues. *International Journal of Production Economics, 169*, 310–332. doi:10.1016/j.ijpe.2015.08.008

O'Reilly, J. (2010). *The Green Rule: Measuring Progress.* Retrieved from http://www.inboundlogistics.com

Odeyale, S. O. (2014). Performance appraisal for green/environmental friendliness of a supply chain department. *Journal of Industrial Engineering and Management, 7*(5), 1316. doi:10.3926/jiem.1057

Olasunkanmi, S., Jelili, A., Olubunmi, E., & Odeyale, L. (2014). Evaluation and selection of an effective green supply chain management strategy: A case study. *International Journal of Research Studies in Management, 3*(1), 27–39.

Olugu, E. U., Wong, K. Y., & Shaharoun, A. M. (2011). Development of key performance measures for the automobile green supply chain. *Resources, Conservation and Recycling, 55*(6), 567–579. doi:10.1016/j.resconrec.2010.06.003

Parisi, C., & Maraghini, M. P. (2010). Operationalising sustainability: How small and medium sized enterprises translate social and environmental issues into practice. In Business Performance Measurement and Management, (pp. 131-147). Springer Berlin Heidelberg.

Saridogan, M. (2012). The impact of green supply chain management on transportation cost reduction in Turkey. *International Review of Management and Marketing, 2*(2), 112.

Sarkis, J. (1998). Evaluating environmentally conscious business practices. *European Journal of Operational Research, 107*(1), 159–174. doi:10.1016/S0377-2217(97)00160-4

Sarkis, J. (1999). A methodological framework for evaluating environmentally conscious manufacturing programs. *Computers & Industrial Engineering, 36*(4), 793–810. doi:10.1016/S0360-8352(99)00166-7

Sarkis, J. (2002). A strategic decision framework for green supply chain management. *Journal of Cleaner Production, 11*(4), 397–409. doi:10.1016/S0959-6526(02)00062-8

Sarkis, J., & Weinrach, J. (2001). Using data envelopment analysis to evaluate environmentally conscious waste treatment technology. *Journal of Cleaner Production, 9*(5), 417–427. doi:10.1016/S0959-6526(00)00084-6

Tanwer, A. K., Prajapati, D. R., & Singh, P. J. (2014). Green Supply Chain Management An Environmental Approach For Manufacturing Industry. *Proceedings of National Conference on Advancements and Futuristic Trends in Mechanical Engineering*. Retrieved from http://www.supply-chain.com/info/faq.html

Theyel, G. (2006). Customer and supplier relations for environmental performance. In Greening the Supply Chain, (pp. 139-149). Springer London.

Tramarico, C. L., Salomon, V. A. P., & Marins, F. A. S. (2016). Multi-criteria assessment of the benefits of a supply chain management training considering green issues. *Journal of Cleaner Production.*

Tsai, W. H., & Hung, S. J. (2009). A fuzzy goal programming approach for green supply chain optimisation under activity-based costing and performance evaluation with a value-chain structure. *International Journal of Production Research, 47*(18), 4991–5017. doi:10.1080/00207540801932498

Tuzkaya, G. (2013). An intuitionistic fuzzy Choquet integral operator based methodology for environmental criteria integrated supplier evaluation process. *International Journal of Environmental Science and Technology, 10*(3), 423–432. doi:10.1007/s13762-013-0180-9

Tuzkaya, G., Ozgen, A., Ozgen, D., & Tuzkaya, U. R. (2009). Environmental performance evaluation of suppliers: A hybrid fuzzy multi-criteria decision approach. *International Journal of Environmental Science and Technology*, *6*(3), 477–490. doi:10.1007/BF03326087

Uygun, O., & Dede, A. (2016). Performance evaluation of green supply chain management using integrated fuzzy multi-criteria decision making techniques. *Computers & Industrial Engineering*, *102*, 502–511. doi:10.1016/j.cie.2016.02.020

Wang, Z., Mathiyazhagan, K., Xu, L., & Diabat, A. (2016). A decision making trial and evaluation laboratory approach to analyze the barriers to Green Supply Chain Management adoption in a food packaging company. *Journal of Cleaner Production*, *117*, 19–28. doi:10.1016/j.jclepro.2015.09.142

Wu, K. J., Tseng, M. L., & Vy, T. (2011). Evaluation the drivers of green supply chain management practices in uncertainty. *Procedia: Social and Behavioral Sciences*, *25*, 384–397. doi:10.1016/j.sbspro.2012.02.049

Xu, L., & Peng, J. (2015). A Green Level Measure Model of Enterprise Supply Chain based on Fuzzy Information. *International Journal of u- and e- Service, Science and Technology*, *8*(2).

Zhang, M., Tse, Y. K., Doherty, B., Li, S., & Akhtar, P. (2016). Sustainable supply chain management: Confirmation of a higher-order model. *Resources, Conservation and Recycling*. doi:10.1016/j.resconrec.2016.06.015

Zhu, Q., Geng, Y., Fujita, T., & Hashimoto, S. (2010). Green supply chain management in leading manufacturers: Case studies in Japanese large companies. *Management Research Review*, *33*(4), 380–392. doi:10.1108/01409171011030471

Zhu, Q., & Sarkis, J. (2004). Relationships between operational practices and performance among early adopters of green supply chain management practices in Chinese manufacturing enterprises. *Journal of Operations Management*, *22*(3), 265–289. doi:10.1016/j.jom.2004.01.005

Zhu, Q., Sarkis, J., & Geng, Y. (2005). Green supply chain management in China: Pressures, practices and performance. *International Journal of Operations & Production Management*, *25*(5), 449–468. doi:10.1108/01443570510593148

Zhu, Q., Sarkis, J., & Lai, K. H. (2007a). Green supply chain management: Pressures, practices and performance within the Chinese automobile industry. *Journal of Cleaner Production*, *15*(11), 1041–1052. doi:10.1016/j.jclepro.2006.05.021

Zhu, Q., Sarkis, J., & Lai, K. H. (2007b). Initiatives and outcomes of green supply chain management implementation by Chinese manufacturers. *Journal of Environmental Management*, *85*(1), 179–189. doi:10.1016/j.jenvman.2006.09.003 PMID:17084502

APPENDIX

As mentioned in the text, Sarkis (2002) proposed four logistics stages for the entire supply chain: "*Inbound*", "*In-plant*", "*Outbound*", and "*Reverse*". Table 17 contains all the studies reviewed in this chapter with relevant logistics stages and classification titles.

Table 17. Classification of the studies within all logistics stages

Author	Logistics Stage	Logistics Sub Stage	Method	Business Decision Category	Sector	Country
Guide et al. (1997)	In-plant & Outbound & Reverse	Remanufacturing	Unknown	All		
Sarkis, J. (1998)	All	All	Analytic Network Process (ANP)	Strategic		
Beamon, B. M. (1999)	All	All	N/A	All		
Sarkis, J. (1999)	In-plant	All	Analytic Network Process (ANP) and Data Envelopment Analysis (DEA)	All		
Sarkis et al. (2001)	In-plant	Waste	Data Envelopment Analysis (DEA)	Tactical	Energy	USA
Knemeyer et al. (2002)	Reverse	All	Qualitative Techniques	Tactical	Computer	USA
Lakhal et al. (2004)	All	All	Unknown	Tactical	Lumber	Canada
Zhu et al. (2004)	All	All	Statistics	All		
Hervani et al. (2005)	All	All	Unknown	All		
Kongar, E. (2005)	Reverse	All	A generic Balanced Scorecard	All	Automotive	Turkey
Zhu et al. (2005)	All	All	Survey	All	Many	China
Kainuma et al. (2006)	All	All	Statistics	All		
Zhu et al. (2007a)	All	All	Statistics	All	Automotive	China
Zhu et al. (2007b)	All	All	Analysis of variance (ANOVA)	All	Many	China
Tuzkaya et al. (2009)	Inbound	Supplier / Purchasing	Hybrid Fuzzy Analytic Network Process (ANP), Fuzzy Preference Ranking Organization Method	Tactical	White Goods	Turkey
O'Reilly, J. (2010)	All	All	N/A	All		

continued on following page

Table 17. Continued

Author	Logistics Stage	Logistics Sub Stage	Method	Business Decision Category	Sector	Country
Parisi et al. (2010)	All	All	Unknown	All	Unknown	Italy
Zhu et al. (2010)	All	All	Statistics	All	Many	Japan, China
Casillejo Tinedo et al. (2011)	Reverse	Packaging	Unknown	All	Automotive	Unknown
Duarte et al. (2011)	All	All	Balance Score Card (BSC)	All		
Li, Y. (2011)	All	All	Statistics	All	Many	China
Olugu et al. (2011)	All	All	Unknown	All	Automotive	Unknown
Wu et al. (2011)	All	All	Fuzzy Decision Making Trial and Evaluation Laboratory (DEMATEL)	All		
Brooks et al. (2012)	In-plant	N/A	N/A	Strategic		
Buyukozkan et al. (2012)	Inbound	Supplier / Purchasing	Hybrid Fuzzy Multiple Criteria Decision Making (MCDM), Fuzzy Decision Making Trial, Evaluation Laboratory Model (DEMATEL), Analytic Network Process (ANP), and Technique for Order Performance by Similarity to Ideal Solution (TOPSIS)	Strategic	Automotive	Turkey
Hassini et al. (2012)	All	All	N/A	All	Electricity	Canada
Miemczyk et al. (2012)	Inbound	All	N/A	All		
Saridogan, M. (2012)	All	Transportation	Statistics	Operational		
Digalwar et al. (2013)	In-plant	All	Statistics	All	Manufacturing	India
Tuzkaya, G. (2013)	Inbound	Supplier / Purchasing	Intuitionistic Fuzzy Choquet integral operator	Tactical		
Luthra et al. (2014)	All	All	N/A	All		
Odeyale, S. O. (2014)	Inbound & In-plant & Outbound	All	Analytic Hierarchy Process (AHP) and Weighing Multiplication and Additive Model (WMAAM)	All		
Olasunkanmi et al. (2014)	All	All	Fuzzy Analytic Hierarchy Process (AHP)	All	Cement Mnfg.	Unknown
Tanwer et al. (2014)	All	All	Unknown	All	Manufacturing	Unknown
Fahimnia et al. (2015)	All	All	N/A	All		

continued on following page

Table 17. Continued

Author	Logistics Stage	Logistics Sub Stage	Method	Business Decision Category	Sector	Country
Govindan et al. (2015)	Inbound	Supplier / Purchasing	N/A	All		
Ntabe et al. (2015)	All	All	N/A	All		
Xu et al. (2015)	All	All	Fuzzy Logic	All		
Ahmad et al. (2016)	Inbound & In-plant & Outbound	All	Statistics	All	Energy	Unknown
Luthra et al. (2016)	All	All	Statistics	All	Automotive	India
Schoggl et al. (2016)	Inbound	All	Statistics	All		
Tramarico et al. (2016)	All	All	Analytic Hierarchy Process (AHP)	All		
Uygun et al. (2016)	All	All	Fuzzy Multiple Criteria Decision Making (MCDM)	All		
Wang et al. (2016)	Outbound	Packaging	Decision Making Trial and Evaluation Laboratory (DEMATEL)	Tactical	Packaging	India
Zhang et al. (2016)	All	All	Exploratory Factor Analysis (EFA) and Confirmatory Factor Analysis (CFA)	All	Unknown	China

Chapter 3
Corporate Social Responsibility-Based Supplier Selection Process in Sustainable Supply Chains

Lei Xu
Tianjin University of Technology, China

Xiaoran Shi
Tianjin University of Technology, China

Yu Xie
Tianjin University of Technology, China

Sang-Bing Tsai
University of Electronic Science and Technology of China (Zhongshan Institute), China & Nankai University, China

ABSTRACT

Due to the rising importance of the global interdependence, decisions related to supplier selection rely on not only corporate operation conditions, but also corporate social responsibility (CSR). Under this circumstance, this chapter focuses on a supplier selection problem concerning CSR. Particularly, seven criteria and their corresponding importance for supplier selection are analyzed. Meanwhile, some relevant sub-criteria are included as well. Based on existing literature review and managerial interviews obtained from leaders in the southern region of India, these issues are compared and ranked. In this chapter, Analytical Hierarchy Process (AHP) is conducted to identify the importance of these considered criteria and sub-criteria. Finally, the chapter concludes with a presentation of rankings and priorities of the proposed seven criteria. Given the obtained results and recommendations, it would be helpful to establish a useful method for selecting the best supplier based on CSR.

DOI: 10.4018/978-1-5225-3537-9.ch003

INTRODUCTION

A supply chain is defined as "the management and integration of a set of selected key business processes from end users through original suppliers that provides products, services, and information that add value for customers and other stakeholders through the collaborative efforts of supply chain members" (Schwartz et al., 2008). A company's supply chain, then, more or less complicated, generally starts from the supplier, goes to the manufacturer, and ends with the customer or consumer. From the manufacturer's perspective, which acts as an intermediate link of a supply chain, collection and distribution represent raw or recycled materials arriving from the supplier and finished products moving toward the consumer (or customer), respectively (Choi et al., 2013). Clearly, the purchasing manager of the manufacturer is responsible for conducting an optimal decision of supplier selection, which may help the manufacturer maintain a competitive position (Govindan et al., 2012) and implement an efficient supply chain[1]. Considering its importance, we will study a supplier selection problem in this chapter.

Supplier selection has been widely studied for a long time with the consideration of various criteria (e.g., price, quality, delivery, etc.). With the globalization of trade, firms face with more challenges and pressures from their competitors, consumers or even the entire society. Under this circumstance, a rising concept, known as corporate social responsibility (CSR), gains its popularity; and thereby endorsement of CSR practices from NGOs, stakeholders, and customers becomes compulsory (Welford and Frost, 2006; Michael and Brown, 2006; Abreu et al., 2012). Whereas there is no universally agreed-upon definition for CSR (Wafula, 2012), one useful definition is "operating a business in a manner that accounts for the social and environmental impact created by the business" (Payne, 2009).

To be specific, researchers demonstrate that customers expect social responsibility from firms these customers support and consider CSR as a basic quality of any firm (Basil and Weber, 2006; Xu et al., 2013). If a firm or organization is unprogressive or insensitive to CSR issues, it runs the risk of losing its core customers and consumers (Blake and Mansour, 2012). Therefore, to maximize profit and minimize risk and pollution, CSR issues must be integrated with the goals of a company's supply chain (Aliakbaria and Seifbarghy, 2011). A company's handling of CSR issues serves as a prominent factor in comprising the success and image of the company (Khaledabadi and Magnusson, 2008). Moreover, according to Salam (2007), CSR has the ability to increase sales revenue and market share by improving the customer's perception of a company. Given these concerns, we will provide in this chapter a method for ranking or prioritizing some CSR issues in order to select the best supplier.

BACKGROUND

In this section, a review of the prior studies regarding supplier selection is conducted. First of all, the primitive criteria considered in supplier selection and the ways in which rankings were arranged are explained. Second, multi-criteria decision making (MCDM) methodologies for supplier evaluation utilized in these studies are discussed.

Supplier Selection Criteria

Supply chain management (SCM) concepts include collaboration, the preferable selection of suppliers, and the efficient utilization of these suppliers. Establishing supplier selection criteria is a crucial process in SCM. In the present competitive environment, the effective selection of suppliers is important to the success of any firm (Liu et al., 2000). From the manufacturer's perspective, supplier selection is a type of multi-criterion problem where the solutions must be scrutinized. Some researchers attempt to conduct the empirical analysis to investigate the supplier selection process that is integrated with supplier involvement, customer satisfaction level, and the firm's performance. Amongst these studies, various product-based criteria are often chosen, such as quality, delivery, reliability, etc. (Tracey and Tan, 2001).

Particularly, Dickson (1996) was the first to examine the vendor selection process by identifying 23 basic criteria. Even in this early research, social responsibility, ethical responsibility, and firm importance were included. In the study, product price was considered in the sixth position, followed by product quality and delivery (which was in the first and second rank, respectively). After Dickson's initial research, similar studies were conducted. Most studies demonstrated similar rankings. For example, the position of cost remained the same. In addition, these studies paid greater attention towards factors such as technique and production capacity (Weber et al., 1991). Based on 113 articles, Roodhooft and Konings (1996) altered the ranking of vital criteria and showed that improvement was directly proportional to time and that quality, delivery, price, repair service, and technical capability were ranked lower.

As aforementioned, supplier selection was originally a lengthy process in the past twenty year, where the supplier was chosen based only on the cost, quality, and some other quantitative and qualitative factors (Bhuttta and Huq, 2002). More recently, Cheraghi et al. (2004) conducted a more robust analysis of the supplier selection process. In their study, based on reviewing more than 110 research papers, the authors provided an update on critical success factors (CSFs) for supplier selection, which included both traditional factors (i.e., quality, delivery, price, service, etc.) and non-traditional ones (i.e., just-in-time communication, process improvement, supply chain management, etc.). In a similar vein, Pal and Kumar (2008) proposed a model for supplier evaluation that was integrated with SCM's overall performance index. Nowadays, due to the pressure from NGO, stakeholders and customers, environmental concerns have been included and considered as a priority among a firm's basic selection criteria. Here, environmental concern means taking into account any environmentally harmful substances disseminated by a particular supplier. The manufacturing firm expects the selected supplier to fulfill environmental procedures, such as implementation of waste reduction procedures, installation of pollution control mechanisms, and substitution of hazardous raw materials with environment-friendly ones (Rao, 2005).

MCDM Methodologies in Supplier Selection

Multi-criteria decision making methodologies are the most common ones for dealing with supplier evaluation and selection problems since these methodologies provide the opportunity to replicate the right solution. Many researchers have worked in this theme with such methodologies as analytical hierarchy process (AHP), fuzzy AHP, genetic algorithm (GA), linear programming (LP), technique for order performance by similarity to ideal solution (TOPSIS), and decision-making trial and evaluation laboratory (DEMATEL), amongst many others. Details of these aforementioned methodologies are provided below.

- Before AHP, there were two approaches which were categorical and weighted pointed evaluation solutions to deal with supplier selection. Regarding categorical approach, there is no doubt that the approach plays an essential role in structuring the evaluation process in a systematic way. Nevertheless, this solution is not able to figure out the important relationship between each criterion as well as to make a subjective decision through this system. As for pointed evaluation solution, it does help resolve the disadvantages of the categorical approach. However, a major disadvantage of this approach is that it doesn't have the effective ability to take account of qualitative evaluation criterion.

- AHP, a well-known multi-criterion decision making approach, offers a methodology to rank alternative courses relied on the decision makers' judgements with respect to the importance of the criterion. First of all, the progress of AHP is to determine the relative importance of criterion. Then, this approach focuses on comparing the criterion and measuring the extent of the alternative. At last, based on the results of the comparison, AHP aims to synthesize the results of the comparison to calculate the relative importance of the alternatives. It is universally acknowledged that managerial judgements are exerted to acquire the AHP approach. The reason why alternative weights reflect the relative importance of the criteria in meeting the goal of the hierarchy is that AHP employs a ratio scale for human judgements. Given this superiority, AHP facilitates to handle all the disadvantages of the two approaches mentioned above.
- There is no doubt that AHP is a widely-used multi-criteria decision making methodology. However, the major defect of AHP is that decision makers sometimes want to employ fuzzy judgements rather than the crisp judgements required by AHP. This practical situation triggers the development of fuzzy AHP methods. Among the approaches to resolve fuzzy AHP, there is a major discovery that many applications have employed the extent analysis method to deal with fuzzy judgements due to its computational simplicity. To be more specific, triangular fuzzy numbers are first used to get pairwise comparison scale of fuzzy AHP. Then, the extent analysis method is applied for the sake of the synthetic extent value of pairwise comparison.
- Compared with the methods applied to linear weighting models (i.e., categorical approach, AHP, fuzzy AHP, etc.), linear programing (LP) and genetic algorithm (GA) are approaches used to solve mathematical programming models (MP-models). Through MP-models, decision makers present their decision problem into a mathematical objective function, which aims to maximize profits or minimize costs by determining the optimal ordering quantities from different suppliers. Once the model is explicitly built, decision makers could obtain the optimal solution with the assistance of LP. The objective function contains many decision variables and the constraints are complex, making it impossible to find the optimal solution using LP. Then, users need to rely on heuristic algorithms (i.e., genetic algorithm, simulated annealing algorithm, ant colony algorithm, etc.) to solve the problem. The major advantage of MP-models is that it is more objective than linear weighting models since decision makers are forced to explicitly state their concerns. However, it gives a more quantitative solution with the expense of huge computational efforts.
- By using the TOPSIS method, a more simple and systematic progress can be easily reached when we rank the alternatives. Using the TOPSIS methodology, positive and negative ideal alternatives could be found based on their corresponding attribute values. Then, the best solution should be the one closest to the positive ideal alternative while farthest from the negative ideal alternative. By doing so, we have the capability to avoid pairwise comparisons among the criteria, which are employed by many methods, such as AHP and fuzzy AHP. However, its major disadvantage is similar to AHP as it suffers from the limitation of using crisp judgements when selecting the alternatives. To overcome this limitation, the fuzzy TOPSIS method is developed.

- When talking about the DEMATEL method, it is widely acknowledged that it is a useful approach to analyze the influential relationship among complex criteria. Based on digraphs, the DEMATEL method helps exchange the relationship between cause and effect factors into an intelligible structural model. Meanwhile, by using the DEMATEL method, we can get the most important criteria compared with other criteria. Differing from other MCDM methodologies, the DEMATEL method can assist to reduce the number of criteria which are used for evaluating the factor effectiveness, thereby reducing computational complexity. Furthermore, based on the impact of digraph map of the DEMATEL method, it is evident that companies are capable of enhancing the effectiveness of specific factors.

Particularly, in Ordoobadi's study (2009a), an attempt in the supplier selection process is made with the Taguchi method. Given the same problem, the author tried to solve the problem via the fuzzy logic (Ordoobadi, 2009b). Haq and Govindan (2006) designed an integrated supplier selection and multi-echelon distribution inventory model, and obtained optimal results with the support of the fuzzy approach. Atul et al. (2013) explored a vendor managed inventory problem, and conducted the analysis with the assistance of AHP. Chan et al. (2008) focused on a global supplier selection problem, and established an evaluation process via the fuzzy AHP method. Similarly, supplier selection through fuzzy TOPSIS in a case study from Turkey is elucidated by Eleren and Yilmaz (2011). Vijayvagy (2012) proposed a decision framework for supplier evaluation, in which AHP and TOPSIS were integrated to select the supplier based on 18 criteria. Moreover, interviews were conducted with 50 purchasing managers to justify the obtained theoretical results. With the development of industrial environment, supplier selection is required to employ the up-to-date criteria and methodologies by using the DEMATEL and fuzzy DEMATEL methods (Ehsanifar, 2011; Gharakhani, 2012; Wang et al., 2016). Furthermore, in order to help in taking meaningful decision on supply chain coordination schemes (SCCS) selection, Kumar (2015) combined DEMATEL and Maximum Mean De-Entropy (MMDE) algorithm to put forward a structural model, through with the relationships among several relevant factors for the development of Impact Relationship Maps (IRM) were explored.

Regarding the sustainable supply chain and green supplier selection issues, Che (2010) employed a model for solving the multi-period supplier selection problem with an assembly sequence using a hybrid Guided-Pareto genetic algorithm. Considering CSR issues, Baskaran et al. (2011) emphasized the perspective view of supplier selection and investigated criteria such as discrimination, abuse of human rights, underage labor, long working hours, unfair competition and pollution in their study. In addition, Kumar et al. (2014) exerted a validated model to assist to deal with the supplier selection process and identified the most important issues which underlined the selection of a supplier based on CSR. Last but not least, according to CSR practices, Govindan et al. (2016) proposed a decision-making model to select the best supplier with the consideration of multiple criteria. Meanwhile, key actors including Shareholder, Governments, Customers, and Community, who play a crucial role in the decision-making procedure, are identified.

To the best of our knowledge, while several authors have discussed CSR issues in their studies, generally, only basic issues are mentioned. Little attention has been devoted to minor issues. To this end, this study attempts to fill the gap by considering both major and minor issues of CSR criteria in the supplier selection process. Meanwhile, the priority or ranking of CSR issues for selecting the best supplier is demonstrated in detail.

MAIN FOCUS OF THIS CHAPTER

In supply chains, supplier selection is a main problem faced by purchasing managers because material cost plays a crucial role in the total cost of production. Hence, the selection of an efficient supplier is considered a major contribution to the supply chain. According to the firm's strategy, supplier selection can be conducted based on various combinations of criteria. Due to the importance of CSR issues arose along with globalization, this paper addresses the problem of supplier selection based on CSR criteria. Specifically, basic CSR issues as well as some additional considerations in supplier selection are taken into account. Then, a priority of CSR issues for selecting the best supplier is provided. An interview was conducted with a manager at a rubber manufacturing industry situated in the southern part of India. In addition, a questionnaire, which includes seven criteria with relevant sub-criteria, was provided to the purchasing manager of the firm. Then, a pairwise comparison of the entire set of criteria over another was made given the obtained responses of the questionnaire. By this pairwise comparison, the AHP approach was applied in data analysis. Finally, the results obtained from this analysis were used to prioritize the CSR issues for selecting the best supplier.

Human Rights Issues (M1)

Per Theobald and Arkani (2007), CSR has included a growing preoccupation with the issue of human rights. Due to this concern in CSR, the issues of human rights are considered as a main criterion in this study. Based on the definition of human rights, which is described as "to have a right is to have a claim against someone whose valid recognition is called for by some set of governing rules or moral principles" (Brown, 2012), we further divided it into some sub-criteria, namely, death penalty, minority rights, and one-child policy.

Underage Labor (M2)

In accordance with the World Bank, the worldwide forced labor participation rate of underage children reached 50% in 2004 (Braun, 2006). Meanwhile, underage labor is an even worse prevalent problem found in Asia and Sub-Saharan Africa. To eliminate or mitigate pernicious social phenomenon, a report has been submitted to appeal for the increase of the base age of child labor (Ray, 2004). Considering its importance, this paper exhibits underage labor as important factor in CSR with the relevant sub-criteria, which are conventional forms of slavery, minimum age convention, trafficking of children, and debt bondage, respectively.

Long Working Hours (M3)

Overtime working culture exists in many countries throughout the world. Statistics imply that in recent years, over 20% workers have been beset by working long hours. Furthermore, workers in Asia, Korea, Pakistan, and Thailand, work more than 60 hours per week, accounting for 35%-45% of the entire labor force (Ogura, 2009). Therefore, it is quite natural to select long working hours as a criterion when dealing with CSR issues. In this work, sub-criteria in terms of long working hours include regular work days, the practice of allocating overtime to limited workers, and limit and over time.

Pollution (M4)

Waste means material loss or economic loss. The recycling of wastes and recovery of discarded products result in the reduction of waste. Global warming and climate change become a more and more serious problem and gain worldwide attention these years; while carbon and other greenhouse gas emissions, especially CO2 emissions, are the main stimulants (Beske et al., 2008). Therefore, the reduction of CO2 must be severely managed by firms. To this end, in our study, environmental issues are classified as operating risk, recycling, CO2 emissions, and recovery of discarded products.

Safeguarding Mechanisms (M5)

Safeguarding mechanism is the monitoring system which governs CSR within the firm, acting as the auditing of CSR with internal and external members. Safeguarding mechanisms can be classified as direct sanctions, goal congruence, third party interventions, trust, and reputation effects. In particular, direct sanctions imply the sudden cancelation of commitment to members who have not followed CSR. Goal congruence represents the motivation provided for members to access the path which is common in the firm. Trust and reputation are usually considered in the measurement strategy.

Feminist Labor Issue (M6)

In the beginning of the nineteenth century, gender conflicts arose in developed countries. As an outcome, the position of women in the labor markets has become the subject of substantial and vital research (Honeyman and Goodmand, 1991). From then on, a lot of rules were framed against the feminist labor market. However, to what extent those laws discriminate against women has been debated drastically across the Anglo-American legal world (Jamieson, 2004). By this argument, the feminist labor issue has been considered in our paper and is followed by some sub-criteria namely sexual harassment, maternity leave, rights for pregnant women, and the literacy level of women.

Organizational Legal Responsibilities (M7)

Taxes and customs are the backbone of a nation's economy, and big businesses have an explicit social duty of paying taxes. However, these leading companies frequently claim to practice CSR by considering their environmental efforts. Rare attention has been paid on the payment of taxes (Jimena, 2010). As a consequence of an increasing black money market, a major problem occurs in most of the developing countries. In this criteria, we include paying taxes, paying customs, and the observance of local and global rules about staff, to consider organizational legal responsibilities in the supplier selection process.

Table 1 shows the main criteria and sub-criteria considered in this study. Their supporting literature and additional references are provided as well.

Table 1. Criteria and sub-criteria of CSR

Main Criteria	Sub-Criteria	References
Human rights issue (M1)	(M1.1) Death penalty	Buhmann (2006); Carasco et al. (2008); Kobrin (2009); Puncheva et al. (2010); Kemp et al. (2010); Dawkins (2012);
	(M1.2) Minority rights	
	(M1.3) One child policy	
Under age labor (M2)	(M2.1) Conventional forms of slavery	Hobbs et al. (1986); Bonnet (1993); White (1996); Winstanley et al. (2002); Dessy (2003); Ray (2004); Braun (2006); Lee and Kim (2009)
	(M2.2) Minimum age Convention	
	(M2.3) Trafficking of children	
	(M2.4) Debt bondage	
Long working hours (M3)	(M3.1) Regular rest days	Stamos (1980); Golden (2000); Ogura (2009); Dembe (2009); Golden (2009); Baskaran et al. (2011)
	(M3.2) Practice of allocating overtime to limited workers	
	(M3.3) Limit and overtime	
Pollution (M4)	(M4.1) Operating risks	Beske et al. (2008); Andersen et al. (2009)
	(M4.2) Recycling	
	(M4.3) CO_2 emission	
	(M4.4) Recovery of discarded products	
Safeguarding mechanism in CSB (M5)	(M5.1) Direct sanctions	Pedersen and Andersen (2006)
	(M5.2) Goal congruence	
	(M5.3) Third party interventions	
	(M5.4) Trust	
	(M5.5) Reputation effect	
Feminist labor issue (M6)	(M6.1) Sexual harassment	Honeyman and Goodman (1991); Jamieson (2004); Neumayer et al. (2007); Pearson (2007); Prieto-Carron (2008);
	(M6.2) Maternity leave	
	(M6.3) Rights for pregnant women	
	(M6.4) Literacy levels of women	
Organizational legal responsibilities (M7)	(M7.1) Paying tax	Jimena (2010); Aliakbaria et al. (2011)
	(M7.2) Paying Custom	
	(M7.3) Observance of rules about staff	

SOLUTIONS AND RECOMMENDATIONS

Solution Methodology

AHP is a multi-criterion decision making tool, which deals with complex engineering problems. The common approach is to combine different themes into a single theme with the assistance of a weighing scheme (Jiang et al., 2011; Luthra et al., 2016). Corporations have to make decisions when they recognize the complexity and importance of their issues. With the assistance of AHP, decision makers are able to structure these complex problems into simple hierarchies (Nilay et al., 2011). Meanwhile, AHP can help them handle both tangible and intangible factors (Haq & Govindan, 2006). Considering the superiority of AHP, this paper analyzed 12 criteria with sub-criteria and made pairwise comparisons among them

Figure 1. Proposed framework of this study

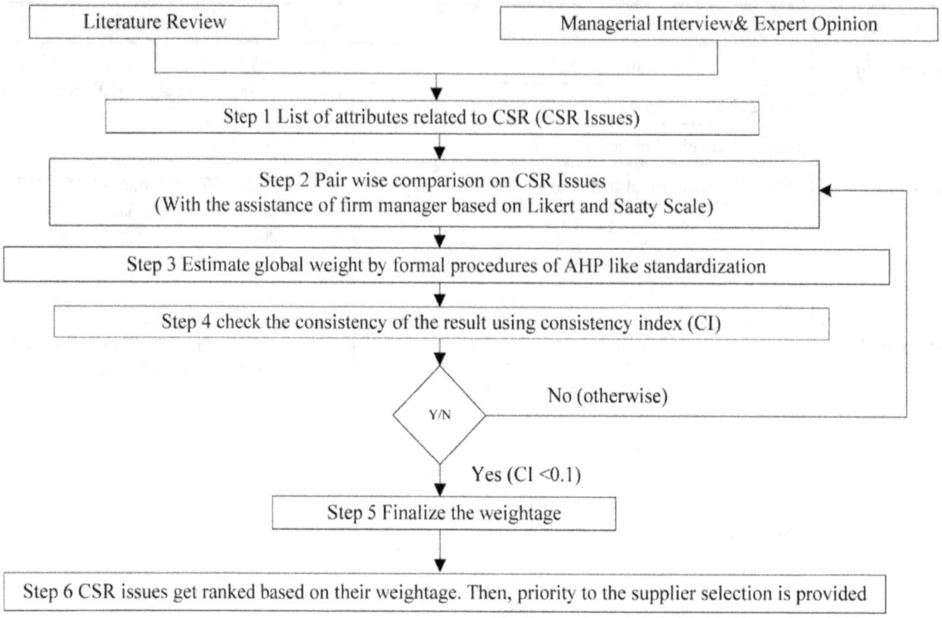

using the AHP methodology. Figure 1 presents the proposed framework for the problem undertaken and also step by step procedures for solving the problem.

Here, the pairwise comparison is conducted on the basis of a Saaty scale (detailed information could be found in Table 2). Saaty scale is the scale which defines the relative importance between one over other criterion (Saaty, 2008). For instance, if criterion A and criterion B are considered to have a pairwise comparison according to Saaty, and if A is nine times more important than B, we denote it as 9; while if B is nine times more important than A, we denote it as 1/9. With the assistance of Saaty scale, each criterion and their relevant sub-criteria were analyzed. Finally, with the application of the AHP tool, the relative weight of each criterion and sub-criterion was calculated.

RESULTS AND DISCUSSION

Table 3 shows the pairwise comparison of the CSR criteria based on managerial interviews. According to the obtained results, organizational legal responsibility takes the first position due to its greatest weight, followed by pollution and human rights issues, respectively. This result can be substantiated by practical situations. First of all, organizational legal responsibility is the most crucial issue for any firm in reality. It has to be addressed as a priori to ensure the legal operation of the firm. Secondly, environmental campaigns take place throughout the world and encourage firms to practice sustainability measures such as reuse, recycle, reduction of emissions, etc. Last but not least, no firm can afford even the perception of profits built on the abuse of human rights.

The fourth and fifth positions are taken by underage issues and long working hours. The underage and long working hour issues have become very strict in recent years and they are severely scrutinized

Table 2. The description of absolute Saaty-scale numbers

Intensity of Importance	Definition	Explanation
1	Equal importance	Two activates contribute equally to the objective
2	Weak or slight	
3	Moderate importance	Experience and judgment slightly favor one activity over another
4	Moderate plus	
5	Strong importance	Experience and judgment strongly favor one activity over another
6	Strong plus	
7	Very strong or demonstrated importance	An activity is favored very strongly over another; its dominance demonstrated in practice
8	Very, very strong	
9	Extreme importance	The evidence favoring one activity over another is of the highest possible order of affirmation
Reciprocals of above	If activity i has one of the above non-zero numbers assigned to it when compared with activity j, j has the reciprocal value when compared with i	A reasonable assumption
1.1–1.9	If activities are very close	Maybe difficult to assign the best value but when compared with other contrasting activities the size of the small numbers would not be too noticeable, yet they can still indicate the relative importance of the activities

by many organizations. The sixth and seventh issues are safeguarding mechanisms and feminist labor issues, respectively.

In Table 4, relative weights of all sub-criteria are presented. Then, their corresponding rankings are demonstrated. In particular, the sub-criteria of organizational legal issues have the relative weights as follows: $M7.2 > M7.1 > M7.3$. In this context, paying customs has the greatest weight, followed by paying tax and legal rules about staff. Regarding pollution criteria, the relative weights are shown as $M3.3 > M3.1 > M3.4 > M3.2$. In accordance of this rank, CO_2 emissions takes the first position among all criteria on pollution issues due to the risk factor, which is the main accountability issue in the sup-

Table 3. Main criteria priority chart with relative weight

	M1	M2	M3	M4	M5	M6	M7	Relative Weights	Rank
M1	1	2	3	0.5	4	5	0.3333	0.158655	3
M2	0.5	1	2	0.3333	3	4	0.25	0.103625	4
M3	0.3333	0.5	1	0.25	2	3	0.2	0.0675646	5
M4	2	3	4	1	5	6	0.5	0.239928	2
M5	0.25	0.3333	0.5	0.2	1	2	0.1667	0.0447693	6
M6	0.2	0.25	0.3333	0.1667	0.5	1	0.1429	0.0311746	7
M7	3	4	5	2	6	7	1	0.354284	1

Table 4. Relative and global weight calculations for CSR issues

Main Criteria	Relative Weights	Sub-Criteria	Relative Weights	Global Weights
Human Rights Issues (M1)	0.158655	(M1.1)	0.539615	0.077088
		(M1.2)	0.296961	0.042423
		(M1.3)	0.163424	0.023346
Under age labor (M2)	0.103625	(M2.1)	0.095435	0.013634
		(M2.2)	0.467296	0.066757
		(M2.3)	0.160088	0.02287
		(M2.4)	0.277181	0.039597
Long working hours (M3)	0.0675646	(M3.1)	0.539615	0.077088
		(M3.2)	0.163424	0.023346
		(M3.3)	0.296961	0.042423
Pollution (M4)	0.239928	(M4.1)	0.277181	0.039597
		(M4.2)	0.095435	0.013634
		(M4.3)	0.467296	0.066757
		(M4.4)	0.160088	0.02287
Safeguarding mechanisms in CSR (M5)	0.0447693	(M5.1)	0.418537	0.059791
		(M5.2)	0.0972509	0.013893
		(M5.3)	0.262522	0.037503
		(M5.4)	0.0617652	0.008824
		(M5.5)	0.159925	0.022846
Feminist labor issues (M6)	0.0311746	(M6.1)	0.467294	0.066756
		(M6.2)	0.160087	0.02287
		(M6.3)	0.0954321	0.013633
		(M6.4)	0.277186	0.039598
Organizational Legal responsibilities (M7)	0.354284	(M7.1)	0.296961	0.042423
		(M7.2)	0.539615	0.077088
		(M7.3)	0.163424	0.023346

plier selection theory. Moreover, sub-criteria of human rights issues are ranked as $\mathrm{M}1.1 > \mathrm{M}1.2 > \mathrm{M}1.3$, which implies that death penalty is the most valuable criterion in supplier selection. The priority order of underage labor sub-criteria is provided as $\mathrm{M}2.2 > \mathrm{M}2.3 > \mathrm{M}2.4 > \mathrm{M}2.1$. This indicates that minimum age convention has higher weightage among all other sub-criteria in this category. For the sub-criteria of long working hours, regular rest days has the highest relative weight over other sub-criteria. Amongst the five sub-criteria of the safeguarding mechanisms category, the order is presented as $\mathrm{M}5.1 > \mathrm{M}5.3 > \mathrm{M}5.5 > \mathrm{M}5.2 > \mathrm{M}5.4$. Last but not least, regarding feminist labor issues, the order of the sub-criteria is given as $M6.1 > M6.4 > M6.2 > M6.3$, and there is no surprise that sexual harassment occupies the first place among the issues. Given these results, it can be straightforward to point out the best sub-criteria and criteria for supplier selection when considering CSR issues.

FUTHER RESEARCH DIRECTIONS

For multinational companies, selecting a supplier selection based on CSR issues is quite prevalent. Conversely, due to the lack of knowledge, inefficiencies of their purchasing manager, amongst many other reasons, small-scale firms and small and medium-scale enterprises (SMEs) do not select their suppliers with the consideration of CSR. The results obtained from this research may provide guidance and theoretical support to select the supplier on a CSR basis for SMEs, which may ease the work of purchasing managers and help them better coordinate with their stakeholders. Meanwhile, in future analyses, this problem can be extended as a supplier evaluation problem through AHP or other MCDM tools that integrate real data obtained from the industry.

CONCLUSION

The paper has examined CSR issues in the supplier selection process by focusing on the priority of CSR to choose the best supplier. Individual weightage of CSR issues has been identified to provide a foundation for firms to meet their legislative commitments with NGOs, stakeholders, and customers. To be specific, this study has considered seven CSR criteria with their relevant sub-criteria, which are human rights issues, underage labor, long working hours, pollution, safeguarding mechanisms in CSR, feminist labor issues, and organizational legal responsibilities, respectively. By conducting a detailed analysis with the assistance of the AHP methodology, the relative weights of all seven criteria and their corresponding sub-criteria are obtained. The decreasing order of the weights of the seven criteria is summarized as M7>M4>M1>M2>M3>M5>M6. The results present a highly coincidence with the existing literature. For any firms, organizational legal responsibility is of the highest priority because firms need their partners to be trustworthy actors. Following this legal issue, pollution and human rights issues are captured at the second and third ranks, respectively. These three criteria are recommended by the stakeholders and other driving factors of CSR. By means of the priority of CSR issues, supplier ranking can be categorized accordingly.

This research derives some interesting managerial implications: (1) different tools can be effectively adopted by firms to monitor their suppliers in terms of CSR (like CSR safeguarding, training programs, etc.); (2) practicing this type of supplier selection strengthens the possibility of a good relationship with NGOs, stakeholders, and customers; and (3) purchasing managers may still adopt future environmental commitment when they select a supplier even while maintaining a focus on CSR issues. This research also exists some limitations. Particularly, selecting suppliers based on CSR issues may be too difficult in SMEs because their suppliers are not able to satisfy all those CSR conditions due to the lack of knowledge and the high cost of initializing CSR activities. Meanwhile, buyers are not able to monitor all CSR activities (except the main issues) of their suppliers when making the selection decision because of lead times, the cost for surveying suppliers, amongst many others. This limitation does not, however, negate the results of this study.

ACKNOWLEDGMENT

This research was supported in part by National Natural Science Foundation of China (Grant No. 71302005, 71672125), major project of National Natural Science Foundation of China (Grant No. 91646117).

REFERENCES

Abreu, M. C. S. D., Castro, F. D., Soares, F. D. A., & Silva Filho, J. C. L. (2012). A comparative understanding of corporate social responsibility of textile firms in Brazil and China. *Journal of Cleaner Production*, *20*(1), 119–126. doi:10.1016/j.jclepro.2011.08.010

Aliakbaria, A., & Seifbarghy, M. (2011). A supplier selection model for social responsible supply chain. *Journal of Optimization in Industrial Engineering*, *8*, 41–53.

Andersen, M., & Skjoett-Larsen, T. (2009). Corporate social responsibility in global supply chains. *Supply Chain Management*, *14*(2), 75–86. doi:10.1108/13598540910941948

Arkani, S., & Theobald, R. (2007). Building civil society: Is there a role for the corporate sector? *Journal of Corporate Citizenship*, 37–50.

Atul, B. B., Govindan, K., & Satish, V. B. (2013). Analytical hierarchy process-based framework for VMI adoption. *International Journal of Production Research*, *51*(4), 963–978. doi:10.1080/00207543.2011.650795

Basil, D. Z., & Weber, D. (2006). Values motivation and concern for appearances: The effect of personality traits on responses to corporate social responsibility. *International Journal of Nonprofit and Voluntary Sector Marketing*, *11*(11), 61–72. doi:10.1002/nvsm.38

Baskaran, V., Nachiappan, S., & Rahman, S. (2011). Supplier assessment based on corporate social responsibility criteria in Indian automotive and textile industry sectors. *International Journal of Sustainable Engineering*, *4*(4), 359–369. doi:10.1080/19397038.2011.579360

Beske, P., Koplin, J., & Seuring, S. (2008). The use of environmental and social standards by German first-tier suppliers of the Volkswagen AG. *Corporate Social Responsibility and Environmental Management*, *15*(2), 63–75. doi:10.1002/csr.136

Bhutta, K. S., & Huq, F. (2002). Supplier selection problem: A comparison of the total cost of ownership and analytic hierarchy process approaches. *Supply Chain Management*, *7*(3), 126–135. doi:10.1108/13598540210436586

Blake, P., & El Mansour, B. (2012). *Exploring Corporate Social Responsibility Leadership*. Retrieved from http://www.ufhrd.co.uk/wordpress/wp-content/uploads/2012/11/UFHRD2012Sust2.pdf

Braun, S. (2006). Core labor standards and FDI: Friends or foes? the case of child labor. *Review of World Economics*, *142*(4), 765–791. doi:10.1007/s10290-006-0092-6

Buhmann, K. (2006). Corporate social responsibility in China: Current issues and their relevance for implementation of law. *Copenhagen Journal of Asian Studies*, *22*(1), 62–91.

Carasco, E. F., & Singh, J. B. (2008). Human rights in global business ethics codes. *Business and Society Review*, *113*(3), 347–374. doi:10.1111/j.1467-8594.2008.00324.x

Chan, F. T. S., Kumar, N., Tiwari, M. K., Lau, H. C. W., & Choy, K. L. (2008). Global supplier selection: A fuzzy-AHP approach. *International Journal of Production Research*, *46*(14), 3825–3857. doi:10.1080/00207540600787200

Che, Z. H. (2010). A genetic algorithm-based model for solving multi-period supplier selection problem with assembly sequence. *International Journal of Production Research*, *48*(15), 4355–4377. doi:10.1080/00207540903049399

Cheraghi, S. H., Dadashzadeh, M., & Subramanian, M. (2011). Critical success factors for supplier selection: An update. *Journal of Applied Business Research*, *20*(2), 91–108. doi:10.19030/jabr.v20i2.2209

Choi, T. M., Li, Y., & Xu, L. (2013). Channel leadership, performance and coordination in closed loop supply chains. *International Journal of Production Economics*, *146*(1), 371–380. doi:10.1016/j.ijpe.2013.08.002

Dawkins, C. E. (2012). Labored relations: Corporate citizenship, labor unions, and freedom of association. *Business Ethics Quarterly*, *22*(3), 473–500. doi:10.5840/beq201222331

Dembe, A. E. (2009). Ethical issues relating to the health effects of long working hours. *Journal of Business Ethics*, *84*(2), 195–208. doi:10.1007/s10551-008-9700-9

Dessy, S. E., & Vencatachellum, D. (2003). Explaining cross-country differences in policy response to child labor. *The Canadian Journal of Economics. Revue Canadienne dEconomique*, *36*(1), 1–20. doi:10.1111/1540-5982.00001

Dickson, G. W. (1996). An analysis of vendor selection systems and decision. *Journal of Purchasing*, *2*(15), 1377–1382.

Ehsanifar, A. (2013). Applying fuzzy DEMATEL method to analyze supplier selection criteria (case study: WagonPars Company). *International Research Journal of Finance & Economics*, *115*, 76–86.

Eleren, A., & Yilmaz, C. (2011). Selection of suppliers by fuzzy TOPSIS model: Sample study from Turkey. *International Journal of Business & Social Science*, *2*(22), 189–200.

Gharakhani, D. (2012). The evaluation of supplier selection criteria by fuzzy DEMATEL method. *Journal of Applied Sciences Research*, *2*(4), 3215–3224.

Golden, L. (2009). A brief history of long work time and the contemporary sources of overwork. *Journal of Business Ethics*, *84*(2), 217–227. doi:10.1007/s10551-008-9698-z

Golden, L., & Figart, D. (2000). Doing something about long hours. *Challenge*, *43*(6), 15–37. doi:10.1080/05775132.2000.11472177

Govindan, K., Khodaverdi, R., & Jafarian, A. (2013). A fuzzy multi criteria approach for measuring sustainability performance of a supplier based on triple bottom line approach. *Journal of Cleaner Production*, *47*(47), 345–354. doi:10.1016/j.jclepro.2012.04.014

Govindan, K., Shankar, M., & Kannan, D. (2016). (in press). Supplier selection based on corporate social responsibility practices. *International Journal of Production Economics*. doi:10.1016/j.ijpe.2016.09.003

Haq, A. N., & Govindan, K. (2006). Design of an integrated supplier selection and multi-echelon distribution inventory model in a built-to-order supply chain environment. *International Journal of Production Research*, *44*(10), 1963–1985. doi:10.1080/00207540500381427

Hobbs, S., & Cornwell, D. (1986). Child labor: An underdeveloped topic in psychology. *International Journal of Psychology*, *21*(1-4), 225–234. doi:10.1080/00207598608247587

Honeyman, K., & Goodman, J. (1991). Womens work, gender conflict, and labor markets in Europe, 15001900. *The Economic History Review*, *44*(4), 608–628. doi:10.2307/2597804

Jamieson, S. (2004). Feminist theory, globalization and comparative labor law: Women workers in Australia and Ireland. *International Journal of Human Resource Management*, *15*(3), 459–465. doi:10.1080/0958519042000181197

Jian Liu, J., Fong-Yuen Ding, F. Y., Vinod Lall, V., Jian, L., Fong-Yuen, D., & Vinod, L. (2000). Using data envelopment analysis to compare suppliers for supplier selection and performance improvement. *Supply Chain Management: An International JournalSupply Chain Management*, *5*(3), 143–150. doi:10.1108/13598540010338893

Jiang, Z., Zhang, H., & Sutherland, J. W. (2011). Development of multi-criteria decision making model for remanufacturing technology portfolio selection. *Journal of Cleaner Production*, *19*(17-18), 1939–1945. doi:10.1016/j.jclepro.2011.07.010

Jimena, J. (2010). Tax is just one more corporate social issue. *Canadian Mining Journal*, *131*(3), 9–9.

Kemp, D., Bond, C. J., Franks, D. M., & Cote, C. (2010). Mining, water and human rights: Making the connection. *Journal of Cleaner Production*, *18*(15), 1553–1562. doi:10.1016/j.jclepro.2010.06.008

Khaledabadi, H. J., & Magnusson, T. (2008). Corporate social responsibility and knowledge management implications in sustainable vehicle innovation and development. *Communications of the Ibima*, *6*(2), 8–14.

Kim, J., Lindgreen, A., & Lee, K. (2009). Current status of CSR in the realm of supply management: The case of the Korean electronics industry. *Supply Chain Management*, *14*(2), 335–340.

Kobrin, S. J. (2009). Private political authority and public responsibility: Transnational politics, transnational firms, and human rights. *Business Ethics Quarterly*, *19*(3), 349–374. doi:10.5840/beq200919321

Kumar, D. T., Palaniappan, M., Kannan, D., & Shankar, K. M. (2014). Analyzing the CSR issues behind the supplier selection process using ISM approach. *Resources, Conservation and Recycling*, *92*, 268–278. doi:10.1016/j.resconrec.2014.02.005

Kumar, P. (2015). Application of DEMATEL and MMDE for analyzing key influencing factors relevant to selection of supply chain coordination schemes. *International Journal of Information Systems and Supply Chain Management*, *8*(2), 51–72.

Luthra S., Mangla S.K., Xu L., & Diabat A. (2016). Using AHP to evaluate barriers in adopting sustainable consumption and production initiatives in a supply chain. *International Journal of Production Economics*, *181*(Part B), 342-349.

Maloni, M. J., & Brown, M. E. (2006). Corporate social responsibility in the supply chain: An application in the food industry. *Journal of Business Ethics*, *68*(1), 35–52. doi:10.1007/s10551-006-9038-0

Neumayer, E., & Soysa, I. D. (2007). Globalisation, womens economic rights and forced labor. *World Economy*, *30*(10), 1510–1535. doi:10.1111/j.1467-9701.2007.01060.x

Ogura, K. (2009). Long Working hours in Japan: An international comparison and research topics. *Journal of Economics*, *36*(2), 23–45.

Ordoobadi, S. (2009a). Application of Taguchi loss functions for supplier selection. *Supply Chain Management*, *14*(1), 22–30. doi:10.1108/13598540910927278

Ordoobadi, S. (2009b). Development of a supplier selection model using fuzzy logic. *Supply Chain Management*, *14*(4), 314–327. doi:10.1108/13598540910970144

Pal, P., & Kumar, B. (2008). 16T: toward a dynamic vendor evaluation model in integrated SCM processes. *Supply Chain Management*, *13*(6), 391–397. doi:10.1108/13598540810905642

Payne, A. (2009). Corporate social responsibility and sustainable development. *Journal of Public Affairs*, *6*(3-4), 286–297. doi:10.1002/pa.230

Pearson, R. (2007). Beyond women workers: Gendering CSR. *Third World Quarterly*, *28*(28), 731–749. doi:10.1080/01436590701336622

Pedersen, E. R., & Andersen, M. (2006). Safeguarding corporate social responsibility (CSR) in global supply chains: How codes of conduct are managed in buyer-supplier relationships. *Journal of Public Affairs*, *6*(3-4), 228–240. doi:10.1002/pa.232

Preuss, L., & Brown, D. (2012). Business policies on human rights: An analysis of their content and Prieto-Carrón M. (2008). Women workers, industrialization, global supply chains and corporate codes of conduct. *Journal of Business Ethics*, *83*(1), 5–17.

Puncheva-Michelotti, P., Michelotti, M., & Gahan, P. (2010). The relationship between individuals recognition of human rights and responses to socially responsible companies: Evidence from Russia and Bulgaria. *Journal of Business Ethics*, *93*(4), 583–605. doi:10.1007/s10551-009-0242-6

Rao, P. (2005). The greening of suppliers—in the South East Asian context. *Journal of Cleaner Production*, *13*(9), 935–945. doi:10.1016/j.jclepro.2004.04.010

Ray, R. (2004). Child labor: A survey of selected Asian countries. *Asian-Pacific Economic Literature*, *18*(2), 1–18. doi:10.1111/j.1467-8411.2004.00148.x

Roodhooft, F., & Konings, J. (1997). Vendor selection and evaluation an activity based costing approach. *European Journal of Operational Research*, *96*(1), 97–102. doi:10.1016/0377-2217(95)00383-5

Salam, M. A. (2007). Social responsibility in purchasing: The case of Thailand. *International Journal of Procurement Management*, *1*(1), 97–116. doi:10.1504/IJPM.2007.015357

Schwartz, K., Tapper, R., & Font, X. (2008). A sustainable supply chain management framework for tour operators. *Journal of Sustainable Tourism*, *16*(3), 298–314. doi:10.1080/09669580802154108

Stamas, G. D. (1980). Percent working long hours shows first post-recession decline. *Monthly Labor Review*, *103*(5), 39–42.

Tracey, M., & Tan, C. L. (2001). Empirical analysis of supplier selection and involvement, customer satisfaction, and firm performance. *Supply Chain Management*, *6*(4), 174–188. doi:10.1108/EUM0000000005709

Vijayvagy, L. (2012). Decision framework for supplier selection through multi criteria evaluation models in supply chain. *International Journal of Management & Innovation*, *4*(2), 16.

Wang, Z., Mathiyazhagan, K., Xu, L., & Diabat, A. (2016). A decision making trial and evaluation laboratory approach to analyze the barriers to Green Supply Chain Management adoption in a food packaging company. *Journal of Cleaner Production*, *117*, 19–28. doi:10.1016/j.jclepro.2015.09.142

Weber, C. A., Current, J. R., & Benton, W. C. (1991). Vendor selection criteria and methods. *European Journal of Operational Research*, *50*(1), 2–18. doi:10.1016/0377-2217(91)90033-R

Welford, R., & Frost, S. (2006). Corporate social responsibility in Asian supply chains. *Corporate Social Responsibility and Environmental Management*, *13*(3), 166–176. doi:10.1002/csr.121

White, B. (1996). Globalization and the child labor problem. *Journal of International Development*, *8*(6), 829–839. doi:10.1002/(SICI)1099-1328(199611)8:6<829::AID-JID404>3.0.CO;2-Z

Winstanley, D., Clark, J., & Leeson, H. (2002). Approaches to child labor in the supply chain. *Business Ethics (Oxford, England)*, *11*(3), 210–223. doi:10.1111/1467-8608.00279

Xu, L., Kumar, D. T., Shankar, K. M., Kannan, D., & Chen, G. (2013). Analyzing criteria and sub-criteria for the corporate social responsibility-based supplier selection process using AHP. *International Journal of Advanced Manufacturing Technology*, *68*(1), 907–916. doi:10.1007/s00170-013-4952-7

Yucenur, G. N., Vayvay, O., & Demirel, N. C. (2011). Supplier selection problem in global supply chain by AHP and ANP approaches under fuzzy environment. *International Journal of Advanced Manufacturing Technology*, *56*(5-8), 823–833. doi:10.1007/s00170-011-3220-y

KEY TERMS AND DEFINITIONS

Analytical Hierarchy Process: A multi-criteria decision making approach that offers a methodology to rank the alternative courses relied on the decision makers' judgement with respect to the importance of the criterion.

Corporate Social Responsibility: Companies self-regulates, monitors, and ensures its business activities to comply with the spirit of the law and ethical standards in terms of their economic, social and environmental impacts.

Green Supplier Selection: Select the supplier based not only on the cost, quality, and some other conventional factors, but also on the environmental concerns.

Multi-Criteria Decision Making Methodology: The methodology assists decision makers to explicitly evaluate multiple conflicting criteria in structuring and solving complex decision-making and planning problems.

Organizational Legal Responsibilities: The organization's legal compliance with the spirit of the law, and national/international norms.

Supply Chain Management: The management and integration of the flow of products, resources (including cash, raw materials, human resources, information, etc.) and services from the point of origin (i.e., raw material suppliers) to the point of consumption (i.e., consumers) through the collaborative efforts of supply chain members. It involves a set of key business activities, including but not limited to the acquisition of the raw materials, preferable selection of suppliers, manufacturing, management of work-in-process inventory and finished goods inventory, transportation, marketing and selling, amongst many others.

Sustainable Supply Chain: The management of social, economic, environmental impacts and following effective governance practices through the entire supply chain. Companies implemented sustainability criteria into their corporate supply chain seek for long-term value creation rather than short-time economic benefits maximization.

ENDNOTE

1 Inefficiencies in a typical supply chain result in up to 25% of a company's operating costs being wasted. In addition, a reduction of 5% wastage throughout the supply chain will double a typical company's profit margin (Dennis et al., 2009).

Chapter 4
Cleaner Production in the Brazilian Sucroenergy Sector

Jéssica Patricia Corrêa Brunhara
University Center Toledo Araçatuba (UNITOLEDO), Brazil

Rosana Pereira Corrêa
University Center Toledo Araçatuba (UNITOLEDO), Brazil

Sergio Ricardo Mazini
University Center Toledo Araçatuba (UNITOLEDO), Brazil

ABSTRACT

This chapter presents a discussion about the real role of companies as social and environmental agents, which is increasingly gaining momentum and timeliness. With industrial expansion and increasing pollution, it has become imperative that companies assume not only the role of producers of goods and services, but also those responsible for the implementation of environmental management systems and their instruments. The Cleaner Production is a process in the production process that helps in the environmental preservation, since it establishes the following order of priority for waste management: elimination - reduction - reuse - treatment - final disposal. In the sugar-energy sector, considered as one of the greatest precursors of environmental degradation by deforestation and burning, Cleaner Production is fundamental for the rational use of natural resources and for minimizing the environmental impacts caused by productive operations.

INTRODUCTION

The environment factor is increasingly being inserted into business strategies. The negative impacts caused by deforestation and pollution, mainly from agricultural and industrial production, are visible. The inclusion of environmental protection among the organization's strategic objectives extends substantially the whole concept of management.

DOI: 10.4018/978-1-5225-3537-9.ch004

The environmental issue is increasingly becoming a mandatory issue on the agendas of the organizations' executives. The internationalization of business and environmental quality standards, along with changes in consumer behavior, require companies to take a responsible position on the impacts of their production processes (DONAIRE, 1999).

Due to the pressure of the clients, of the globalized and highly competitive market, the demands for environmental management processes in organizations grow, which is reaching even the micro and small companies that supply national and multinational corporations (SEIFFERT, 2011).

Cleaner Production is a preventive environmental strategy applied in production processes, products and services to minimize the impact on the environment. This production model has been developed by UNEP and the United Nations Industrial Development Organization (UNIDO) since the 1980s (BARBIERI, 2007).

In the sugar-energy sector Cleaner Production is considered as an environmental protection, with greater attention to strategies and practices focused on the rational use of natural resources and minimizing the environmental impacts caused by their productive operations.

After the bibliographic research on environmental management, cleaner production and the sugar-energy sector in Brazil, the following questions were raised:

- What are waste generated by the sugar-energy sector and its environmental impacts?
- What are the cleaner production practices adopted by the Brazilian sugar and ethanol industry?

This chapter presents an analysis of the environmental impacts caused by waste from the productive activities of the Brazilian sugar and ethanol industry and Cleaner Production in the sector. The survey of the plants installed in Brazil was carried out by means of a survey on the websites of the entities and bodies of the sector, such as UNICA - Union of the Sugarcane Industry, UDOP - Union of Producers of Bioenergy, CONAB - Companhia Nacional de Abastecimento, CETESB - Environmental Company of the State of São Paulo.

The collection of data and information was done through searches on the websites of the groups of companies in the sector and in their sustainability reports published on the internet.

BACKGROUND

Environmental Management

The new consumption demands generated by population growth pressure the increase in the production of goods and services, promoting changes and environmental impacts resulting from residues arising from economic activities and human consumption. On the other hand, climate change, a decrease in the ozone layer, loss of biodiversity, contamination of soil and air, show that the planet is no longer able to absorb human-generated pollution. This situation is compromising the future of the environment, humanity and all living beings, and may affect future generations.

The expansion of the collective consciousness, with respect to the environment and the complexity of the current social and environmental demands that the community passes on to the organizations, induces a new positioning on the part of the entrepreneurs and executives in face of such questions (TACHIZAWA, 2011)

The economic development of a country and a society is, among other factors, related to the productive capacity and wealth generation of its companies, and these, in turn, have their consolidation, expansion and development increasingly linked to the issues of Social and environmental responsibility.

Corporate environmental management is a relatively new topic and is directly related to the capacity of companies to interact with the environment in which they operate, establishing relationships of equity, respect and responsibility with society, government and markets. Companies and the market are undergoing a process of repositioning and raising awareness about the importance of promoting systematic changes in production processes that minimize the damages generated to the environment and society (BARBIERI, 2007). Environmental damage generated by companies can be understood from three different approaches (Table 1).

- **Pollution Control:** This approach is characterized by the adoption of practices to prevent the effects of pollution generated by a given production process. It is usually accomplished through localized actions with little articulation between them and almost always result from a reactive stance of the company in the face of the problem and the pressures exerted by the society's command organs. Usually, the forms of control do not alter the productive process and the technologies employed are remediation, which seek to solve a problem that has already occurred;
- **Pollution Prevention:** This type of business approach to the preservation and care of the environment aims primarily at preventing the generation of pollution through actions that improve the production process, and encompasses changes in the production process or product and aiming at Reduce or eliminate tailings, before reaching the environment;
- **Incorporating These Issues Into Business Strategy:** This type of environmental business approach addresses environmental issues as a strategic issue, where pollution control and prevention practices are treated as market opportunities or as ways to minimize future environmental problems.

Table 1. The approaches of corporate environmental management (Barbieri, 2007)

Characteristics	Approaches		
	POLLUTION CONTROL	POLLUTION PREVENTION	STRATEGY
Basic Concern	Compliance with legislation and responses to community pressures.	Efficient use of inputs	Competitiveness
Typical Posture	Reactive	Reactive and proactive	Reactive and proactive
Typical actions	Corrective. Use of end-of-pipe remediation and control technologies. Application of safety standards.	Corrective and preventive Conservation and replacement of inputs. Use of technologies.	Corrective, preventive and anticipatory. Anticipating problems and capturing opportunities using medium and long term solutions. Use clean technologies.
Perception of employees	Additional charge	Reduce cost and increase productivity	Competitive advantages
Involvement of top management	Sporadic	Newspaper	Permanent and systematic
Involved areas	Environmental actions confined to pollution-generating areas	Growing involvement of other areas such as production, purchasing, product development and marketing	Environmental activities made by the organization. Expansion of environmental actions for the entire production chain

Currently, the concept of management has been used to include, in addition to the public management of the environment, the action programs developed by companies and non-governmental institutions to manage their activities within the modern principles of environmental protection (PHILIPPI JR, 2014).

There is also an increasing organizational concern to implement environmental management instruments in its production processes, to link its image to social and environmental responsibility actions, with the objective of gaining recognition and acceptance of society in order to increase its competitiveness and consolidate itself in the market.

Cleaner Production

Cleaner Production is a strategy of reducing (or eliminating) waste (or pollutants) in the generating source by developing actions that promote the reduction of waste, the conservation of natural resources, the reduction or elimination of toxic substances (present in Raw materials or ancillary products), reducing the amount of waste generated by processes and products, and consequently reducing pollutants released into air, soil and water (SEIFFERT, 2011).

The Cleaner Production model consists of 3 levels with the purpose of minimizing or reusing waste and emissions (Figure 1). Level 1 has a higher priority, involves changes in products and processes to reduce emissions and waste at source, as well as eliminate or reduce their toxicity. Tier 2 indicates that emissions and waste that are still being generated should be reused internally. Level 3 occurs when the production unit is not able to take advantage of the emissions and waste generated by it, with alternatives such as: external recycling, sale or donation of reusable waste, and adequate disposal of reusable waste (BARBIERI, 2007).

The Cleaner Production methodology aims to identify and quantify the generation of solid wastes, effluents and atmospheric emissions, from the analysis of the transformation process (input / process-

Figure 1. Levels of performance of cleaner production (cebds.org, 2016, p.38)

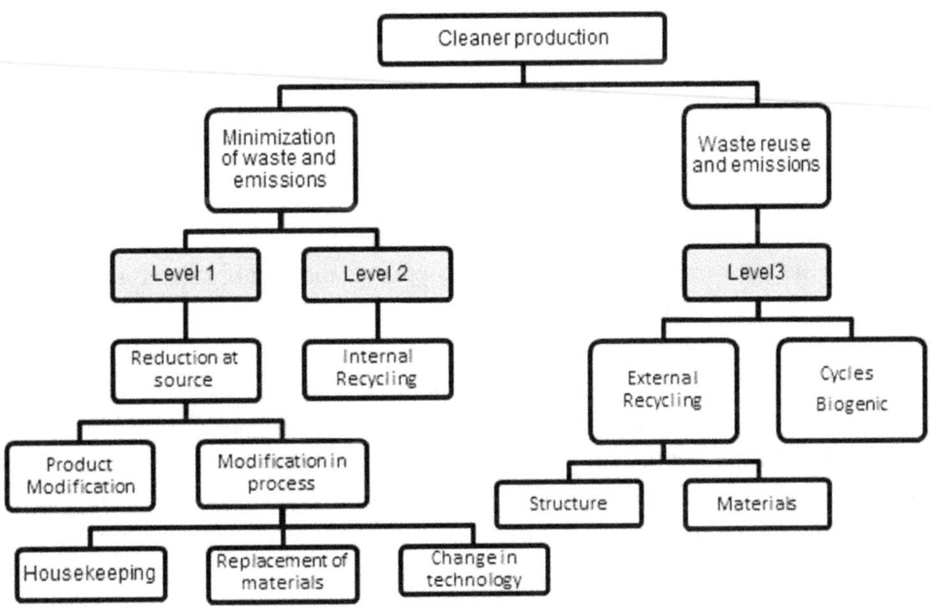

ing / exit), in order to find alternatives to minimize or avoid the consumption of Raw materials, energy and water, also the internal or external reuse of waste or by-products of the production process. It is a precautionary approach to avoid excessive use of a given resource or to make it more efficient, instead of adopting only measures to control waste, effluent and emissions generated.

The basic principle of the Cleaner Production methodology is to eliminate pollution during the production process, not in the end. In addition to environmental issues, it is important to consider that all the waste that the company generates cost you money because they were bought at the price of raw material and consumed resources such as water and energy. Once generated, they continue to cost money, either in the form of treatment and storage costs, or in the form of fines for lack of such care, or even damage to the company's image and reputation. For the implementation of the P + L methodology the company will go through 18 stages (MORE CLEAN PRODUCTION GUIDE, 2012):

- **Task 01:** Commitment of the company management
- **Task 02:** Raising employee awareness
- **Task 03:** Formation of ECOTIME
- **Task 04:** Presentation of the methodology
- **Task 05:** Pre-evaluation
- **Task 06:** Drafting of flowcharts
- **Task 07:** Quantitative Tables
- **Task 08:** Definition of indicators
- **Task 09:** Evaluation of the data collected
- **Task 10:** Barriers
- **Task 11:** Selection of the focus of evaluation and prioritization
- **Task 12:** Mass and energy balances
- **Task 13:** Assessment of the causes of waste generation
- **Task 14:** Generating the Cleaner Production options
- **Task 15:** Technical, environmental and economic evaluation
- **Task 16:** Selecting the option
- **Task 17:** Implementation
- **Task 18:** Monitoring and Continuity Plan

Jabour and Jabour (2013) explain the four steps for implementing the P + L methodology proposed by the Cleaner Production Guide.

Step 1: In preparing the work team, top management support is the first big step for an organization that wants to implement P & L. It is the top management that supports and sensitizes the work team, directs resources and establishes deadlines for the execution of data surveys for later proposition of solutions. With the support of top management, it is important to form a multidisciplinary group (employees from different functional areas) to compose the so-called Eco-Time. It is this team that will disseminate the P + L proposal in order to facilitate the process of data collection in the different micro-operations of the organization and that will be reported to the top management in order to transmit the advances and the barriers of the elaboration of the diagnosis.

Step 2: In the data collection phase, Eco-Time, in a first moment, needs to know and evaluate the macro-operation in terms of the environmental impact generated, that is, emissions, effluents and waste. In a second moment, the Eco-Time will assemble the global flowchart (macro-operation) and intermediate (micro-operation), that is, it will design the global and intermediate transformation model, in order to identify the inputs (water, energy And inputs), processing and outputs (emissions, effluents and waste). Knowing the main inputs and outputs, the next step is to quantify, in terms of unit of measure (consumption), the cost of obtaining and eventual destination (fines, transportation, etc.), over a year, for Each resource and identify which micro-operation is the most critical in terms of environmental and economic impacts.

Step 3: In the critical analysis stage, there are three possible solutions: minimization of waste, emissions and effluents, either at source (level 1) or by internal recycling (level 2) or reuse of waste, emissions and effluents (level 3) With a focus on external recycling. Minimizing waste, emissions and effluent at source (level 1) means adopting changes in the production process or product design. In this way, it is sought to make adjustments in the process through the use of cleaner technology (example: less energy consumption), substitution of raw material and / or actions of organization / cleaning of the production environment. Regarding the product, a proposal is to develop a more environmentally appropriate product design (eg change of composition of inputs) or simply to acquire the inputs in the required quantity or exact specifications (to avoid wastage). The minimization of waste, emissions and effluents by internal recycling (level 2) means taking advantage of these wastes as inputs to the production process, for example: using sugarcane bagasse to make cogeneration of electricity, or redirecting process vapors Thermometers for heating or cooling other process.

Currently Brazil has 371 units of active sugar-energy plants (table 2), the sector generates more than 900 thousand direct formal jobs and benefits 70 thousand independent rural sugar cane producers. The gross value handled by the sugarcane chain4 exceeds US $ 100 billion, with a GDP of approximately US $ 43 billion, equivalent to about 2% of the Brazilian GDP (UNICA 2017).

In the 2015/2016 crop, Brazil produced 665,586.2 million tons of sugarcane, with the highest amount produced, 397,587.6 million tons, concentrated in the state of São Paulo and the lowest amount produced, 48,816.7 million tons in the North-Northeast Region. The state of São Paulo is also the largest producer of sugar, 21,302.7 million tons, and ethanol, approximately 15 billion liters.

The activities of the sugarcane industry bring economic benefits to the country generating direct and indirect jobs, provides technological development, but land use causes environmental impacts.

MAIN FOCUS OF THE CHAPTER

Impacts Caused by the Sugar-Energy Sector

There are several environmental impacts and, in the same way, several factors contribute to these impacts, degrading natural resources that are already considered to be scarce and harming the population. First, the local population and, later, the global population, that is, the community and society in general. The main environmental impacts are presented in Table 3.

Figure 2. Map of the production of sugarcane in Brazil (UNICA, 2017)

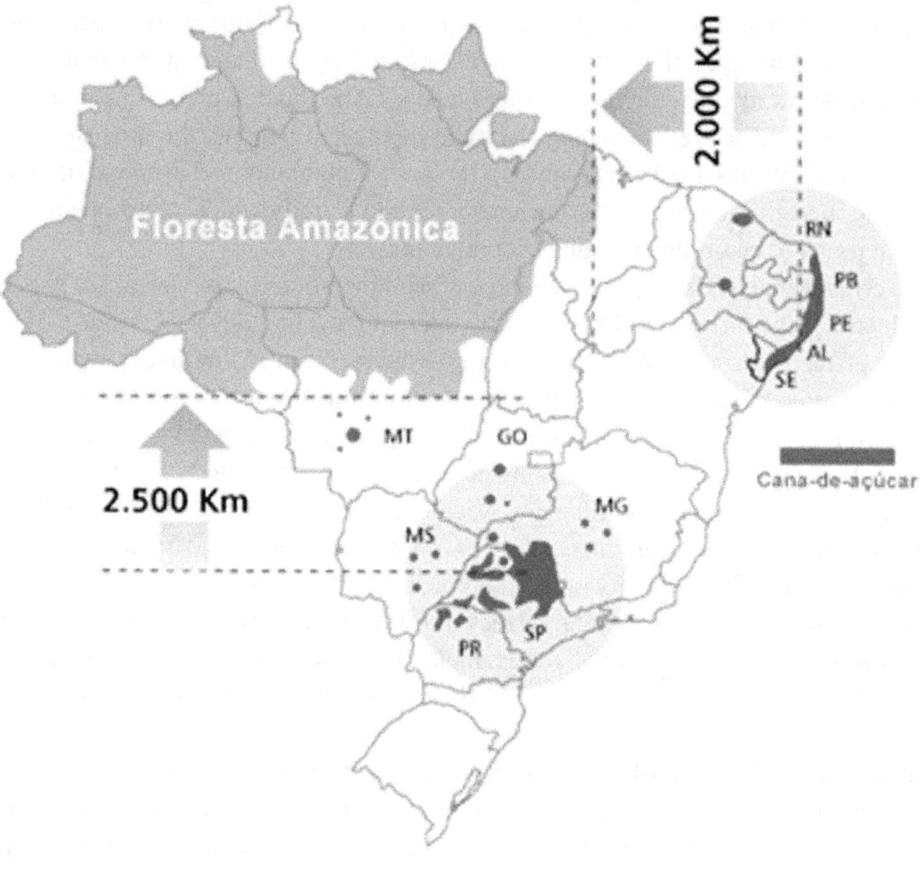

The impacts caused by the sugarcane industry may include deforestation, soil compaction due to the use of agricultural machinery, burning, water pollution, soil contamination and improper disposal of waste from the production process.

Vinasse, sugarcane bagasse and filter cake are the main by-products generated from the processing of sugarcane and cause irreversible damage to the environment if not managed correctly.

- Vinhaça: is the residue of the distillation process, source rich in potassium, has calcium, magnesium sulfur and micronutrients. The way to reduce this residue is to use in the fertigation process. Due to the chemical composition, the vinasse is nutritionally utilized by the cultivated sugar cane, replacing the use of chemical fertilizers.
- Bagasse: it is the result of the milling of the sugarcane to extract the broth that is raw material for the production of sugar and ethanol. The way to eliminate this residue is to use in the process of cogeneration of electric energy from the burning of the bagasse of the sugar cane. For own consumption and sale of surplus.
- Filter cake: it comes from the filtration of the broth extracted from the mills in the rotary filter. This residue is used as fertilizer for crops and feed for livestock. It is necessary to go through process due to the high degree of fermentation.

Table 2. 2015/2016 harvest

States	Sugar Cane (Thousand ton)	Sugar (Thousand ton)	Ethanol (Thousand liters)
Norte	**3.541,90**	**34,6**	**276.973,4**
Rondônia	191.0	0	12,908.8
Acre	86.1	0	4,511.5
Amazonas	216.3	12.4	5,802.3
Pará	682.3	22.2	40,932.3
Tocantins	2,366.2	0	212,816.6
Nordeste	**45,274.8**	**2,574.0**	**1,717,860.9**
Maranhão	2,455.1	12.5	187,297.2
Piauí	967.4	66.9	32,669.8
Ceará	208.6	0	14,594.5
Rio Grande do Norte	2,467.7	137.7	89,428.3
Paraíba	5,532.5	129.1	339,748.6
Pernambuco	11,349.0	822.3	347,727.3
Alagoas	16,193.4	1,213.2	373,106.5
Sergipe	2,264.7	105,4	111.896,9
Bahia	3,816.4	86,8	221.391,8
Centro-Oeste	**139,026.4**	**3.554,4**	**8.824.842,0**
Mato Grosso	17,150.5	337,1	1.326.604,0
Mato Grosso do Sul	48,685.4	1.325,1	2.820.217,8
Goiás	73,190.5	1.892,2	4.676.020,2
Sudeste	**439,395.8**	**24.623,0**	**18,061,102.5**
Minas Gerais	64,932.4	3,249.4	3,083,943.4
Espírito Santo	2,809.6	70.9	150,836.6
Rio de Janeiro	1,066.2	0	58,655.6
São Paulo	367,587.6	21,302.7	14,767,665.9
Sul	**41,816.7**	**2,703.0**	**1,580,745.7**
Paraná	41,266.1	2,703.0	1,576.962.0
Rio Grande do Sul	61.2	0	3,783.7
Norte/Nordeste	**48,816.7**	**2,608.6**	**1,994,834.3**
Centro-Sul	**616,769.5**	**30,880.5**	**28,466,690.2**
Brazil	**665,586.2**	**33,489.1**	**30,461,524.5**

Source: (prepared by the authors based on the Conab 2016 data)

It is important that the company adopt actions to minimize or eliminate the impacts generated by the production of raw material, by the production process and by-products / wastes from its industrial activities. Thus, you can produce your products, using the natural resources in a responsible way, without degrading the environment.

Table 3. The main environmental impacts

CATEGORY	ENVIRONMENTAL IMPACT
Air	Air Quality Change Reduction of the ozone layer Contribution to global warming Odors Generation of acid rain
Water Ground	Alteration of surface water quality Alteration of groundwater quality Sedimentation or alteration of watercourses Radiation contamination Alteration of soil quality Radiation contamination
Community	Noise Vibration Radioactivity Heat Damage to health Uncomfortable
natural Resources	Exhaustion or reduction of the availability of natural resources
Animals and Plants	Damage to animals Damage to plants

Source: Moura (2008, p.115).

Environmental Actions in the Brazilian Sugar and Ethanol Sector

Since the 1990s, the preservation of the environment has become one of the most influential factors in business. Companies are beginning to come up with solutions to achieve sustainable development while at the same time increasing the profitability of their business. To do so, companies can use environmental indicators to verify their organizational behavior to the needs and requirements of the new business world.

With the increase in sugarcane production in the State of São Paulo, there was also an increase in the burning of sugarcane crops, which had an impact on society and the environment, and again the State had to seek alternatives to These impacts.

Technological evolution is therefore gradual, requiring the development of realistic policies for the recycling and reuse of manpower and the monitoring of environmental impacts related to erosion and the proliferation of pests from mechanization.

Main Decrees and State Laws that regulate the practice of burning in sugarcane plantations in the state of São Paulo:

- State Decree 41.719 of April 16, 1997. It prohibits the burning of sugar cane and determines the gradual reduction of this practice until its extinction in 8 years, in the areas defined as mechanizable, and in 15 years in the areas defined as non-mechanizable;
- State Law 10,547 of May 2, 2000. It maintains the ban on the burning of sugar cane, but changes the term of its extinction to 20 years, that is, in the year 2020 and regulates the practice of burning during this period;

- State Law 11,241 of September 19, 2002. It maintains the prohibition of fires and differentiates the deadlines, considering the mechanizable and non-mechanizable areas, determining its extinction in the year 2021 for the mechanizable areas and in the year 2031 in the non-mechanizable areas;
- Decree 47,700, dated March 11, 2003. Regulates State Law 11,241 / 2002 and establishes the deadlines and percentages for the gradual elimination of sugarcane fires in the State of São Paulo; and
- Agro-environmental Protocol of June 4, 2007. Establishes the anticipation of the elimination of the burning of sugarcane in the State of São Paulo.

The state government authorities have created laws aimed at mitigating these consequences by prohibiting actions or even by raising awareness, as an example of the case of the Agro-Environmental Protocol, which has voluntary adhesion of sugarcane producers, mills and distilleries Who grow their own cane.

The Agro-Environmental Protocol of the Sugar and Alcohol Sector was signed on June 4, 2007, by UNICA (2017), representative of the sugar cane industry and the Government of the State of São Paulo.

According to UNICA (2017), the adhesion to the Agro-Environmental Protocol aims to:

- Reduce the concern of environmentalists with the expansion of sugarcane for the production of biofuels;
- To offer a differentiated treatment in the production of cane and ethanol;
- Demarcar areas to be protected;
- Recover and build biodiversity corridors;
- Protect water sources and combat erosion;
- Encourage the mechanization of sugarcane cutting and eradication of fires;
- Reduce water consumption and air pollution by power plants;
- Ensuring good labor and social practices.

Brazil is recognized worldwide by the strong renewable component of its energy matrix. Currently, more than 40% of all energy used in the country comes from renewable sources. The sugar-energy sector plays a key role in this scenario: sugarcane, the raw material used to produce ethanol and bioelectricity, is the country's second largest source of renewable energy, accounting for almost 16% of the country's energy matrix. (UNICA, 2017).

The consumption of ethanol hydrated by flex fuel automobiles, called flex, combined with the current mandatory mix of 27% anhydrous ethanol in gasoline, has reduced greenhouse gas (GHG) emissions by more than 300 million tons of CO2eq since the launch date In Brazil, from March 2003 to May 2015. In addition to being clean and renewable, bioelectricity is generated close to the electricity consuming centers, which significantly reduces system losses and the need for investments in transmission, and is Complementary to hydroelectric generation, the potential for bioelectricity generation by the plants in the Center-South of the Country is concentrated between April and November, months of harvest of sugarcane and corresponding exactly to the driest period of the year. (UNICA, 2017).

Table 4. presents the rejects generated from sugarcane milling and the Cleaner Production measures that can be taken for each reject.

Table 4. Cleaner Production Measures in the sugar-energy sector

Reject	Source	Composition	Examples of Cleaner Production Measurement	
			Reduction	Reuse / Recycling
Cane Washing Water	Washing the cane before grinding	High sucrose content, especially in the case of despalha with fire; - Vegetable matter, soil and adhered adhesives;	- Recycling in the process of imbibition (allows recovery of part of the diluted sucrose); - Recycling in the washing process (treatment required for removal of coarse solids and sedimentable residues, and possibly for removal of soluble organic substances)	- Recycling in the process of imbibition (allows recovery of part of the diluted sucrose); - Recycling in the washing process (treatment required for removal of coarse solids and sedimentable residues, and possibly for removal of soluble organic substances)
Water from the barometric condensers and condensed water on the evaporators	Washing the cane before grinding	• Reduction of syrup loss: - reduction of flow velocity; -reduction of condensation water temperature; • Recovery of the syrup: - use of obstacles that reduce the drag (separators and stoves); - increasing the height of the evaporators;	• Reduction of syrup loss: - reduction of flow velocity; -reduction of condensation water temperature; • Recovery of the syrup: - use of obstacles that reduce the drag (separators and stoves); - increasing the height of the evaporators;	• Recycling of water in the process itself (careful with sugar content); • Recycling in the process, but in another stage, such as: - sugar cane imbibition; - washing the honey after crystallization of the sugar; - generation of steam; - washing filters; - preparation of solution for chalking (in clarification);
Bagasse	Cane milling and broth extraction	Cellulose, with a moisture content of 40 to 60%	-	Electric power cogeneration; - Composting (use as fertilizer); - Animal feed production; - Production of agglomerates; - Production of cellulose;
Filter cake	Filtration of the sludge generated in the clarification	- liming wastes; - Rich in phosphates	-	- Use as soil conditioner; - Production of animal feed
Vinhaça	- Residues from the distillation of fermented molasses to obtain alcohol.	High DBO e DQO	-	- Use as fertilizer (important use it at the appropriate application rate, otherwise there is contamination)
Dornas Wash Water	- Washing of the fermentation containers, for obtaining alcohol (reduced volume);	- Similar to vinhoto, but much more diluted (about 20% of vinhoto);	-	
Molasses	- Manufacture of sugar	High DBO	Practically all used in the production of alcohol.	- Production of alcohol - Manufacture of yeast
Cane tip	- Cane cutter for milling	-	-	Animal food

(CETESB, 2002).

Cleaner Production in Brazil's Sugar-Energy Sector

The Sucroenergetic Sector, as one of the great economic forces and production of goods of the Country, has sought to improve and optimize its productive process.

The sugar and alcohol mills have long been developing and implementing measures that minimize the environmental impacts resulting from their productive activity. Many of these measures are part of the P & L concept (CETESB, 2002).

The Cleaner Production model aims to reduce or eliminate waste or pollutants in the generating source (BARBIERI, 2007). The actions carried out by the sector are the conscious use of raw materials, the treatment and reuse of water, the production of electric energy from sugarcane bagasse for its own consumption during the harvest, the management of solid wastes and effluents from the production of Sugar and ethanol from sugarcane.

Ten groups were surveyed in the Brazilian sugarcane industry with the objective of analyzing the Cleaner Production practices adopted by the mills.

AGROVALE

The AGROVALE plant was founded in 1932 in the São Francisco Valley - Bahia, the industry was established in a dry and atypical soil for production, but with investment in irrigation, AGROVALE overcame the challenges and stands out in the national scenario for its most Higher than the national average. AGROVALE is recognized as one of the largest producers of sugar and alcohol in the Northeast. In 2005, a new distillery was inaugurated in the industry, increasing production capacity from 120,000 liters to 450,000 liters per day, an increase that will further increase the production of alcohol from Bahia, generating more wealth in agriculture and industry.

The Environmental Management System is present in the company's strategy, the fibrous waste resulting from the extraction of sugarcane juice is mainly used as fuel in boilers for power and heat generation. AGROVALE markets sugarcane bagasse in the following ways:

Table 5. Group/State (Authors, 2017)

Group	Unit Amount	Present in the STATE
Agrovale	1	**BA**
Biosev	11	**SP, MG, MS, RN, PB**
Bunge	8	**SP, MG, MS, TO**
Carlos Lyra	4	**AL, SP**
Clealco	4	**SP**
Colorado	2	**GO, SP**
Corprodia	1	**MT**
Coruripe	5	**AL, MG**
Olival Tenório	2	**AL**
Raízen	22	**GO, SP, MS**

Table 6. Type of Bagasse (Authors, 2017)

In-Nature Bagasse	Used by farmers to maintain soil moisture, it minimizes the effects of drought and as a soil fertilizer.
Hydrolyzed Bagasse	Obtained through bagasse hydrolysis, the product is used in animal husbandry as an animal feed, with high gustatory taste by cattle.
Packaged Bagasse	Substrate composed of the mixture between sugarcane bagasse and filter cake, the product is used as organic fertilizer of great value fertilizer for plants.

The enriched bagasse marketed by AGROVALE has less moisture, which guarantees a more consistent product

AGOVALE generates its own electricity through the burning of sugarcane bagasse, a clean and renewable energy. The surplus is sold to third parties.

It realizes practices of rational use of soil and water, through the implantation of underground drip systems in its fields of production, this system provides the extension of the canevial renewal, ensures agricultural productivity above our current averages increasing the production of greater volume of sugarcane Within the same production area, this differential has many positive environmental impacts, one of them is the reduction of deforestation and consequently the maintenance of the biodiversity of our biome, lower water consumption, lower degradation of cultivated soils when compared with the traditional irrigation methods Used, with higher productivity and long life, injection of organic matter directly into the root area of the crop, among other economic advantages and environmental gains.

BIOSEV

Biosev is one of the world leaders in the sugar and alcohol sector, with 13 units located in the main producing regions of the country.

The company founded in 2009, as a result of the merger of LDC Bioenergia, of the Louis Dreyfus Commodities group, with Santelisa Vale, one of the largest national companies in the production and processing of sugarcane.

Biosev's units produce electricity by burning biomass (bagasse and straw) generated from the production of sugar and ethanol. The surplus energy is marketed to third parties.

The products produced by Biosev serve large companies related to the animal nutrition sector and also suppliers and partners of the sugarcane production activity. The source of raw material are the byproducts of the processing of sugarcane.

The company adopts the strategy of elimination and reduction of waste in the generating source. In this way, the vinasse and the wastewater generated in the industrial processes are reused and applied in the growing areas as a complement to the fertilizers, as they also contribute to the improvement of soil conditions.

The plant conducts water consumption management aimed at reducing consumption, with actions that involved closing of circuits, installation, revision and calibration of flow meters.

The monitoring of emissions of particulate matter and NOx (nitrogen oxides) is carried out by specialized companies, whose analytical certificates subsidize maintenance activities and periodic adjustments in the systems for collecting particulate matter and gas washing equipment from the production processes.

Table 7. Processing of SugarCane (Authors, 2017)

Dry Yeast and Yeast Cell Wall	Obtained by means of alcoholic fermentation in the alcohol manufacturing process and subsequent drying; Dry yeast is an additive used in the basic composition of feed, concentrate and supplementation recommended for all animal species.
Molasses Powder	Ingredient used to feed different animal species (cattle, horses, pigs, poultry and fish) as energy source, palatabilizing and flavoring
Molasses Powder - MBS	Organic, non-toxic and biodegradable binder used in the refractory and ceramic materials industry in general.
Animal Food	Produced from co-products of the industry (raw bagasse, hydrolyzed bagasse, yeast and molasses), it is intended for the feeding of various categories of ruminants and horses.
Fuselage Oil	By-product of the alcohol distillation process, it consists mainly of isoamyl alcohol and is used in the chemical industry for the production of solvents, varnishes and fixative in perfumery, among others.

BUNGE

Bunge, present in Brazil since 1905, is one of the main agribusiness and food companies in the country. Since 2006 it has also been active in the sugar and bioenergy segment. Present in 16 states in all regions of Brazil, Bunge now has more than 20,000 employees and about 150 units, among industries, distribution centers, silos and port facilities. Bunge Brasil belongs to the holding company Bunge Limited, founded in 1818, with headquarters in White Plains, New York, USA.

The Bunge Group adopts the Cleaner Production model in its production processes. The company prioritizes areas that have already been degraded and associated with the use of technologies that increase productivity, that is, that enable an increasing production without the need to open new areas to meet the growing global demand for food and energy. The Group establishes strict quality rules for all suppliers and constantly performs compliance checks. The monitoring of the use of natural resources is carried out daily in order to be used as necessary.

It produces clean and renewable energy, obtained from the burning of sugarcane bagasse. Of the eight units, six are able to export energy to the National Interconnected System (SIN) and, therefore, to market the electric energy produced. In addition, three Bunge plants have the Bonsucro certification, granted by an international organization, allowing part of the ethanol and sugar produced to be marketed in restricted markets, such as the European market, or can make the raw material of sustainable chains such as Chemical industry and green plastics.

Bunge Limited has a Global Non-Deforestation Policy. Committed to eliminating deforestation from its agricultural supply chains around the world, employing proven methodologies incorporating carbon and biodiversity protection, respecting the rights of local and indigenous communities, and applying improved traceability and transparency of supply chains- And to make public the progress made.

CARLOS LYRA

The history of the Carlos Lyra Group began in 1951, when businessman Carlos Benigno Pereira de Lyra Neto took over Algodoeira Lagense S / A. The company, based in the municipality of São José da Lage (AL), was active in the purchase, processing and commercialization of cotton, corn, castor beans and coffee. It is currently composed of four industries producing sugar, ethanol and Bioelectricity, the Carlos Lyra Group is present in the States of Alagoas and São Paulo. In addition to the sugar-energy sector, it also operates in the textile, cattle raising, taxi-riding and broadcasting segments, generating more than 14,000 direct jobs for the country.

For the Carlos Lyra Group, the adoption of environmentally sound management is an example to other companies and future generations. Usina Caeté S / A, Matrix is in the process of certification of ISO 14001 Version 2004. The units of Alagoas have 8,150 ha, destined for permanent preservation and legal reserve. Approximately 500,000 seedlings of native trees have already been planted. Several environmental education projects were implemented in partnership with non-governmental institutions and public agencies.

The Carlos Lyra Group created two Private Reserves of Natural Patrimony (RPPN's), duly registered at ICMBio: the Salvador Lyra Reserve (Caeté S / A - Marituba Unit), with 1,392 ha, considered the largest RPPN in the Alagoas restinga and the second And the Reserva Senador Carlos Lyra (Usina Caeté S / A - Cachoeira Unit), with 449 ha, the largest RPPN in Maceió (AL).

At the Paulicéia Unit (SP), environmental education actions are being developed involving employees, public school students and the community in general. Since 2010, the company has the title of Green Ethanol, a certificate that is being renewed every year.

Another important initiative is the creation of environmental reserve areas in the Paulicéia Unit, where some 400,000 native tree seedlings of the region have already been planted, aiming at the recovery of riparian forests.

The Group invests in its units for environmental preservation and in the implementation of the Cleaner Production model. Such as the closed effluent circuit, thermal boiler and gas scrubbers. In the field, the use of filter cake and vinasse, contribute to the reduction of chemical fertilization.

All the units produce their own electric energy from the burning of sugarcane bagasse.

COLORADO

The first distillery of the Colorado Group was founded in 1980 from a contract signed with Proálcool by Oswaldo Ribeiro de Mendonça and his brother Geraldo Ribeiro de Mendonça in the municipality of Guaíra - São Paulo.

The Colorado Group's Environmental Management System seeks to reduce the consumption of inputs such as water, electricity and materials, extending this commitment to the entire production chain. The production of ethanol and electricity from sugarcane reflects the commitment to environmental issues.

All energy used in the industrial process of the production of ethanol and sugar is generated within the own mills from the burning of sugarcane bagasse.

Grupo Colorado has a continuous plan to reduce water use in its production stages. The main factor contributing to the reduction of water consumption was the evolution of the mechanical harvesting of raw cane, which eliminated the need to wash the cane.

CORPRODIA

COPRODIA was founded in 1980 in the state of Mato Grosso, in a region with no infrastructure, made up of producers and their families from the south of the country, driven by their longings, Bet on the strength of work and cooperation and turned the dream into reality, with great courage and determination faced the difficulties and today reap results that make this region the world's largest arable land.

Corprodia uses the Cleaner Production model in its production processes. The industrial system seeks to minimize the use of natural resources where processes that use water are operated in closed circuit, where water is recycled, reused and capture is given only to replace losses. It performs monitoring of the quality of surface and groundwater in the areas that are influenced by the activities of the sugar and ethanol production process. The byproducts generated such as filter cake and boiler ash are used in the fertilization of cane growing areas, which reduces the consumption of chemical fertilizers and minimizes impacts to the environment. Vinasse is used in the fertigation process. There is a dequation of the pollution control systems, consisting of gas scrubbers that guarantee a significant reduction of the emission of pollutants by the boilers and annual monitoring of the gas emissions, which allows better management of the productive activities. More than 98% of the sugarcane harvest is mechanized, this system maintains on average 12 tons of straw / hectare, increasing the organic matter content, improving the physical properties of the soil. In addition, it reduces the environmental impacts once caused by the practice of burning the cane fields. It performs the biological control of pests to avoid to the maximum the use of agricultural pesticides in the crops. COPRODIA produces all the electric energy it consumes being self-sufficient in its production with the use of part of the bagasse generated in the cane milling process. Cane bagasse is also used to feed cattle in a feedlot system. It has a program of preservation of the Amazon Forest. CORPRODIA performs modern techniques such as varietal management, precision agriculture resources that minimize input use, crop rotation with legume use, no-till, and complement the cleaner production process.

CORURIPE

The Coruripe Mill was founded in 1925, by José de Castro Azevedo, from the union of several sugarcane mills, in the municipality of Coruripe - Alagoas. Over the years with its development has transformed a small sugar mill into one of the most important business corporations in Brazil. Currently, the Coruripe Matriz Power Plant in Alagoas is part of the Tércio Wanderley Group, which also has four industrial units in Minas Gerais: Iturama, Campo Florido, Limeira do Oeste and Carneirinho, as well as Coruripe Energética. Together they produce the most sugar and alcohol in the North and Northeast of Brazil.

The concern and preservation of nature and caring for the human being are also priorities in the company.

It is self-sufficient because it produces its own electric energy from the burning of sugarcane bagasse. The surplus is sold to utilities.

The by-products of sugarcane processing are used. Fusel Oil, a byproduct of ethanol distillation, is used for pharmaceutical and chemical industries. A yeast by-product of sugar and ethanol, serves as a complement to animal feed. Cane bagasse hydrolyzed by-product of the sugar cane used for feed. Product hydrated with steam at high pressure to facilitate animal digestion. Vinasa is a byproduct of the manufacture of ethanol. It is used as fertilizer in the cane field. A Pie by-product of the manufacture of sugar and ethanol. It is used as fertilizer.

In addition, Coruripe has been recovering riparian forests and replacing cane planted in hillside areas by reforestation. The company also implemented a sowing of Atlantic Forest trees that annually produces more than 50 thousand seedlings of native species.

Respect for nature led Usina Coruripe to allocate 17,000 hectares of land for preservation. More than nine thousand hectares have already been transformed into RPPNs - Private Reserves of Natural Heritage (RPPN).

CLEALCO

The Clealco Group was founded in the 70's by politicians, cattle ranchers and farmers, due to the prices of oil barrels constantly rising in Brazil, the government launched a sugarcane plantation project with the objective of leaving the country less dependent Of feedstock imports and make ethanol a new fuel option. Currently, it has three plants, located in the cities of Clementina, Queiroz and Penápolis.

The company maintains its sustainability policy as a strong policy. The company seeks to integrate the social, economic and environmental pillar in the relationship with all stakeholders, in order to realize the productive processes in order to reduce environmental impacts, preserve the health and safety of its employees, and contribute continuously to the development Social and economic development of the more than 26 municipalities where it operates.

In June 2011 the building of Ceretas (Center for Recovery and Screening of Wild Animals) was inaugurated at UNESP of Araçatuba (State University of São Paulo). The space has 695 meters of constructed area and cost R $ 800 thousand, was built with investments of companies of the sugar-energy sector, among them Clealco. The center is a specialized hospital to care for and study the genetics of animals, with a surgical center, ICU, cage complex, rooms for emergency care, X-ray and ultrasound. In order to accelerate the process of reforestation of the region's riparian forests, the company has a universe of seedlings located on the left bank of the highway SP 463, height of KM 07, in the district Martelli in Clementina. The site produces seedlings of more than 80 species of native trees, which after being completely developed are planted in the preservation areas and distributed in schools and institutions of the region, contributed to afforestation of the municipalities.

The company has a system of cane washing in closed circuit, with which it is possible to reuse the water used in the industrial processes through a sedimentation tank. Among the processes of water treatment are the correction of pH, which facilitates the decantation of impurities and prevents the proliferation of microorganisms. This system is integrated with the cooling water of the sugar factory, distillery, fermentation and gas washes of the boiler. Successfully performs selective collection, procedures take place in all sectors of the company.

OLIVAL TENÓRIO

It is inevitable to speak of the history of the Olival Tenório Group without mentioning the life trajectory of the businessman Olival Tenório Costa, since the two are confused practically in one, the son of José da Costa Tenório and Isabel Tenório Cavalcante, "Dr. Olival "was born on November 8, 1922, in the city of Atalaia - AL, he graduated in Industrial Chemistry from the Federal University of Pernambuco, beginning his professional career as a Chemist at the Uruba Power Plant in Atalaia - AL, in 1958, Acquired in partnership with his father - in - law José Tenório de Albuquerque, Engenho Ilhota, in Boca da Mata - AL, current Usina Triunfo, in 1971, still in partnership with his father - in - law, acquired Usina Puerto Rico, located in the municipality of Colônia Leopoldina - AL and then transferred to the municipality of Campo Alegre - AL, after that acquisition, at the same time, decided to end the partnership, being the businessman Olival Tenório with the Usina Puerto Rico, an entrepreneur, Olival Tenório reactivated the unit of Cologne Leopoldina, Later denominated as Autonomous Distillery Porto Alegre.

The Group has an Environmental Management System in its processes. The reduction of the use of chemical additives in the fight against pests and diseases of the cane, being priority the biological or mechanical control. It takes advantage of the byproducts generated in the processing of sugarcane as the filter cake used for fertilizer, the vinasse in the fertigation system, the sugarcane bagasse in the cogeneration of electric energy for the own consumption and commercialization of the surplus. Responsible use of natural resources in the processing of sugarcane. It has programs for the preservation of the Atlantic Forest in partnership with The non-governmental organization IPMA - Institute for Preservation of the Atlantic Forest.

RAÍZEN

The Raízen group stands out as one of the most competitive energy companies in the world. Betting on the innovation, technology and talent of around 30 thousand employees to generate sustainable solutions that contribute to the economic development of the country in an ethical way, with respect to people and the environment, following the values and standards of conduct that are in keeping with our Position.

Built as part of the Shell and Cosan business, it has grown into a unique company with experience and wide market penetration. It implements a fully integrated process of the sugar-energy sector, with action in all stages: sugarcane cultivation, sugar and ethanol production, internal and export logistics, distribution and marketing.

With energetic alternatives of renewable matrices, with constant investment in technology, research and development. And so, for example, the production of bioelectricity from the co-products of sugarcane, as well as reuse a part of such co-products in the production of second generation ethanol. All of this is part of one of the biggest goals as a company: to offer the energy of the future today.

The group has environmental certifications such as Selo energia verde, Bonsucro, and actively participates in the international multistakeholder participation forum - counting on the participation of NGOs, consumer industry, traders, producers - that presents an exclusive certification model for sugarcane and considers In its standard the various aspects of sustainability. Producing units are audited by independent companies and, when they meet the criteria, they receive a stamp.

The Agro-environmental Zoning prepared by the Secretariats of Environment and Agriculture and Supply was announced in September 2009 by the then governor of São Paulo, José Serra. The Zoning was created to guide the sustainable planning of the growth of the sugar-energy sector in the State. In the survey, hydrographic, physical, topographic and climatic aspects were considered. In addition to the areas, rules were established for the licensing of new ethanol plants in São Paulo.

CONCLUSION

The groups surveyed have an Environmental Management System, with the objective of reducing the impacts of their productive activities on the environment. The adoption of the Cleaner Production model allows the policy to eliminate or reduce waste and emission of pollutants and reduce the waste of natural resources.

The management of the residues generated from the processing of sugarcane, are carried out efficiently by all groups of sugarcane plants researched. The by-products of the cane milling process are used in the process itself.

The sugarcane bagasse is used in energy cogeneration that causes positive impacts, since there is a use of the residue and does not overload the country's electricity grid.

The use of the filter cake as fertilizer in the cane fields, avoids the use of pesticides that cause degradation to the environment.

The practice of fertigation with vinasse, aims at greater agricultural productivity and reduction in the use of chemical fertilizers.

It is worth mentioning that the application of vinasse and filter cake should be carried out under qualified professional guidance, which will indicate the quantity and periodicity appropriate to the type of soil.

The closed-loop cane washing system allows the water used in the process to be treated and reused. Thus reducing the use of natural resources.

In the State of São Paulo, 90% of the sugarcane harvest is mechanized. In the Groups surveyed, this percentage is 100%, which benefits the environment, since it eliminates the burning and maintains organic matter in the soil, contributing For carbon sequestration.

All groups have a reforestation program, with their own nurseries rich in native seedlings. Most Groups allocate part of their own land to conservation areas.

The Groups seek continuous improvement in the processes to produce more and more in a sustainable way and preserving the environment.

However, there is still much to be done, considering that many actions are recent and others depend on changes in the organizational culture.

In this sense, there is a need for several other researches on the subject in the sugar-energy sector and other sectors of the economy, as well as involving its stakeholders.

REFERENCES

Alves Filho, M. (2007). Colhedora de cana preserva emprego e reduz desperdício. *Jornal da Unicamp*. Retrieved from http://www.unicamp.br/unicamp/unicamp_hoje/ju/julho2007/ju365pag05.html

Barbieri, J. C. (2007). *Gestão Ambiental Empresarial: conceitos, modelos instrumentos*. São Paulo, Brasil: Saraiva.

CETESB - Companhia Ambiental do Estado de São Paulo. (n.d.). *A Produção Mais Limpa (P+L) no Setor Sucroalcooleiro – informações gerais*. Retrieved from http://www.agencia.cnptia.embrapa.br/Repositorio/Adubacao_organica_producao_ mais_limpaID-37HFh1RpEg.pdf

CONAB. (n.d.). Retrieved from http://www.conab.gov.br/conabweb/download/safra/1graos_08.09.pdf

Donaire, D. (1999). *Gestão ambiental na empresa*. São Paulo, Brasil: Atlas.

Jabour, A. B. L. S., & Jabour, C. J. C. (2013). *Gestão ambiental nas organizações; fundamentos e tendências*. São Paulo, Brasil: Atlas.

Moura, L. A. A. (2008). *Qualidade e gestão ambiental: Sustentabilidade e implantação da ISO 14.001* (5th ed.). São Paulo, Brasil: Juarez de Oliveira.

Ronquim, C. C. (2010). *Queimada na colheita de cana-de-açúcar: impactos ambientais, sociais e econômicos*. Campinas: Embrapa Monitoramento por Satélite. Retrieved from https://www.cnpm.embrapa.br/projetos/carbcana/download/ser_doc_77_queima_cana.pdf

Seiffert, M. E. B. (2011). *Gestão ambiental: Instrumento, esferas de ação e educação ambiental*. São Paulo, Brasil: Atlas.

Shikida, P. F. A. (2013). Expansão canavieira no Centro-Oeste: Limites e potencialidades. Revista de Política Agrícola, 22(2), 122-137.

Silva, V. L. M. M., Gomes, W. C., & Alsina, O. L. S. (2007). *Utilização do bagaço de cana-de-açúcar como biomassa adsorvente na adsorção de poluentes orgânicos. In Revista Eletrônica de Materiais e Processos* (Vol. 2, pp. 27–32). Retrieved from www2.ufcg.edu.br/revista-remap/index.php/REMAP/article/download/28/64

Tachizawa, T. (2011). *Gestão ambiental e responsabilidade social corporativa: estratégia de negócios focadas na realidade brasileira*. São Paulo, Brasil: Atlas.

UDOP. (n.d.). *Subprodutos da cana se tornaram fonte de receita*. Retrieved from http://www.udop.com.br

UNICA. (n.d.). *Bagaço de cana pode ganhar valor substituindo areia na construção civil*. Retrieved from http://www.unica.com.br/noticias/show.asp?nwsCode=%7B4E794FD3-7EC2-403C-A66D-325016046000%7D

KEY TERMS AND DEFINITIONS

Agro-Environmental Zoning: Is also known as Ecological-Economic Zoning. It is an instrument of environmental and urban legislation that establishes measures and standards of protection for the environment, aiming to ensure the conservation of biodiversity, environmental quality, water resources and soil, At the same time, the sustainable development of the economy and the improvement of the quality of life of the population.

Cleaner Production: Is a management strategy focused on the efficiency of raw materials, energy, water and inputs, with the aim of minimizing the environmental impacts of processes, services, or products.

Engenho: Is the whole agro-industrial establishment specializing in the transformation of sugarcane into sugar, molasses, sugarcane brandy and ethanol.

Environmental Management System: Is an interrelated set of policies, practices and organizational, technical and administrative procedures of a company that aims to obtain better environmental performance, as well as, control and reduction of its environmental impacts.

Sucroalcooleiro: Is the sector that produces sugar and alcohol. It comprises from the agricultural part to the industrial.

Sugarcane: The sugarcane industry in Brazil comprises all agricultural and industrial activities related to the production of sugar, bioethanol and bioelectricity.

Sustainability: Promote the exploration of areas or the use of natural resources, natural or otherwise, in a way that does not harm the balance between the environment and human communities and the entire biosphere that depend on it to exist.

Chapter 5
Sensitization of Sustainable Manufacturing Strategies to Benefit Indian SMEs

Vimal Kek
Sri Venkateswara College of Engineering, India

Jayakrishna Kandasamy
VIT University, India

ABSTRACT

In recent times, globalization policy eliminates the restriction of market boundaries, enabling all manufacturers to exploit the markets of developing countries, due to which, the environmental regulations in the developing countries are evolving to secure their nature with stringent regulation. In this view, this article is written to impart knowledge on sustainable manufacturing to all manufacturers. However, for successful implementation of sustainable manufacturing practices, formulation of strategies is found to be essential. In this view, critical insights have been presented by reviewing existing practices of sustainable manufacturing, how to deploy them and their correlation with the environmental benefits are mapped to support practicing managers. The practices discussed in the chapter include process optimization in view of the environment energy modeling, waste minimization options, better end-of-life policies, material efficiency and need to employee training.

INTRODUCTION

In the past, lack of availability of natural resources forced the manufacturing organizations to improve labor efficiency through various manufacturing strategies by utilizing abundant natural resources. However, increase in population and over consumption of natural resources redefines manufacturing strategy. The evolution of manufacturing system starts with Craft Production System (1890 – 1910) in Automobile manufacturing and in the year 1910 Ford Motor company introduced Mass Production System which is based on economies of scale. After the Second World War, an advanced production system called Toyota Production System (TPS) (1950 onwards) in Toyota Motor Company, then in 1980,

DOI: 10.4018/978-1-5225-3537-9.ch005

lean manufacturing evolved by integrating advanced tools from world-class manufacturing system and TPS which also includes Supply Chain Management. In 1987, Brundtland Commission emphasized the importance of sustainable concepts and defined sustainable development as "Sustainable development is the development that meets the needs of the present without compromising the ability of future generations to meet their own needs".

The sustainability concepts have been further flourished by many researchers in the past decade (Vinodh, 2011). Sustainability can be viewed in three orientations namely product, process, and material. In India, manufacturing organization contributes to 18% of total energy consumption (Central Statistics Office, 2013). Manufacturing processes account for a major proportion of energy consumed by manufacturing organizations. India ranks 5th in Generation of Greenhouse Gases (GhG) and these non-specific processing industries contribute approximately 134 million tonnes of CO2 equivalent. Thus, the interest has been shifted towards manufacturing processes in the recent times. For successful implementation of sustainable manufacturing practices, identification and implementation of strategies form critically. Thus, literature search strategy for this research was developed by first identifying the likely relevant databases, conferences, and journals and using keywords (such as sustainable strategies, sustainable manufacturing practices, green manufacturing practices, environmental conscious processes etc.). The insights from the literature review are presented in the following sections.

SUSTAINABLE MANUFACTURING PRACTICES

In the past, green competitive pressure and government legislations were not same for developing and developed countries. But, scarcity of material and energy, the improved environmental consciousness of customers and globalization turned the scenario (Nidumolu et al. 2009). These form the major reason for restructuring organizations stake on sustainable development. Altogether, these reasons force organizations to shift from pollution control strategy to prevention strategy.

Also, based on analysis of process flow diagram (Figure 1) of the manufacturing process, the identified wastes are released to technosphere and ecosphere. Kleindorfer et al. (2005) suggested employee involvement, waste reduction, energy conservation, and emission control are important internal strategies to enable sustainability of organizations. Also, Rashid et al. (2008) pointed out, for enabling sustainable material unitization: waste minimization, material efficiency, resource efficiency and eco-efficiency as the critical strategy. However, sustainable manufacturing practices have three perspectives namely material, system and process (Vimal et al. 2014). Sustainable manufacturing practices focus on input materials, processes, Transportation, packaging and disposal after use phase, among others. Of the life cycle stages, material and manufacturing processes are found to be significant as far as the environment is considered.

Considering the process flow diagram of manufacturing processes and after through literature review, five sustainable manufacturing strategies which gain huge interest among researchers and practitioners are Energy modeling and optimization studies, Employee skill training and involvement program, process emission studies, process parameters optimization considering sustainability and waste minimization and disposal strategies.

Figure 1. Input-Output analysis of manufacturing process

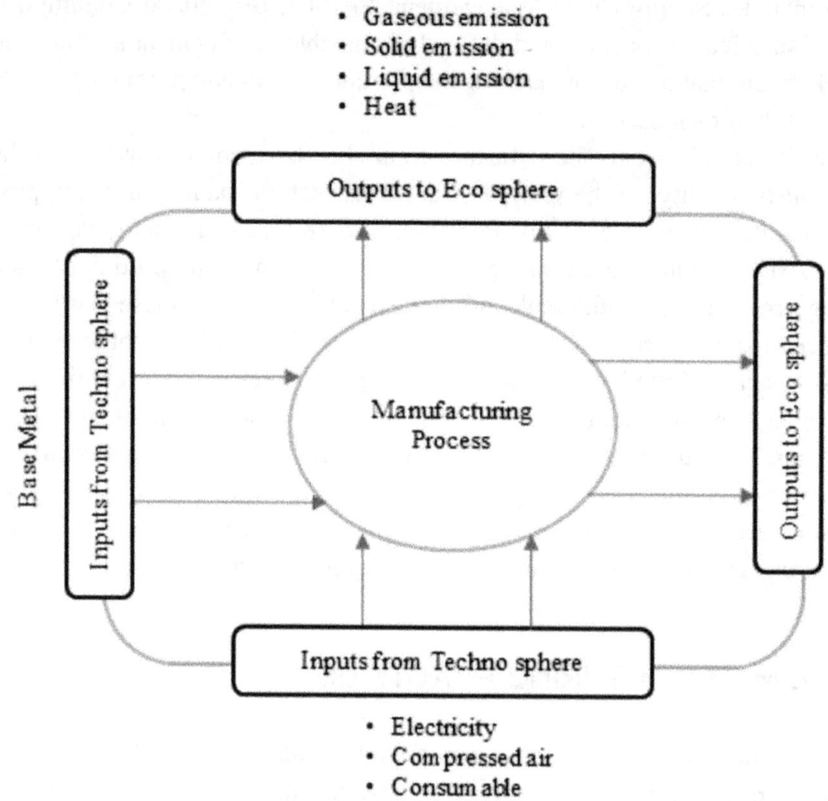

WASTE MINIMIZATION

Due to global warming potential and material scarcity, nowadays term 'waste' refers to underutilization or incomplete utilization of material (Rothenberg et al. 2001). Initiative to minimize waste is viewed as conservation of natural resources. Apart from this, waste minimization concerned with reduced water consumption and effluent generation; etc. The reduction of waste at source point conserves natural resources which again minimize the energy consumption due to recessing, transportation etc. Thus, reuse, recycling, and recovery are also included in waste minimization which was not considered in definition by Organization for Economic Co-Operation and Development (OCED) (1998). This change in viewpoint is large because of natural resource scarcity.

Gupta et al (1995) stated that, through adaptation of environmental friendlier waste disposal operations, green image of the product can be improved. Anastas and Zimmerman (2006) stated recycling and waste minimization are means of approaching sustainability. Foolmaun and Ramjeawon (2013) used Life Cycle Assessment (LCA) methodology to determine environmentally friendlier disposal options for used PET bottles. Thus, the focus of Waste minimization and disposal strategy is to conserve material through waste minimization and proper adaptation of disposal scenarios.

EDUCATION AND EMPLOYEE TRAINING

Employee training and involvement are identified as an important strategy while deploying advanced manufacturing system. People around the world recognize that current economic development trends are not sustainable and public awareness, education, and training are key strategies to move society toward sustainability. Earlier few studies pinpoint the importance of employee training and involvement (Vimal and Vinodh, 2012; Muduli et al, 2013). Klassen and Whybark (1999) identified employee training as one of the important tools to incorporate environmental strategies. Shen et al (2007) used staff factor as important criteria for deploying sustainability in a manufacturing environment. Training has been seen as a solution for environmental problems by many researchers (Carter and Dresner, 2001; Walker et al., 2013; Muduli et al., 2013).

Education, public awareness, and training are core issues of sustainable development. The training scheme planned for the current study is of two modules: basic training includes reducing waste, avoid spills, energy conservation etc. and advanced training includes awareness of sustainability manufacturing. Education is an essential tool for achieving sustainability.

ENERGY EFFICIENCY

For metal removal process, 90% of the environmental impact is due to electrical energy consumption. Another option for improving the environmental performance of manufacturing process is through improving energy efficiency during the operational phase. Initiation of measures to improve the energy consumption of machine tools provides substantial leverage to reduce associated environmental impacts in use phase. Newman et al. (2012) identified possible ways for improving the energy efficiency of the manufacturing process are through energy consumption mathematical modeling, energy consciousness metal cutting etc. Gutowski et al. (2006) further studied various electrical energy requirements for manufacturing process through analyzing specific energy requirements. The energy studies can be viewed in the process, machine, line and factory levels.

PROCESS EMISSION STUDIES

The percentage of Greenhouse Gases (GhGs) emitted indicates the contribution of the organization to Global Warming, Ozone Depletion and Eutrophication (Azzone and Noci, 1998). Thus steps to be taken towards reducing the emission of GhGs. Rashid et al (2008) stated reducing the cost of environmental damage can be improved by reducing emissions. Azzone and Noci (1998) stated that, the process efficiency indices aimed at assessing environmental performance which concerned air emissions. Also, Sustainable Production Indicators (SPI) proposed by LCSP (1998), included GhG intensity and air emission as an important indicator for assessing manufacturing process.

MATERIAL AND RESOURCE EFFICIENCY

Lovins (2003) suggested that from 100% of natural material extracted to make a product only 7% of materials people as useful product Out of 7%, 1% becomes durable and 6% becomes waste from customer first use. Going further, of this 1% consumer durables, only 0.02% is recycled or remanufactured and the balance of 0.98% becomes persistent waste from disposal (typically landfill). Thus, to increase efficiency, efforts have been made towards the recovery of products or materials at the end of their useful life which includes re-use, remanufacturing, and recycling. Efficiency is the ratio of output to the input of any system. Thus, material efficiency can be stated as 'the ratio of the output of products to the input of raw materials'. Worrell et al. (2011) defined material efficiency as 'reducing the consumption of primary materials without affecting the level of human activities qualitatively'. Young et al. (1994) suggested the need for an efficient material economy as the next efficiency revolution, claiming that it will require more than technological improvements alone, and calls for a shift of culture and ethics.

On the other hand, resource efficiency is the strategy that 'strives for efficient use, reduction of flow and consumption of resources drawn from nature' (Schmidt-Bleek 1996). Resource efficiency, also called resource productivity, has received much attention recently because it is focused on managing resources as a whole rather than being limited to one stage of the resource's life, such as waste minimization and pollution prevention which centered the focus on waste (cause) and its polluting effect. The strategy of resource efficiency defines itself as concerned with a wider scope of impacts on the environment including reducing the generation of waste, using fewer resources and using resources, production processes and extracting processes that have less impact towards the environment.

The above five strategies are the most frequently found in the literature evidence. However, the implementation of these strategies in SMEs was found to be very limited. The evidence of sustainable manufacturing practices in SMEs is discussed in the following section.

EVIDENCE OF SUSTAINABLE MANUFACTURING PRACTICES IN SMES

Millar and Russel (2011) evaluated the adaptation of sustainable manufacturing practices in Caribbean industries. However, the literature review suggests that 80% of the organization in the Caribbean are SMEs. Of the assessment of 76 companies, 87% of the companies are SMEs. They surveyed the understanding of sustainable concepts, tools by the organization. Further, they assessed current practices followed by the organization which was mainly focused on improving material efficiency and EoL disposal strategies. Also, non-investment for sustainable manufacturing practices has been identified as a principal barrier and various reasons have been discussed. Seidal et al (2007) conducted a study on sustainable manufacturing practices in SMEs. The application of tools such as life cycle analysis, stakeholder analysis, and SWOT analysis was discussed, and the roles of ecolabeling, environmental standards, national and international certification programs and government agencies were also explored. The benefits of collaboration with local universities and utilization of student projects as a low-cost development mechanism have been highlighted. Seidal et al (2006) discussed the establishment of sustainable practices in medium-scale furniture industries in New Zealand. With the experience, they explained various environmental policies, strategies required for successful implementation of SMEs. However, from the above review, there exists the potential for sustainable manufacturing practices in Indian SMEs. Thus, the implementation of these strategies may fetch better results as the scope is higher.

CONCLUSION

It can be concluded that from the survey, that sustainable manufacturing practices are the need of the hour. However, past literature suggests that successful implementation of sustainable manufacturing practice depends on the adaptation of correct strategies. From the literature, widely used five strategies, and their scope and advantages are detailed. However, evidence of sustainable manufacturing practices in Indian manufacturing organization was not recorded in the past. From the viewpoint of benefits, implementation of these practices can be largely expected in the near future.

REFERENCES

Abdul Rashid, S. H., Evans, S., & Longhurst, P. (2008). A comparison of four sustainable manufacturing strategies. *International Journal of Sustainable Engineering*, *1*(3), 214–229. doi:10.1080/19397030802513836

Anastas, P. T., & Zimmerman, J. B. (2006). The twelve principles of green engineering as a foundation for sustainability. *Sustainability Science and Engineering: Defining Principles*, 11-31.

Azzone, G., & Noci, G. (1998). Identifying effective PMSs for the deployment of green manufacturing strategies. *International Journal of Operations & Production Management*, *18*(4), 308–335. doi:10.1108/01443579810199711

Central Statistics Office. (2013). *Energy Statistics 2013*. Retrieved from http://mospiold.nic.in/Mospi_New/upload/Energy_Statistics_2013.pdf

Foolmaun, R. K., & Ramjeawon, T. (2008). Life Cycle Assessment (LCA) of PET bottles and comparative LCA of three disposal options in Mauritius. *International Journal of Environment and Waste Management*, *2*(1), 125–138. doi:10.1504/IJEWM.2008.016997

Gupta, M. C. (1995). Environmental management and its impact on the operations function. *International Journal of Operations & Production Management*, *15*(8), 34–51. doi:10.1108/01443579510094071

Gutowski, T., Dahmus, J., & Thiriez, A. (2006, May). Electrical energy requirements for manufacturing processes. *13th CIRP International Conference on Life Cycle Engineering*, 31.

Kleindorfer, P. R., Singhal, K., & Wassenhove, L. N. (2005). Sustainable operations management. *Production and Operations Management*, *14*(4), 482–492. doi:10.1111/j.1937-5956.2005.tb00235.x

Lovins, A. (2003). *Natural capitalism and the new North Carolina economy*. Rocky Mountain Institute. Available from: http://www.rmi.org/images/ other/Energy/E03-11_NapCapNewNCEcon.pdf

Lowell Center for Sustainable Production. (1998). *Sustainable production: A working definition*. Informal meeting of the committee members. Author.

Millar, H. H., & Russell, S. N. (2011). The adoption of sustainable manufacturing practices in the Caribbean. *Business Strategy and the Environment*, *20*(8), 512–526. doi:10.1002/bse.707

Muduli, K., Govindan, K., Barve, A., Kannan, D., & Geng, Y. (2013). Role of behavioural factors in green supply chain management implementation in Indian mining industries. *Resources, Conservation and Recycling*, *76*, 50–60. doi:10.1016/j.resconrec.2013.03.006

Newman, S. T., Nassehi, A., Imani-Asrai, R., & Dhokia, V. (2012). Energy efficient process planning for CNC machining. *CIRP Journal of Manufacturing Science and Technology*, *5*(2), 127–136. doi:10.1016/j.cirpj.2012.03.007

Nidumolu, R., Prahalad, C. K., & Rangaswami, M. R. (2009). Why sustainability is now the key driver of innovation. *Harvard Business Review*, *87*(9), 56–64.

OECD. (1998). *Waste minimisation profiles of OECD member countries*. Organisation for Economic Co-operation and Development.

Posselt, G., Kellens, K., Thiede, S., Herrmann, C., Dewulf, W., & Duflou, J. R. (2013). Combining Machine Tool Builder and Operator Perspective towards Energy and Resource Efficiency in Manufacturing. In *Re-engineering Manufacturing for Sustainability* (pp. 209–214). Springer Singapore. doi:10.1007/978-981-4451-48-2_34

Section 2
Green Environment and Management

Chapter 6
Green Environment and Management:
Environmental Management System (EMS)

Fatma Ince
Mersin University, Turkey

ABSTRACT

This chapter on environmental management system (EMS) addresses the relationship between organizational factors and environmental goals. Because the increasing awareness about environment affects the related groups and forces the firms have an environmental approach. The global competition, regulations and other pressures can change the tools of the sustainability. So, the firms start to consider an environmental management system as an innovative instrument as well as an adaptation strategy. Because, environmental standards and policies give the firm an opportunity to improve the business performance. From this viewpoint, this chapter provides an overview of the context, organizational drivers and implementation of the EMS.

INTRODUCTION

Firms start to embrace environmental practices as part of their competitive strategy. The pressure of the laws, customers, competitors, investors and employees accelerate the trend toward proactive environmental management systems. EMS is not an obligation; it helps the firm provide integrated system which includes economic, social and environmental considerations in all aspects of sustainability. Because, the firms are directly and indirectly in charge of the implementation of environmental systems. In this line, the firms can improve their capability to gain environmental goals and performance. Because of this, in this chapter, such dimensions of EMS are defined first and their relationships with organizational dimensions are explained afterwards.

For that purpose, the chapter is organized as three parts. In the first section of the chapter a review of literature on the Environmental Management System, its importance, its evolution, its scope and approaches are given. And the second part of the chapter is about organizational drivers of the EMS

DOI: 10.4018/978-1-5225-3537-9.ch006

and includes the relationship between business strategies, competitiveness, innovation, sustainable development, business performance and EMS. The remainder and also the last part of the chapter give information about the implementation of the EMS and the standards of the ISO 14001 and EMAS which are used for the sustainability environmental performance.

WHAT IS AN ENVIRONMENTAL MANAGEMENT SYSTEM (EMS)?

Environment face varied issues such as global warming, nuclear hazards, desertification, genetically modified organisms, deforestation. To face problems head on and to the preservation of the resources for the benefit of the future generations, a number of approaches emerge, such as maintaining ecosystem services, biodiversity conservation, sustainable livelihoods, human well-being, ecosystem health, integrity and sustainability (Laberge, 2007). Moreover, the interconnectedness of ecological systems and human leads to environment management, because of the growing awareness of nature. The economy and human health can be affected by pollution, deforestation, flooding and wetland loss. So these cascading effects need to new approach about the environment and human activities.

Various perspectives are available about environmental management because of the concept of subject. Saving the nature is one of the most narrow-minded viewpoints, because environmental management includes the effects and impacts of various interactions or activities that human societies have on the environment. The main point of the term is about being beneficial to the whole humanity (Sulphey and Safeer, 2015). Luke (2007) emphasizes the firms should define the managerial goals with considering the relations ecosystem and goods or services. So, EM should plan and operate with the protection and conservation of the politic and economic interest that surround the natural resources. Because modern resource management approach starts to see the nature as an economic and political asset.

More comprehensive and systematic approaches also emerge because of the rapid growth in industrial and urban development. The growing needs of population like agriculture, recreation and travel, protected areas, water resources, costs and other lands intense pressure on nature. From this point of view, the systems about human and environment became a necessity and need to planning and managing the relationship between human and nature. But this management style should include more comprehensive, participatory and integrated approach by the help of the multi-disciplinary applications.

The researches and opinions about using and conservation of natural resources, protecting of habitats, control of hazards, measuring the field of applied ecology are subjects of the environmental management. This management style is not only about industries or companies, but also about individuals that undertake to regulate and protect the health of the nature. Because the main idea behind the environmental management is to control human impact on the natural world and interaction with the nature in order to preserve the resources. This perspective comes from the aim of the improvement of human welfare for the present and future generations in the long run.

Some serious issues about nature, such as wasting resources, the sustainable development of raw materials and emission affect the business to change equipment and procedures to meet imposed standards. And consequently, the business communities try to new proactive methods to reduce the risk of noncompliance and improve the health or safety practices of society and workers. An environmental management system which is a set of processes or practices that enable the reducing damages and increasing the efficiency is one of the adaptation strategies of the business.

Sheldon and Yoxon (2012) explain the environmental management as "if financial management is managing the finances of a company and quality management is managing the quality of its products and process, then it stands to reason that environmental management is managing the environment that the company operates in". From this viewpoint Pataki and Crooty posited that (2017), "an Environmental Management System is a set of management tools and principles designed to create the administrative procedures that an organization needs to integrate environmental concerns into its daily business practices".

The basis of EMS base on a proactive management approach which help reduce the risks while improve the safety and health. Also, this system helps the organization satisfy the regulatory demands with cost-effective manner. It is not only the manner of following the environmental rules; it is also about the source efficiency and the stronger operational control. Because the system helps the employees expand awareness and the operational control. Other basic elements of an EMS include the following (EPA, 2017):

- Goal settings about environmental issues,
- Operations in full compliance with the environmental rules and requirements,
- Analyzing and reducing the environmental impacts
- Programs to gain the environmental aims
- Controlling and measuring the progress of success
- Creating and ensuring employee awareness and competence
- Reviewing progress of the EMS to make improvements

It is clearly seen that EMS has a holistic approach about environmental issues which includes all the activities of the business. From source efficiency to a change adaptation, EMS gives the firms an opportunity to reduce damages and increasing the efficiency. After the explaining of the EMS; the importance, the evolution, the scope and the approaches of the term will be given in this part.

The Importance of Environmental Management System

The wide scope of environmental management involves the complex relations that all creatures have their habitat. The environment has biotic and abiotic components as well as the human environment with the social cultural and economic systems. These social systems include civil rights, women and tribal issues, while the environmental systems include the endangered species and the protection of areas, water and lands (Sulphey and Safeer, 2015). Increasing natural disasters and other requirements, let experts and scientists highlight the need for a more careful, protective and integrated environmental approaches for living beings. However, the old and limited viewpoint that includes the protecting endangered species, improving human life quality, reducing the negative effects of the projects and enhanced water and air quality, should be expanded to really benefit.

Enhanced compliance, image with the public, regulators, lenders, investors, pollution prevention, and resource conservation are some of the social benefits of EMS, while increasing efficiency or reduced costs, new markets, improved environmental performance, employee awareness of environmental issues and responsibilities, and enhanced employee morale is contributing to the firms directly. On the other hand all employees spend time to learn and practice the environmental methods. And it has a cost for firms as well as receiving consultancy service on external resources or outside training (EPA, 2017). It will be a profitable investment in the long run at these initial costs. The answer of the question whether the EMS should be adopted or it is unnecessary for the firm is about the following questions (Pataki and Crotty, 2017):

- Does a lack of time or resources, prevent your business from taking charge of its environmental obligations?
- Is the state of your business' environmental affairs a significant liability?
- Are you looking at ways to improve your overall environmental performance?
- Does your business know how its environmental objectives relate to its business objectives?

If the answer to any question is yes, EMS gives the firms benefit and can help various important ways of using environmental approach and practices. Of course, there are some costs about setting capabilities, resources and ownership structure of the firm, but EMS gives opportunities savings, identifying and eliminating the environmental problems. Because making a product quality first time is cheaper than reworking later. Also, being proactive has less cost than compensating it. So, it can be said that prevent practices are most cost-effective than managing the pollution after exiting. Such environmental sensitive activities attract the customers and help the firm's sustainability.

Moreover, EMS has economic and global outcomes in the long term. Like all activities of the firms which are a part of the economic system, this kind of management systems affects the future of the humans and next generations. So, firms should consider the resource effectiveness and other environmental activities not only for individuals but also for the society. From this point of view, industrializing nations like Japan and Europe adopt environmental protection as part of their international competitive strategy. These environmental strategies take place through the firms due to the press from customers, competitors, employees and governments. Because the relationship between nature and firms' performance. So, the governments start to pressure the firms to embrace proactive environmental systems to reduce natural disasters. Lastly, it is obviously seen that EMS is a shield both external factors such as the laws, the competitive and financial institutions and internal factors like efficiency. So, EMS is not only an adaptation strategy, both also it is a proactive approach as sectoral leadership and exploring newer markets. Establishing an EMS can be a good investment to improve organizational performance and quality by reducing the risks.

The Evolution of Environmental Management System

European Committee for Standardization defined the EMS with the management approach as "the part of the overall management system that includes organizational structures, planning activities, responsibilities, practices, procedures, processes and resources for developing implementing, achieving, reviewing and maintaining the environmental policy". This dedication has a holistic approach because it characterizes the EMS as a systems management strategy considering the environment. As far as this point, there are various views on the environment from the 1960s to today. Based on this EMS development stage can be summarized as follows (Weiß and Bentlage, 2006):

- Environmental awareness (1960s)
- End of pipe approach (1970s)
- Process-integrated approach (1975s)
- Environmental co-ordination (1980s)

- Environmental management: EMS and Auditing (1985s)
- EMS standards (BS7750, ISO 14001) (1992s)
- EMAS-Reporting Certification (2000s)
- Sustainability Approach (2000s)

As seen from the items, the evolution of the management has a long process from the environmental awareness to the EMS and sustainability. Firstly, the awareness about the relationship between substances that hazardous to the nature and firms emerge, then various integrated approaches are considered, lastly the standards are set about the environment by the committees. The reasons behind the growing interest of the local authorities and other committees arise from four main factors (UNEP, 2017):

- **Complexity of Environmental Challenges:** Instead of avoiding the problem, the inadequacy of compensating practices as end of pipe approach has begun to be noticed. Because urban issues need to more holistic aspects included both socioeconomic and environmental outcomes.
- **Pressure From Citizens:** Social responsibility groups as citizens, cities and their representatives and officials force the local authorities to consider their environmental performance like the private sector, for better environmental quality.
- **Rapid Population and Urban Growth:** The cities are growing with increasing quality of life demand because of the needs of houses, offices, shops, factories, roads and other services.
- **International and National Political Pressure:** According to the better understanding of the influences of natural disasters, national and international agreements and laws increase the pressure on the local authorities to assure a cleaner environment.

Because of these reasons EMS applied to local and international authorities, committees and firms provide a systematic approach to develop practices, policies and practices that can respond to the social, economic and environmental challenges ahead.

The Scope of Environmental Management System

The scope of an environmental management system is up to the environmental policies and the possibilities of the business. Intrinsically the environmental perspective of the business is a function of the entire management system. The vision and other main targets of the organization can affect the environmental activities and set out the boundaries of the future applies to.

For the financing of environmental investment and designing of instruments to enable raising of funds, some deciding making principles are emerging. These principles shape the way people interact and help guide with the environment as nations develop. According to Sulphey and Safeer (2015) the five principles of environmental management can be summarized as follows:

- **The Polluter Pays Principle (PPP):** In the Rio Conference of 1992, the polluter pays principle endorsed by the public authorities. The underlying aspect behind this principle is about making more efficient production process through promotion of economic efficiency of pollution control policies and reducing potential trade distortions. According to the using of resources and the damages caused to the society, the responsible foundations pay the price or tax.

- **Users Pays Principle (UPP):** This principle is based on the full cost payment of losses for the future generations. Users pay principle is similar with the polluter pays principle because of the focusing the equality and achieving efficiency. The costs of using the resources and its related service should be paid by the beneficiaries or customers.
- **Precautionary Principle (PP):** The aim of this principle is to ease environmental stress before conclusive evidence of damage exists and adopt policy when the clear evidence is available. Unfortunately, there is a lack of scientific certainty about the threats or damages. But this gap should not use for postponing of the cost and prevent environmental degradation.
- **Subsidiary Principle (SP):** Despite this principle is developed for political decisions, it is used for applying the UPP, PP and the PPP. These political standards and processes are based on the decisions which are made by the public authority dependants with effective action.
- **Intergenerational Equity Principle (IEP):** The main concern of the principle is sustainable development according to the Brundtland Report. So, the basic point of the principle is the environmental accounting measures of sustainable income which includes the trilogy of economic, environmental and social outcomes.

These five principles are used to guide the sectoral policies and budgeting process for the purpose of minimizing the potential distortions and resource efficiency. The environmental management system helps companies their regulatory demands, reducing their environmental impacts and increasing their operating efficiency. Also the energy conservation and operational control can be designed based on the environmental management system more easily. There are some requirements in the organization for the efficiency of the system. These requirements and also the basic elements of EMS can be defined as follows (GIIRS, 2017):

- Environmental goals, reviewing
- Environmental impact and legal requirement analyze
- Environmental goals setting
- The target of reducing environmental impact and comply with legal requirements
- The program is established for the targets and goals
- The aim of monitoring and measuring
- The environmental awareness and ability of employees
- Monitoring the progression of environmental management system
- Improving the system and practices

The system has internal and external potential costs such as employee time, potential consulting assistance and outside training of personnel while it has lots of benefits. First of all, the system improves environmental performance and increase efficiency, while it reduces the costs of environmental issues and responsibilities. Then, the system helps the organization enhance the compliances and employee morale with a good image towards the public, regulators and investors. Also, the pollution prevention, resource conservation, the opportunities of new markets are some of the benefits of the environmental management system (Poveda and Young, 2015). Finally, the long term performance efficiency can be achieved with this proactive system, because of the facilitating effects of this management style.

The scope of environmental management system follows a repeating cycle, because it is an endless process with five stages. For the environmental performance, the system starts the *environmental policy* which is used for the planning of the targets. Whether the targets are being met or not is considered in the *implementation* stage. If there is a gap between existing and planning outcomes, corrective action is taken not a moment too soon. This feedback helps the manager review the *evolution* and whether or not the EMS is working according to the plan (Weiß and Bentlage, 2006). So the *revised* plan is the final stage of this cycle as well as it is the beginning of the next cycle of the EMS. The revolution of the EMS cycle gives the organization the opportunity of improving.

Approaches of Environmental Management System

In some situations, the reason of events may be more important than the events. So, environment and overall performance need to systematic identification and correction of system deficiencies. This means that environmental management is only available with continuous improvement and conscious of quality. Therefore, the requirements of an efficient EMS can explain with the basic of Total Quality Management (TQM) and the Deming Cycle. The basis of the TQM is based on the togetherness of management and employees to solve problems and improve the process. From top manager to front worker, the organization has active employee involvement to focus on the system.

The basic dimensions of the EMS are built on TQM concepts. So, the stages of the putting the EMS into effect and the principles of it, can explain with the Deming Cycle named PDCA cycle. According to the Deming Quality Management Model, the elements of an EMS can define as follows (Sammalisto and Brorson, 2008):

- **Plan:** A proactive plan, which is considering the environmental policy and aspects, legal requirements and organizational targets, starts the cycle.
- **Do:** Employees need training about environmental awareness and behaviors. Also EMS is a team work, so self-responsibility is needed for effective communication between all level employees. Organizational structure should help the process, documentation, operation and control of the activities.
- **Check and Correct:** the progress should monitor and measure to fulfil the plan effectually. The Conformance, corrective and preventive records, audits are so important to achieving the environmental goals. The gap between the present results and plan should evaluate to assure success.
- **Act and Improve:** The results of the checking are also an input of the improving process. The endless cycle always repeats itself due to the management review.

These steps also provide a basis the ISO 14001 Environmental Management standard which is explained in the last section of the chapter. This is a complete environmental model which includes the all level employee participations and commitment. The adaptation degree of the firms can change due to the strategic decision for determining whether to move ahead with complete or partial EMS. According to Rendell and McGinty (2004), the decision about implementation of EMS requires a systematic process. It is recommended that a municipality follows a systematic process when deciding whether it should implement an EMS. It is all too easy to decide that an EMS is not required simply because the task seems difficult or one is not familiar with the terminology. The decision making process has a step-by-step approach to support the decision and include the followings (Rendell and McGinty, 2004):

- **Gather Information:** The managers or decision-makers are in need of information gathering and classification about environment to decide if an EMS makes sense for the firm.
- **Assess Drivers:** To progress in the decision process, internal and external drivers should discuss and determine.
- **Estimate Cost and Benefits:** Estimated benefits and costs of the EMS should assess for operational effectiveness.
- **Set Goals:** Strategic and operational goals, establish to take concrete steps in this stage. Also, all aspects of the implementation are determined and publicize to assure a better environmental position.

The degree of the embracement of EMS is also about fulfilling the expectations of the stakeholders, regulators, local residents and other commissioner. Because, while the firm is adopting a long term environmental approach, it has also satisfied the needs of pressure or interest groups.

ORGANIZATIONAL DRIVERS FOR ENVIRONMENTAL MANAGEMENT SYSTEM

The convenience of organizational structure, management style and environmental requirements is so important to carry into effect. According to the results of the Pennsylvania EMS pilot project, there are some key success factors. The support of the management is the first factor which can affect the operational process easily. So, a management commitment should be established to monitor and review. EMS is a team work, so a clear project team and a dedicated project leader or manager have critical roles. The skills of the team and meeting management can provide the employee commitment. EMS has a strategic perspective, so the commitment of local authority is needed (Rendell and McGinty, 2004). The integration of municipal system and organizational process should gain success in this endless journey.

All environmentally-conscious approach supports the organizational goals and implementation with considering the future generations. So, this broad perspective includes not only strategic management and resource management, but also organizational competitiveness and business performance.

Business Strategies and EMS

EMS is seen as a tool of business strategy development because of the holistic content of the term. Strategic aspects start from the top management which is a driving force of environmental change. Firstly, top managers decide the level of the participation of environmental management. When the business strategy is defined, environmental intention should be concretized clearly. Full cost accounting, total cost assessment and life-cycle costs are some of the tools of the estimating. But the focus of the cost is not enough to embrace the environmental approach. Environmental accounting which provides normative models of conduct is used for the measuring environmental performance exceeding regulatory standards.

According to Tinsley (2002) study, four levels EMS model can see in the electronic and telecommunications organizations. This model is about the degree of integration of the EMS and business strategy which is summarized as following (Tinsley, 2002):

- **Devoid EMS Model:** In this model, there is no accreditation of EMS. It just includes the implementation of EMS with the lower level environmental risks. The aim of this process is to reduce the operational disadvantages of the organizations. In the model, the firm cannot see an important reason to make enough effort for the EMS. So, it is an avoidance approach which provides energy efficiency and reducing the waste of resources.
- **Isolated EMS Model:** In this model, the firm has an accredited system, but it is limited to one or two units of the organizations. So, the environmental issues are a problem for only one small team. The main goals of the model are to follow the environmental rules and avoid legal sanctions.
- **Devolved EMS Model:** In this model, the firm has an EMS accreditation system and this system is seen as a part of daily operational activities. A responsible manager assures the progress and with periodic reports and training programs. Also, monitoring the energy efficiency and waste recycling is important for the organizations.
- **Integrated EMS Model:** This model has a systemic approach to the EMS and includes both environmental health and safety factors and quality system. EMS is the key operational elements of the whole operations. As a proactive approach, integrated EMS Model provides the employee commitment and self-regulation.

Transfer of models to application based on the organizational success factors such as communication, awareness, participation and training. All models have some barriers and some advantages for the organizations. After the support of the management, EMS is need of policy integration. The business policy and environmental policy should be in a harmony because they are part of each other, so they cannot be considered separately. From the strategic level to the tactics, environmental principles should be considered for all activities. So, the mission, goal, objective and strategy are established in the framework of the EMS. Then, implementation, monitoring and controlling of the process perform due to the environmental plans and programs (Yakhou and Dorweiler, 2004).

Innovation, Competitiveness and EMS

Innovation is one of the strong competitiveness tools for the firm which wants to be a sector leader. Also, environmental aspect gives the firms an opportunity to explore or transfer an innovation to the environmental process or product. Moreover, firms can choose to environmental innovation as a business strategy for sustainability. If the direction of the technological change can be defined with environmental innovation, it means that the firm has ecologically specified sustainability targets. The kind of innovation can emphasize the degree of environmental adaptation of the firm.

According to the Wagner (2008) study, EMS is associated with the process innovations, while it is not associated with product innovations. The study finds that information about consumers and eco-labelling activities have a positive association with EMS without regardless of the firm size. In another study dealing with innovation from a different side, EMS is considered as an improving tool of environmental performance. According to the study, a firm's voluntary adaptation of the EMS and its environmental innovation behavior can affect each other (Frondel et al., 2008). The results of the investigation show that innovation behavior is correlated to the stringency of environmental policy. The scope of the environmental innovation is about the new and modified process, practices, techniques and products that avoid undesirable environmental impacts. The solution can be both technical and organizational according to the business strategy. If the innovation can lead the organizational change, then it can be thought as a competitiveness instrument.

According to the resource-based view, critical resources provide a firm a competitive advantage. From this viewpoint, EMS can be a part of the value chain management, because it has more than cost efficiency. For this reason, EMS should be seen not only a cost oriented method, but also a strategic opportunity for the competitiveness. The awareness of environmental benefits needs a long term environmental mindset. Becoming a supply chain members and government adaptation are important as well as corporate social responsibility, profitability and cost effectiveness. According to Sroufe et al. (1998), EMS has a lot of benefit for the firms, but there are some requirements to utilize these benefits. Firstly, the firm thinks the process strategically and consider the external and internal risks together. The aim of reducing costs and taking the opportunities is the part of environmental mindset. Secondly, the firm should think about technological reality and determine the internal risks. After this process, the firm has an awareness of environmental benefits so; it takes the decision about the degree of EMS implantation. At the end of this process, the impact of the EMS leads the firm to a better environmental position in a global competition.

Business Performance, Sustainable Development and EMS

Each firm has own environmental patterns and they can provide different performance improvements. Therefore, a better understanding of the influence of EMS should include the different types of performances such as finance, manufacturing and operational performances. Because, there are different type of environmental practices due to the participation level of the EMS. Normally, the effects of the EMS can be varied to the organizations.

According to González-Benito and González-Benito (2005), environmental productivity is a source of strategic resources and capabilities. They emphasize that environmental approach has a potential to generate sustainability. The authors choose to consider environmental management practices as planning and organizing, product related operational, process related operational and communications practices. The results of the study show that operational performance objectives such as reliability, quality, and volume flexibility can be improved with more ecological supply chain management and with adequate recycling and reverse logistics systems.

The relationship between the sustainable development and EMS can explain with the resource based view. For the sustainability, intangible resources such as human resources and reputation are important as much as tangible or financial resources. From this point of view, Russo and Fouts (1997) notice that resources need to analysis to consider the firm organizational capabilities which include three dimensions. First one is the physical assets and the technologies and skills required to use them, the second one is human resources and organizational capabilities, which include culture, commitment, and capabilities for integration and communication, last one is the intangible resources of reputation and political acumen.

An investigation about the oil and gas industry shows that the business and environmental issues were associated with the emergence of unique organizational capabilities. The proactive strategies can help the act in environmental uncertainties due to the capabilities. In the research, Sharma and Vredenburg (1998) identified three key capabilities derived from environmental commitment. The capability for the capabilities of influencing stakeholders or mitigating their pressures; the capability for higher-order learning which include the exploring new interpretations of the existing procedures or new alternatives; the capability for continuous innovation. These capabilities are seen to have implications for firm competitiveness.

Sustainable product development is only a part of environmental innovation and competitiveness. According to the Maxwell and Van Der Vorst (2003), the initiatives of the eco-design support the firms develop more sustainable products. Environmentally Superior Product (ESP) of Ireland is one of the most important examples of initiatives that support industrial companies to incorporate a more sustainable approach to the development of products and services.

Therefore, multinational enterprises start to engage in corporate citizenship programs to promote sustainable development. It is a tool of external relationships with stakeholders to address social problems, while it has flexible environmental practices to pursue sustainable development objectives (Rondinelli and Berry, 2000). Research conducted by Darnall et al. (2008) also supports these results. They notice that the complementary resources and capabilities of the firms, which have motivation to adopt the EMS, have greater overall facility-level business performance.

Fülöp and Pelcz (2010) posited that corporate sustainability needs the four elements. First one is sustainable development and it determines the environmental matters and common societal goals. Second one is corporate social responsibility and it is about the reasons of the environmental approaches. Stakeholder theory is the third one and it is about the business arguments. Lastly, corporate accountability theory includes the ethical arguments in reports of sustainable performance. On the other hand, environmental approaches are also important for the countries' sustainable development and economies as well as firms.

THE IMPLEMENTATION OF EMS

EMS needs to be structured by the roles and responsibilities at the organizational level. Firstly, the process of the EMS should define clearly to enable to act in line with plan. After the preparation phase, organizational structure is brought into compliance with the roles and responsibilities. Finally, all stages of the EMS should be documented and communicated, after the determination of the structure and responsibilities. So, the support and monitoring of the top management are so important in the implementation of EMS. Because EMS is an extensive approach that requires full participation.

For organizational efficiency, there are some complimentary management tools and methods for improving the environmental performance, such as Pareto analysis, process mapping, brain writing, root cause analysis and bubble up and down. Such systems approaches help to understand the key dimensions of the processes, operations and management issues. The concept of these management tools can be summarized as follows (Rendell and McGinty, 2004):

- **Pareto Analysis:** "Vital few" and "trivial many" can explore with this method which include that 80% of results are produced by 20% of actions. Known as the 80/20 rule gives information about the importance and effects of the behaviors and activities.
- **Process Mapping:** The processes are needed to be identified detail to complete a task or project timely. So, inputs and outputs of the process are determined though mapping to see the progress clearly.
- **Brain Writing:** This brain storming approach is aimed at aiding EMS processes by stimulating creativity. The principles of the method based on the identifying a range of alternative solutions about problems and process.

- **Root Cause Analysis:** Fishbone, Failure Mode and Effect Analysis and Cause and Effect methods are some of the root cause analyses. These problem solving and decoding methods give the employees an opportunity to understand the issues in detail.

If the participant of the employees is a necessity for the organization, these deciding and problem solving technics help the increasing the organizational attention. As a quality approach, EMS is established by the detailed implementation plan which includes the followings (Rendell and McGinty, 2004; Weiß and Bentlage, 2006):

- **Policy Development and Management Commitment:** This phase embraces the development processes such as train core team, secure management commitment, resources and publicize, define the EMS boundary, develop or review energy and environmental policy, perform a gap assessment, develop an implementation plan and schedule.
- **Planning and Information Management:** Information management needs all of the planning requirements such as development objectives and targets, develop or revise energy and environmental management programs, perform aspects and impacts analysis, identify key stakeholders and requirements.
- **EMS Manual Preparation, Training and Implementation:** This phase is about the developing procedures and documents about EMS such as developing training modules, procedures for operational control, EMS auditing procedures, monitoring and measurement procedures, procedures for implementing corrective and preventive actions, management review procedure, emergency preparedness and response procedures, any necessary additional documentation. Also, it is about to assemble existing documentation and EMS manual as well as delivering training to staff. Finally, EMS implementation can start as planning.
- **Initiate Continuous Improvement of the EMS:** The last phase includes the self-audit, EMS and EMS management review cycle performance. Also, one of the EMS certificates can be taken if requested in this section.

This long process should be restricted for a certain period to achieve the environmental management in time. The preparations for the implementation in other words the first three phases of the implementation plan require the most effort and costs. These processes have been systematized and certified by various management systems. The most well-known of EMS model is ISO 14001 Environmental Management System as an international quality system. Secondly, more common EMS model European Eco-Management and Audit Scheme (EMAS). Also there are some other EMS models such as Responsible Care Model developed by the American Chemical Council (ACC), EPA National Enforcement Investigation Center (NEIC) Compliance Focused EMS, US Department of Justice (DOJ) Seven Key Compliance Program Elements, National Biosolids Partnership EMS, DEP College & University EMS and Best Practices Manual, The Occupational Health and Safety Assessment Series OHSAS 18001, The British guidance standard BS 8800, The Security Certificate Contractors (SCC), Standard on Quality Management ISO 9001:2000. The most common standards will explain in the next two titles of the chapter.

The ISO 14001 Environmental Management System

The ISO 14000 series has two major groups as organization-based and product based standards. Firstly, standards of organization, evaluation include Environmental Management Systems (ISO 14001 and ISO 14004), Environmental Performance Evaluation (ISO 14031 and ISO/TR 14032), and Environmental Auditing (ISO 19011 and ISO 14015). Secondly, product based standards are Life Cycle Assessment (ISO 14040, ISO 14041, ISO 14042, ISO 14043, ISO/TR 14047, ISO/TS 14048 and ISO/TR 14049), Environmental Labelling (ISO 14020, ISO 14021, ISO 14024 and ISO/TR 14025), Environmental Aspects in Product Standards (ISO 14062).

The aim of the organization based standards are to establish, maintain and evaluate an EMS, while the aim of the product based ones is to determinate the environmental impacts of product and services by collecting information and communicating with interested groups. For EMS, the most important standard is ISO 14001 because it is an environmental management standard, it is not only about the outputs or performance. This standard is not regulated by the law, it means that it is voluntary, so it can be applied to any country in the world. The main idea of the standard is to improve resource efficiency and control environmental impacts. Continuous improvement diagram based on the Deming cycle and give information about principles of the standard. The diagrams include five stages as following (iso14001.com, 2017):

- **Environmental Policy:** Top management is committed to regulatory compliance as pollution prevention and continuous improvement.
- **Planning:** In this stage, environmental interactions, significant aspects, legal and other requirements are identified and environmental objectives, targets and programs are developed.
- **Implementation and Operation:** Structure and responsibilities are defined; training needs are identified, and all procedures and plans are established in this stage. After the identification, implementation takes places according to the plans.
- **Checking:** Environmental interactions are monitored and measured; compliance is evaluated; corrective action and preventative action of the system are established; periodic internal audits of the EMS are performed at this stage.
- **Management Review:** Environmental performance, policy, priorities and objectives are reviewed and improvements are recommended by the top managers at this stage.

After the reviewing, the results of the practices are used the inputs of the first stage, so this cycle is always continuing to improve the efficiency of environmental management. Plan, do, check and act as known Deming cycle is basic of the standard and it is followed as environmental policy and planning, implementation of the EMS, checking and reviewing, EMS continuous improvement (Sammalisto and Brorson (2008). Environmental impacts of the standards emphasize the air and resource pollution, resource depletion and noise. For the success of the standard, the process needs to be documented. Because of the importance of the information and monitoring, relevant information is collected and classified both on paper and in the computer file. The operational control system is only available by recording the all process. Both the requirements of the laws, implementing and checking and corrective action need to establish a regular measuring. The results of the measuring are also a decision making tool for the top manager and it helps the employee self-improvement about environmental goals and organizational effectiveness. In this process, maintenance has a critical role for the implement of the EMS. The EMS is reviewed consistently to ensure sustainability and adequacy.

To accessibility and be able to adapt, EMS has general principles. This flexibility gives the firms an opportunity to design and progress the organizational structure. But the scope, defining is important to clarify the boundaries of the firms. Because organizational boundaries state the credibility of the environmental management system. There may be an incompatible unit, and it may hinder the EMS process. So, it is suggested that the scope should be defined firstly, and then impacts and legal requirements should be considered.

EU Eco-Management and Audit Scheme (EMAS)

European Commission defined the EMAS as "EMAS is a premium management instrument developed by the European Commission for companies and other organizations to evaluate, report, and improve their environmental performance". EMAS has a flexible characteristic for every type of organization that wants to improve the environmental performance. EMAS support three main dimensions of the environmental management. These dimensions can explain as following (ec.europa.eu, 2017):

- **Credibility:** EMAS registration process needs the third party verification to guarantee the independence of the process.
- **Performance:** The aim of this section is to support the organizations to find the right environmental tools. This voluntary process emphasizes not only the evaluating, but also reducing the firms environmental impact.
- **Transparency:** Both the employees and external related groups should inform the environmental involvement. The transparency and giving information to the public is important for the EMAS success.

European commission explains the EMAS process with ten steps and four key principles to becoming an environmental leader or achieving sustainable improvement. Contact the component body and an initial environmental review are first to step of the process, while the Deming cycle is the next four steps. Do, check, act and plan cycle is enlarged as implementation the environmental management system, internal environmental audit, continuous environmental performance improvements, environmental policy and programs. In consequence of the Deming cycle, an environmental report is obtained to use for external process which include the last three steps as verification and validation, registration, promote. It is clearly seen from the steps that the purposes of the EMAS to help evaluation, impact reduction, reputation, environmental policy, culture change, improvement, pollution prevention, legal compliance, corporate responsibility, resource management. Moreover the specific and main benefits of the EMAS can be explained as following (nqa.com, 2017):

- **Enhanced Credibility and Recognition:** The external audit and launching make reliability and accuracy image for the firm. So, the information transparency gives the public correct information and help people feel safety.
- **EMAS Register:** EMAS regulation and other environmental and organizational requirements take part in the registration lists. So, all needs can be seen clearly.
- **EMAS Logo:** To commitment and to promote the activities, a sharing EMAS logo is used. It gives also brand reliability to external stakeholders.

As can be understood from the explanations the benefits of the EMAS and ISO 14001 are similar, but EMAS also include the publicly available environmental statement and verified results.

To gain EMAS registration has six stages which start with ISO 14001. The firm firstly has to meet the requirements of the ISO 14001 and then can continue the EMAS registration. Secondly, the firm prepare an environmental statement. If this statement can pass the verification step, validation starts. In this step, the firm can demonstrate the ability to fulfil the requirements of EMAS, it gains acceptance to application step. After the application, register number is given and the firm can use the EMAS logo finally. Therefore, it can be said that EMAS may take a long time, but it is a strong tool of management both for internal and external shareholders.

The overall framework of quality systems for environmental management can see the requirements of the ISO 14001 and EMAS standards. This framework can give some information about the other standards and their implementation.

CONCLUSION

Increasingly scarce resources such as wasting water, raw materials and energy affect the way the firms work and change it permanently. To adapt to this change is not only a willing process but also it is a legal necessity. Because laws require firms to change equipment and procedures to meet imposed standards and to reduce the wasting resources. As individual awareness increases because of concerns about the future, consumer preferences change as well as the efforts of some voluntary associations. Environmental issues are becoming increasingly important and attract the attention of businesses. So, firms start to use the protective and proactive environmental measures and change content of some practices to reach new target markets with environmentally friendly policy.

EMS and the standards about it have total quality characteristics. But first practices about EMS are more limited such as environmental awareness, end of pipe approach. In 1960s, environmental issues are seen as cost enhancing factors. Since the standard commissions and some councils have established, the systematic and overall environmental viewpoint start to improve such as various EMS standards, EMAS and sustainability approaches. The EMS adaptation will be a profitable investment in the long run at these initial costs. But the degree of the participation an EMS depends on the strategy of the firm. The firm can choose the devoid EMS model for just environmental implementation without any accreditations. The models of EMS, from devoid to integrated model show the level of accreditation the system and business strategy of the firm.

Innovative perspective gives some environmental outputs like environmental product or services. Environmental innovation as a competitive tool gives the firm strong strategic and market position. So, business performance can improve the various environmental implementations. To sustainable development, the EMS should always manage with the technics and analysis like pareto, process mapping and brain writing. Because, the measurements are important for the review and maintenance of the EMS cycle. For planning, implementation and improvement of the EMS some standards can be used. The ISO 14000 standard series and Eco-Management and Audit Scheme (EMAS) are the most common standards for the sustainable development with EMS.

Lastly, it can be said that it is the responsibility of the firms as well as everyone to protect the resources that are exhausted for future generations. And, the firms should create value by offering new methods or products for environmentally sensitive practices or consumptions.

REFERENCES

Abaza, H., Bisset, R., & Sadler, B. (2004). *Environmental Impact Assessment and Strategic Environmental Assessment: Towards an Integrated Approach*. Geneva: UNEP.

Darnall, N., Henriques, I., & Sadorsky, P. (2008). Do environmental management systems improve business performance in an international setting?. *Journal of International Management*, *14*(4), 364–376. doi:10.1016/j.intman.2007.09.006

EPA. (2007). *Learn About Environmental Management Systems*. Retrieved April 3, 2017, from https: // www.epa. gov/ems /learn-about-environmental-management-systems

Frondel, M., Horbach, J., & Rennings, K. (2008). What triggers environmental management and innovation? Empirical evidence for Germany. *Ecological Economics*, *66*(1), 153–160. doi:10.1016/j.ecolecon.2007.08.016

Fülöp, G., & Pelcz, I. G. (2010). Concept of environmentally-conscious strategic management. *European Integration Studies*, *8*(1), 143-172.

GIIRS. (2017). *Emerging Market Assessment Resource Guide: Implementing an Environmental Management System*. Retrieved April 17, 2017, from http: //www. bcorporation. net/sites/ default/files /documents /bestpractices /em_implementing_ environmental_management_system.pdf

González-Benito, J., & González-Benito, Ó. (2005). Environmental proactivity and business performance: An empirical analysis. *Omega*, *33*(1), 1–15. doi:10.1016/j.omega.2004.03.002

ISO 14001. (2017). *ISO 14001 Requirements*. Retrieved April 16, 2017 from http: //www.iso14001.com. au /iso-14001-requirements.html

Laberge, Y. (2007). Ecomanagerialism. In P. Robbins (Ed.), *Encyclopedia of Environment and Society* (pp. 593–594). Sage Publications.

Luke, T. (2007). Ecomanagerialism. In P. Robbins (Ed.), *Encyclopedia of Environment and Society* (pp. 524–525). Sage Publications.

Maxwell, D., & Van der Vorst, R. (2003). Developing sustainable products and services. *Journal of Cleaner Production*, *11*(8), 883–895. doi:10.1016/S0959-6526(02)00164-6

Nqa. (2017). *EMAS: EU Eco-Management and Audit Scheme*. Retrieved April 16, 2017 from https: // www.nqa. com/en-gb/ certification/ standards/emas

Pataki, G. E., & Crotty, E. M. (2017). *Understanding and Implementing an Environmental Management System: A step by step guide*. New York State Department of Environmental Conversation Pollution Prevention Unit. Retrieved April 3, 2017, from http: //www.dec. ny.gov/ docs/permits_ej _operations_pdf/ p2emsstep1.pdf

Poveda, C. A., & Young, R. (2015). Potential benefits of developing and implementing environmental and sustainability rating systems: Making the case for the need of diversification. *International Journal of Sustainable Built Environment*, *4*(1), 1–11. doi:10.1016/j.ijsbe.2014.12.003

Rendell, E. G., & McGinty, K. A. (2004). *Environmental Management Systems: A Guidebook for Improving Energy and Environmental Performance in Local Government*. Five Winds International. Retrieved April 16, 2017, from http: //www. lgean.org/documents /EMSGuidebookPA.pdf

Rondinelli, D. A., & Berry, M. A. (2000). Environmental citizenship in multinational corporations: Social responsibility and sustainable development. *European Management Journal*, *18*(1), 70–84. doi:10.1016/S0263-2373(99)00070-5

Russo, M. V., & Fouts, P. A. (1997). A resource-based perspective on corporate environmental performance and profitability. *Academy of Management Journal*, *40*(3), 534–559. doi:10.2307/257052

Sammalisto, K., & Brorson, T. (2008). Training and communication in the implementation of environmental management systems (ISO 14001): A case study at the University of Gävle, Sweden. *Journal of Cleaner Production*, *16*(3), 299–309. doi:10.1016/j.jclepro.2006.07.029

Satpathy, I. (2005). *Environment Management*. New Delhi: Excel Books.

Sharma, S., & Vredenburg, H. (1998). Proactive corporate environmental strategy and the development of competitively valuable organizational capabilities. *Strategic Management Journal*, *19*(8), 729–753. doi:10.1002/(SICI)1097-0266(199808)19:8<729::AID-SMJ967>3.0.CO;2-4

Sheldon, C., & Yoxon, M. (2006). *Environmental management systems: a step-by-step guide to implementation and maintenance* (3rd ed.). Routledge.

Sroufe, R. P., Melnyk, S. A., & Vastag, G. (1998). *Environmental Management Systems as a source of competitive advantage*. Michigan State University-Department of Marketing and Supply Chain Management.

Sulphey, M. M., & Safeer, M. M. (2015). *Introduction to Environment Management* (3rd ed.). Delhi: PHI Learning Pvt. Ltd.

Tinsley, S. (2002). EMS models for business strategy development. *Business Strategy and the Environment*, *11*(6), 376–390. doi:10.1002/bse.340

UNEP. (2017). *UNEP / FIDIC / ICLEI Urban Environmental Management: Environmental Management Training Resources Kit*. Retrieved April 17, 2017, from http: //www. unep.or.jp /ietc/Announcements / EMSkit_launch.asp

Wagner, M. (2008). Empirical influence of environmental management on innovation: Evidence from Europe. *Ecological Economics*, *66*(2), 392–402. doi:10.1016/j.ecolecon.2007.10.001

Weiß, P., & Bentlage, J. (2006). Environmental Management Systems and Certification. Baltic University Press.

Yakhou, M., & Dorweiler, V. P. (2004). Environmental accounting: An essential component of business strategy. *Business Strategy and the Environment*, *13*(2), 65–77. doi:10.1002/bse.395

Chapter 7
A Research of the Influence on the Consumer Behavior From CSR Information Embedded in the Process of Online Purchasing

Li Yu
Qilu University of Technology, China & Shandong University, China

Sang-Bing Tsai
University of Electronic Science and Technology of China (Zhongshan Institute), China & Civil Aviation University of China, China

Dakai Li
Shandong University of Finance and Economics, China

Limei Teng
Beijing Normal University, China

Guodong Li
Civil Aviation University of China, China

ABSTRACT

As the online purchasing is becoming prevalent, more and more corporates started to show their CSR information in the processes of online retail. For example: some CSR information could be showed in the "COMMODITY DETAILS". Could this kind of information affect consumer behavior? By recognizing three types of embedded CSRC information, according to the consumer behavior experiments, this paper will show CSR's influences on the consumer behavior such as evaluation of the company, purchasing intention, and purchasing intention. By the comparison of the results after three types' CSR communication, we found that CSR information could significantly affect consumer behavior and the consumer response will differ from different types of CSR information embedded.

DOI: 10.4018/978-1-5225-3537-9.ch007

INTRODUCTION

It is one of the most important blasting fuses that consumers' awareness of corporate social responsibility(CSR) which led to the movement of CSR. An increasing number of enterprises not only paid more attention to their CSR behaviors but also to the CSR communications(CSRC), especially when the Internet age came with the high-speed and large extension of information exchange. It is found that lots of enterprises tried to lead a CSRC in the process of online retail. For example, some CSR information was included in a description of online products in the e-business, which we call it "Embedded CSRC in E-business" (E-ECSRC).

More tremendous number of works focused on the CSR-consumer binomial have mainly concentrated on understanding how CSR perceptions influence the cognitive, affective and behavioral responses of consumers to the firm (Alvarado-Herrera A et al., 2015). For example, CSR perceptions have been shown to have a positive impact on various types of responses, such as the consumer's purchase intention (Trudel and Cotte, 2009) and attitude towards the firm (Brown and Dacin, 1997; Du et al., 2011; Sen and Bhattacharya, 2001). But few researches focused in the influence of CSR information in the specific process of online retail. Can E-ECSRC touch their consumers so that company can realize the objective of CSRC? Can E-ECSRC have an affection on the purchasing intention? If the answer is "yes", what is the deep reason? In addition, how many kinds of E-ECSRC were used now and is there a more effective way based on these good effectiveness?

To seek for the answer of these questions, we conducted a research on the E-CSRC and found two types of CSR information that were embedded in "COMMODITY DETAILS": (1) a screenshot of the CSR news from the third party of media (Type I of the company's E-ECSRC information); (2) a summary or short introduction of the CSR report (Type II of the company's E-ECSRC information). For the reason that the consumer online star-rated system was used in most of the e-commerce platform which could affect the consumer purchasing intention, can an online star-rated system for CSR be designed to promote the affection of CSRC (Type III of the company's E-ECSRC information)?

In this research, we led a consumer experiment to confirm that if the three types of E-ECSRC can effect consumers' perception of CSR, company evaluation and purchasing attention, and if the perception of CSR and company evaluation can affect consumer purchasing attention. The conclusion can help corporations make decisions on their CSRC or improve their sale in e-Business.

HYPOTHESIS AND CONCEPTUAL MODEL

Effect of CSR on Consumer Evaluations of the Company

Having a pro-social agenda is thought to be a powerful marketing tool upon which to build a company' s reputation and gain a competitive advantage through differentiation (Bronn and Vrioni 2001). A majority of investigations attested to the positive effects of people's perceptions of CSR on their evaluations of the company. Based on the experiences of a company's CSRC, consumers establish an inventory in their memory with a combination of attributes, benefits, attitudes, and beliefs about the company, which could be called corporate associations (Brown and Dacin, 1997). Consumers' corporate associations include two major categories and one of them is CSR associations, which deal with the "character of the company, usually with regard to important societal issues". CSR associations will be the important reference for consumers to evaluate the company and direct their purchasing behaviors (Brown & Dacin, 1997).

On the base of their conclusion, marketing scholars tested, developed or modified their models by applying it to various marketplace context. Sen (2001) found that CSR's initiatives would improve the evaluations and this relationship would be moderated by consumers' support of the CSR domain. Berens et al. (2005) used artificial advertisements of a real financial service provider as stimulus in experiment to test the relationship between corporate associations and consumer product responses. Salmones et al. (2005) studied the CSR in telecommunication service industry, concluding that corporates' ethic-law and philanthropic responsibilities had significant effect on consumers' general evaluation and in turn, affected consumers' loyalty. Then, Biehal and Sheinin (2007) held a laboratory experiments using artificial advertisements to test the relationship between corporate message and company's portfolio. Kim (2011) studied three corporate communication strategies (i.e., company ability, CSR, and hybrid) in the model, and found that when a company is well-known to consumers, a CSR strategy is more effective in influencing both consumer corporate ability (CA) and CSR associations and in turn, company/product evaluations. Drawing on social identity theory, Marin et al. (2009) developed a model of influence of CSR on loyalty and tested it using a sample of real consumers. Results demonstrate that CSR initiatives are linked to stronger loyalty both because the consumer develops a more positive company evaluation, and because one identifies more strongly with the company. Lii (2012) investigated the efficacy of three corporate social responsibility (CSR) initiatives-sponsorship, cause-related marketing (CRM), and philanthropy- on consumer-company identification (C–C identification) and brand attitude and, in turn, consumer citizenship behaviors. CSR reputation is proposed as the moderating variable that affects the relationship between CSR initiatives, C–C identification, and brand attitude. Tao and Wilson's (2016) experiment examines how the three corporate communication strategies (Kim,2011) positively affect consumers' evaluation of a company.

This study argues that the existed three types of positive E-ECSRC information have different positive effects on consumer evaluations of the company. Because the positive CSR information can be exposure directly to consumer in the process of online purchasing, consequently, consumers may consider the embedded message as more useful for evaluating the company. Therefore, based on the reviewed literature, this study hypothesizes that:

H1: A company's E-ECSRC information will enhance consumer evaluations of the company.
- **H1a:** Type I of the company's E-ECSRC information will enhance consumers evaluations of the company.
- **H1b:** Type II of the company's E-ECSRC information will enhance consumers' evaluations of the company.
- **H1c:** Type III of the company's E-ECSRC information will enhance consumers' evaluations of the company.

Consumer's Perception of CSR

As the component of the enterprises' competitive advantages, it is very important to improve the perception of CSR through an effective corporate social responsibility communication. There are several traditional approaches for CSR communication such as CSR report, newspaper, website, mass media, and some informal approaches such as word-of mouth etc. More and more scholars tried to understand the consumer's perception of CSR. Despite such an expanding literature exploring consumers' response to CSR, it remains implicit that how consumers perceive CSR (Oberseder et al. 2014).

In this paper, we are inclined to understand that if the three kinds of E-CSRC have same effect on consumer's perception of CSR.

For Type II of the company's E-ECSRC information came from CSR report which included comprehensive CSR information about the enterprise. For both Type Iand Type III of the company's E-ECSRC information are online word-of mouth. Researchers try to explore why consumers rely on online word-of mouth (Cheng-Hao Chen, 2015). Researchers found eight key reasons as to why they do: to reduce risk; because others are using it; to identify lower prices; to get information easily; by accident (unplanned); because it is "cool"; because they are stimulated by off-line inputs such as television; and to obtain prepurchase information (Goldsmith and Horowitz,2006). Hennig- Thurau & Walsh (2003) also found that consumers relied on online word-of mouth to save time and make better decisions. Gretzel, Yoo, &Purifoy (2007) found that consumers relied on online word-of mouth to receive up-to-date information (Gretzel, Yoo, & Purifoy, 2007). In addition, online word- of mouth can help consumer to enjoy a reliable search process (Arsal et al., 2010; Xie et al., 2011).

In consideration of the content of the three kinds of CSR information, based on these research, we conduct a hypothesis that consumer can perceive CSR by online word-of-mouth and different kinds of CSR information will affect their perception.

H2: The perception of CSR will be differed from different type of a company's E-ECSRC information.
H2a: The rank of the perception of CSR after CSRC is: Type III>Type II>Type I

Consumer Purchasing Intention

Researchers and companies are paying increasing attention to the reaction of consumers to CSR such as purchasing intention. Most of researches reflected a positive correlation between these constructs. Jin (2006) and Xie and Zhou (2009), who both found that CSR can positively influence consumers' purchase intention. Romani et al. (2013) conduct a field experiment finding that CSR has an effect on consumers' intentions to (1) say positive things about the company, and (2) participate in advocacy actions benefiting the company. Ferreira and Ribeiro (2017) supports that both the moderation hypothesis of country of origin in the relationship between CSR/CSIR and willingness to pay and between CSR/CSIR and purchase intention.

Therefore, based on the existing researches, the authors propose the following hypothesis:

H3: The consumer purchasing intention will be differed from different types of a company's E-ECSRC information.

H3a: The rank of the consumer purchasing intention after CSRC is: Type III>Type II>Type I

In addition, gender is a common variable in these kinds of research. So, we also paid attention to the CSR's effect by gender to consumer purchasing intention.

H4: The consumer evaluation of a company after different type of a company's E-ECSRC will be differed by gender.

H4a: The consumer evaluation of a company after Type I of the company's E-ECSRC will be differed by gender.

H4b: The consumer evaluation of a company after Type II of the company's E-ECSRC will be differed by gender.

H4c: The consumer evaluation of a company after Type III of the company's E-ECSRC be differed by gender.

H5: The consumer perception of CSR after different type of a company's E-ECSRC information will be differed by gender.

H5a: The consumer perception of CSR after Type I of the company's E-ECSRC information will be differed by gender.

H5b: The consumer perception of CSR after Type II of the company's E-ECSRC information will be differed by gender.

H5c: The consumer perception of CSR after Type III of the company's E-ECSRC information will be differed by gender.

H6: The consumer purchasing intention after different type of a company's E-ECSRC information will be differed by gender.

H6a: The consumer purchasing intention after Type I of the company's E-ECSRC information will be differed by gender.

H6b: The consumer purchasing intention after Type II of the company's E-ECSRC information will be differed by gender.

H6c: The consumer purchasing intention after Type III of the company's E-ECSRC information will be differed by gender.

Experiments studies have generally found a positive impact of CSR. Lafferty and Goldsmith (1999) conducted an experiment and found that corporate credibility had a significant impact on attitudes toward the brand and purchase intent. Bhattacharya and Sen (2003) made a similar argument for how non product aspects of a company, such as CSR, can lead to customer loyalty and other positive post purchase outcomes. Mohr and Webb (2005) conducted an experiment to examine the influence of CSR and price on consumer response. In addition, results from a national sample of adults indicate that CSR in both domains had a positive impact on evaluation of company and purchase intent. Base on the conclusion, we propose hypothesis 7 and hypothesis 8:

H7: The consumer evaluation of a company will promote consumer purchasing intention after E-ECSRC.

H7a: The consumer evaluation of a company will promote consumer purchasing intention after E-ECSRC of Type I.

H7b: The consumer evaluation of a company will promote consumer purchasing intention after E-ECSRC of Type II.

H7c: The consumer evaluation of a company will promote consumer purchasing intention after E-ECSRC of Type III.

H8: The consumer perception of CSR will promote consumer purchasing intention after E-ECSRC.

H8a: The consumer perception of CSR will promote consumer purchasing intention after E-ECSRC of Type I.

Figure 1. Conceptual model

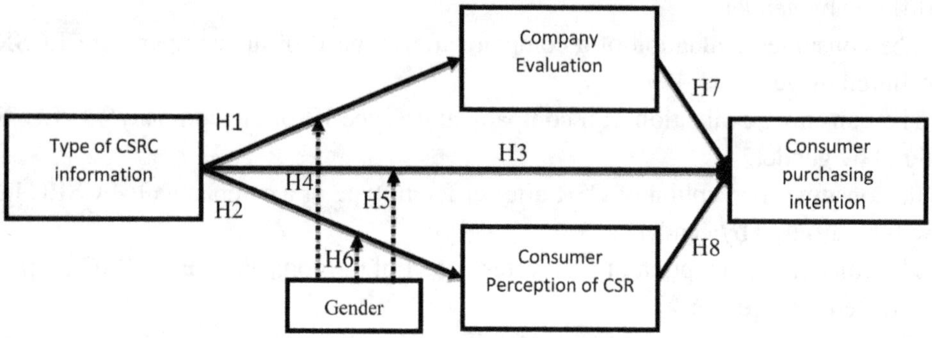

H8b: The consumer perception of CSR will promote consumer purchasing intention after E-ECSRC of Type II.

H8c: The consumer perception of CSR will promote consumer purchasing intention after E-ECSRC of Type III.

METHOD

Design

To test the model, we examined the effects of online embedded CSRC information on subjects' evaluations of the company, perceived CSR, purchasing intent and purchasing decision by using a 3 (type of online embedded CSRC information) x 2 (gender) between-subjects design. There are three types of online embedded CSRC information (1= a summary of the CSR report, 2= a screenshot of the online news from the third party media,3=a result of the dynamic evaluation system provided by authority).

Dependent Variables

Dependent variables. This study has three dependent variables: (1) company evaluation, (2) consumer perception of CSR, and (3) purchasing intention.

Consumer Company Evaluation

Subjects' evaluation of the company (Company Evaluation) was measured as their overall impression ratings of the company (1 = "very unfavorable," 7 = "very favorable") based on Sankar Sen and C. B. Bhattacharya (2001). We obtained these company evaluations both before and after subjects' exposure to online embedded CSRC information so that we could test the effect of the CSRC on the subjects' evaluation of the company.

Perception of CSR

Subjects' perception of CSR was measured by 17 items, 7 Likert scale (1= "strongly disagreed", 7 = "strongly agreed") based on Fatma M et al.(2016).

Consumer Purchasing Intention

Fishbein & Ajzen (1975) suggest that the person's intention is a function of the person's attitude towards the behavior. So in this research, we mainly research the consumer purchasing intention. Subjects' purchasing intention was measured through their product purchasing intention (1 = "not at all likely to buy", 7="very likely to buy") based on Sankar Sen and C. B. Bhattacharya (2001).

Stimuli

To enhance external validity, subjects were exposed to the real CSR information about company A. According to Sankar Sen and C. B. Bhattacharya (2001), this paper choosed diversity issues as our CSR domain for four reasons: (1) diversity is not only one of the primary domains of CSR activity among companies today (Hopkins 1999) but also one of the most active ones (Kinder, Lydenberg, Domini & Co. Inc. 1999); (2) consumers' support for this domain is likely to be varied, particularly compared with other domains, which enabled us to implement the CSR support construction meaningfully; (3) in line with Brown and Dacin (1997), this domain's irrelevance to the company's ability to produce quality products enabled us to minimize any confounding of CSR with CA associations; and (4) subjects' familiarity with the focal company's actions in the diversity domain is likely to be lower compared with other, better publicized CSR domains (e.g., support of social causes), which enabled us to minimize the confounding effects of prior knowledge in subjects' reactions to CSR information.

By the information collection and research, we found that there are three main types of the online embedded CSRC information could be selected as showed in Figure 2, the first picture is: Type Iis a summary of the CSR report of company A including four aspects such as consumer, green value chain, employees and The public interest activity; Type IIis a screenshot of the online news about company A from the third party media which was about the award because its effort on CSR; type III is a result of the dynamic evaluation system provided by authority and the modality was imitation from the dynamic evaluation system of the online satisfaction degree on Taobao.com. All of these three pictures were embedded at the top of a description of milk copied from the real product description of company A.

We obtained information about the company's diversity- related CSR record from its CSR report (2016) and the news from a website(http://www.ce.cn/) which has significance influence in China. The result of the dynamic evaluation system was designed according to the scale of perception of CSR based on Fatma M et al. (2016), and the score was estimated by its CSR report. All of these information is positive because we only intended to research the effect on consumer sof the positive CSR activities to make out if a good CSR performance and active communication can improve consumer behavior.

Figure 2. Three Types of the Online Embedded CSRC Information

Type I Type II Type III

Procedure

To make sure the validity of the experiments and subjects could have a sensitive understanding to the experimental documents, a dairy enterprise was selected as the stimuli. In detail, CSR information is about a dairy enterprise A which engages in dairy products (including infant formula powder) manufacturing, food and beverage processing, livestock and poultry, automobile cargo transportation, catering services etc. The process of the experiments was illustrated as below:

A total of 1189 subjects complete the online study after they received the invitation which was shared from his social network members from WeChat or QQ. They were firstly asked to finish the company evaluation of the company A according to their personal perception. Then they showed a milk product's description which included an embedded CSRC information from one of the three types randomly in at least 20 seconds.

Then the subjects were told that they are assumed to purchase milk products and finish a questionnaire which include three dependent variables: (1) Consumer company evaluation: subjects were asked to rate the company(1 = "very unfavorable," 7 = "very favorable"); (2) perception of CSR: subjects were asked to rate the perception of the company on a 17 items, 7 Likert scale(1= "strongly disagreed", 7 = "strongly agreed" ; Cronbach's α is 0.968); λ(3) Consumer purchasing intention: subjects were asked to rate their purchasing intention on a 3 items, 7 Likert scale (1 = "not at all likely to buy", 7="very likely to buy").

Finally, the subjects were asked for the demographic information including gender, age, education etc. Every subject has a single code.

Result

According to these hypothesizes, it is expected that the rate on consumer evaluation after the online embedded CSRC will be differed from the rate before it. The assumption is tested with the help of the paired-sample T test. Because the subjects were asked to answer the question twice before and after, the online embedded CSRC information was showed. As Table 1 showed: (1) for Type I of the company's online embedded CSRC information, the mean *score* of consumer evaluation of company A before the CSRC is 5.77, which was increased to 5.85 after the CSRC significantly ($t=-1.438$, $p=0.000<0.01$); (2) for Type II of the company's online embedded CSRC information, the mean score of consumer evaluation of company A before the CSRC is 6.08, which was increased to 6.18 after the CSRC significantly (t=-2.129, p=0.000<0.01); (3) for Type III of the company's online embedded CSRC information, the mean score of consumer evaluation of company A before the CSRC is 6.37, which was increased to 6.40 after the CSRC significantly (t=-0.952, p=0.000<0.01). Based on the results, it can be concluded that hypothesis1 is accepted.

It is expected that different types of a company's E-ECSRC information exert different perception of CSR. The Cronbach's α of perception of CSR scale is 0.968.

To test hypothesis 2, a one-way between ANOVA test was conducted. The dependent variables are perception of CSR. Table 2 shows the results that the difference between types of E-ECSRC information is statistically significant (F=40.905, p=0.000). Based on the results, it can be concluded that hypothesis 2 is accepted.

To test hypothesis 2a, a post hoc multiple comparisons was conducted (Table 3). By the way of Least Significant difference (LSD), the mean score of perception of CSR after CSRC Type III is the highest which could produce better effect than CSRC Type I(p=0.000) and CSRC Type II(p=0.000). The difference between the mean score of perception of CSR after CSRC Type Iand CSRC Type II was not statistically significant(p=0.606). Based on the results, it can be concluded that hypothesis 2a is just partly supported.

It is expected that different types of a company's E-ECSRC information exert different consumer purchasing intention. To test hypothesis 3, a one-way between ANOVA test was conducted. The dependent variables are consumer purchasing intention. Table 4 shows the results that the difference between type of E-ECSRC information is statistically significant (F=27.890, p=0.000). Based on the results, it can be concluded that hypothesis 3 is accepted.

To test hypothesis 3a, a post hoc multiple comparisons was conducted (Table 5). By the way of Least Significant difference (LSD), the mean score of perception of CSR after CSRC Type III is the highest which could produce better effect than CSRC Type II(p=0.006), and the mean score of perception of

Table 1. Results hypothesis1

Test Variable	Group Variable	Mean	Std.	T-Test	P-Value
Consumer Evaluation	Before CSRC Type I	5.77	1.352	-1.438	.000
	After CSRC Type I	5.85	1.264		
	Before CSRC Type II	6.08	1.167	-2.129	.000
	After CSRC Type II	6.18	1.006		
	Before CSRC Type III	6.37	1.153	-.952	.000
	After CSRC Type III	6.40	1.056		

Table 2. Results hypothesis 2

Test Variable	Group Variable	Mean	F
perception of CSR	CSRC Type I	5.8604	40.905***
	CSRC Type II	6.2779	
	CSRC Type III	6.3063	

***p<0.001

Table 3. Results hypothesis 2a

Multiple Comparisons						
Dependent Variable: Consumer Purchasing Intention						
LSD(L)						
					95% Confidence Interval	
(I) CSRC type	(J) CSRC type	Mean Difference (I-J)	Std. Error	Sig.	Lower Bound	Upper Bound
CSRC Type II	CSRC Type I	.41745*	.05514	.000	.3093	.5256
	CSRC Type III	-.02841	.05503	.606	-.1364	.0796
CSRC Type I	CSRC Type II	-.41745*	.05514	.000	-.5256	-.3093
	CSRC Type III	-.44586*	.05517	.000	-.5541	-.3376
CSRC Type III	CSRC Type II	.02841	.05503	.606	-.0796	.1364
	CSRC Type I	.44586*	.05517	.000	.3376	.5541

*. The mean difference is significant at the 0.05 level.

Table 4. Results hypothesis 3

Test Variable	Group Variable	Mean	F
consumer purchasing intention	CSRC Type I	5.5880	27.890***
	CSRC Type II	5.9564	
	CSRC Type III	6.1755	

***p<0.001

CSR after CSRC Type II could produce better effect than CSRC Type I(p=0.000). Based on the results, it can be concluded that hypothesis 3a is supported.

According to these hypothesizes, it is expected that the consumer evaluation of a company after the E-ECSRC will be differed from the different gender. The assumption is tested with the independent-sample T test. As Table 6 showed: (1) for Type I of the company's E-ECSRC information, the difference of between male and female of the mean score of consumer evaluation of company is not statistically significant (t=-0.997, p=0.320>0.05), hypothesis 4a is not accept; (2) for Type II of the company's E-ECSRC information, the difference between male and female of the mean score of consumer evaluation is not statistically significant (t=-1.065, p=0.288>0.05), hypothesis 4b is not accept; (3) for Type III of the company's E-ECSRC information, the difference between male and female of the mean score of consumer evaluation is not statistically significant (t=-1.419, p=0.154>0.05), hypothesis 4c is not accept. Based on the results, it can be concluded that hypothesis 4 is not supported.

According to these hypothesizes, it is expected that the rate on consumer perception of CSR after the E-ECSRC will be differed from the different gender. The assumption is tested with the independent-sample T test. As Table 7 showed: (1) for Type I of the company's E-ECSRC information, the difference between male and female of the mean score of consumer perception of CSR of company is not statistically significant (t=-0.998, p=0.324>0.05), hypothesis 5a is not accept; (2) for Type II of the company's E-ECSRC information, the difference between male and female of the mean score of consumer perception

Table 5. Results of multiple comparisons for hypothesis 3a

Dependent Variable: Perception of CSR						
LSD(L)						
					95% Confidence Interval	
(I) CSRC type	(J) CSRC type	Mean Difference (I-J)	Std. Error	Sig.	Lower Bound	Upper Bound
CSRC Type II	CSRC Type I	.36846*	.07943	.000	.2126	.5243
	CSRC Type III	-.21903*	.07928	.006	-.3746	-.0635
CSRC Type I	CSRC Type II	-.36846*	.07943	.000	-.5243	-.2126
	CSRC Type III	-.58750*	.07948	.000	-.7434	-.4316
CSRC Type III	CSRC Type II	.21903*	.07928	.006	.0635	.3746
	CSRC Type I	.58750*	.07948	.000	.4316	.7434

*. The mean difference is significant at the 0.05 level.

Table 6. Results hypothesis4

Test Variable	Group Variable	Mean	Std.	T-test	P-value
consumer evaluation of a company	Male in CSRC Type I	5.91	1.310	-.997	.320
	Female in CSRC Type I	5.78	1.216		
	Male in CSRC Type II	6.13	1.127	-1.065	.288
	Female in CSRC Type II	6.24	.866		
	Male in CSRC Type III	6.33	1.052	-1.419	.157
	Female in CSRC Type III	6.48	1.057		

Table 7. Results hypothesis5

Test Variable	Group Variable	Mean	Std.	T-test	P-value
consumer perception of CSR	Male in CSRC Type I	5.8158	.96791	-.998	.324
	Female in CSRC Type I	5.9050	.81896		
	Male in CSRC Type II	6.2229	.63688	-1.959	.050
	Female in CSRC Type II	6.3333	.47641		
	Male in CSRC Type III	6.1357	.84927	-4.268	.000
	Female in CSRC Type III	6.4830	.76802		

of CSR of company is statistically significant (t=-1.959, p=0.05), hypothesis 5b is accept; (3) for Type III of the company's E-ECSRC information, the difference between male and female of the mean score of consumer perception of CSR of company is statistically significant (t=-4.268, p=0.000<0.001), hypothesis 5c is not accept. Based on the results, it can be concluded that hypothesis 5 is just partly supported.

According to these hypotheses, it is expected that the rate on purchasing intention after the E-ECSRC will be differed from the different gender. The assumption is tested with the independent-sample T test.

As Table 8 showed: (1) for Type I of the company's E-ECSRC information, the difference between male and female of the mean score of consumer purchasing intention of company is not statistically significant (t=-0.212, p=0.832>0.05); (2) for Type II of the company's E-ECSRC information, the difference between male and female of the mean score of consumer purchasing intention of company is not statistically significant (t=-1.731, p=0.084>0.05); (3) for Type III of the company's E-ECSRC information,, the difference between male and female of the mean score of consumer purchasing intention of company is not statistically significant (t=-1.013, p=0.312>0.05). Based on the results, it can be concluded that hypothesis 5 is not supported.

As Table 9 shows, for the three types of E-ECSRC, all of the variables have a significant correlations relationship, on the base of which we can test H7 and H8 by linear-regression analysis.

For the E-ECSRC of Type I, as Table 10 shows coefficient, significance and collinearity statistics, consumer evaluation of the company(β=0.470 , $p = 0.000 < 0.001$) and consumer perception of CSR(β=0.384 , $p = 0.000 < 0.001$) have significantly positive affection to consumer purchasing intention.

Table 8. Results hypothesis5

Test Variable	Group Variable	Mean	Std.	T-test	P-value
consumer purchasing intention	Male in CSRC Type I	5.5753	1.25790	-.212	.832
	Female in CSRC Type I	5.6007	1.11810		
	Male in CSRC Type II	5.8683	1.13711	-1.731	.084
	Female in CSRC Type II	6.0455	.89029		
	Male in CSRC Type III	6.1188	1.10887	-1.013	.312
	Female in CSRC Type III	6.2342	1.16082		

Table 9. Results of correlations for hypothesis7 and hypothesis8

	Consumer Evaluation	Consumer Perception of CSR	Purchasing Intention
	E-ECSRC of Type I:		
consumer evaluation	1	.555**	.682**
consumer perception of CSR	.555**	1	.644**
Purchasing intention	.682**	.644**	1
	E-ECSRC of Type II:		
consumer evaluation	1	.340**	.710**
consumer perception of CSR	.340**	1	.387**
Purchasing intention	.710**	.387**	1
	E-ECSRC of Type III:		
consumer evaluation	1	.575**	.647**
consumer perception of CSR	.575**	1	.611**
Purchasing intention	.647**	.611**	1

**. Correlation is significant at the 0.01 level(2-tailed).

The adjusted R square is 0.565, which means that the model can explain 56.5% of the change of consumer purchasing intention after E-ECSRC.

For the E-ECSRC of Type II, as Table 10 shows coefficient, significance and collinearity statistics, consumer evaluation of the company(β=0.654 , $p = 0.000 < 0.001$) and consumer perception of CSR(β=0.164 , $p = 0.000 < 0.001$) have significantly positive affection to consumer purchasing intention. The adjusted R square is 0.526, which means that the model can explain 52.6% of the change of consumer purchasing intention after E-ECSRC.

For the E-ECSRC of Type III, as Table 10 shows coefficient, significance and collinearity statistics, consumer evaluation of the company(β=0.441 , $p = 0.000 < 0.001$) and consumer perception of CSR(β=0.358 , $p = 0.000 < 0.001$) have significantly positive affection to consumer purchasing intention. The adjusted R square is 0.501, which means that the model can explain 50.1% of the change of consumer purchasing intention after E-ECSRC.

In conclusion, H7 and H8 were accept.

A summary of the study results is presented in Table 11.

CONCLUSION

In this paper, we conduct an experiment to test the differences of consumer behavior after three types of "Embedded CSRC in E-business"(E-ECSRC).

Firstly, all of these three kinds of company's E-ECSRC can enhance consumer evaluations of the company. Secondly, the perception of CSR will be different from different types of a company's E-ECSRC information. We also researched the rank of the perception of CSR which was affected by the three types of E-ECSRC information: Type II and type III can significantly affect consumer perception of CSR much more than Type I. We can conclude that more dimension of the CSR information and rate

Table 10. Results hypothesis7 and hypothesis8

Type of E-ECSRC		Unstandardized Coefficients		Standardized Coefficients	t	Sig.	Collinearity Statistics	
		B	Std. Error	Beta			Tolerance	VIF
Type I	(Constant)	.026	.266		.097	.923		
	consumer evaluation	.442	.038	.470	11.748	.000	.692	1.445
	consumer perception of CSR	.509	.053	.384	9.595	.000	.692	1.445
Type II	(Constant)	-.033	.405		-.081	.935		
	consumer evaluation	.667	.037	.654	17.808	.000	.884	1.131
	consumer perception of CSR	.298	.067	.164	4.464	.000	.884	1.131
Type III	(Constant)	.050	.320		.157	.875		
	consumer evaluation	.474	.047	.441	10.164	.000	.669	1.495
	consumer perception of CSR	.490	.059	.358	8.244	.000	.669	1.495

a. Dependent Variable: consumer purchasing intention

Table 11. Summary of study findings

Hypothesis	Results	Short Interpretation
H1	Accepted	A company's E-ECSRC information will enhance consumer evaluations of the company.
H1a	Accepted	Type Iof the company's E-ECSRC information will enhance consumers evaluations of the company.
H1b	Accepted	Type IIof the company's E-ECSRC information will enhance consumers evaluations of the company.
H1c	Accepted	Type IIIof the company's E-ECSRC information will enhance consumers evaluations of the company.
H2	Accepted	The perception of CSR will be differed from different type of a company's E-ECSRC information.
H2a	Partly Accepted	The rank of the perception of CSR after CSRC is: Type III>Type II>Type I
H3	Accepted	The consumer purchasing intention will be differed from different type of a company's E-ECSRC information.
H3a	Accepted	The rank of the consumer purchasing intention after CSRC is: Type III>Type II>Type I
H4	Rejected	The consumer evaluation of a company after different type of a company's E-ECSRC will be differed by gender.
H4a	Rejected	The consumer evaluation of a company after Type Iof the company's E-ECSRC will be differed by gender.
H4b	Rejected	The consumer evaluation of a company after Type IIof the company's E-ECSRC will be differed by gender.
H4c	Rejected	The consumer evaluation of a company after Type IIIof the company's E-ECSRC be differed by gender.
H5	Partly Accepted	The consumer perception of CSR after different type of a company's E-ECSRC information will be differed by gender.
H5a	Rejected	The consumer perception of CSR after Type Iof the company's E-ECSRC information will be differed by gender.
H5b	Accepted	The consumer perception of CSR after Type IIof the company's E-ECSRC information will be differed by gender.
H5c	Accepted	The consumer perception of CSR after Type IIIof the company's E-ECSRC information will be differed by gender.
H6	Rejected	The consumer purchasing intention after different type of a company's E-ECSRC information will be differed by gender.
H6a	Rejected	The consumer purchasing intention after Type Iof the company's E-ECSRC information will be differed by gender.
H6b	Rejected	The consumer purchasing intention after Type IIof the company's E-ECSRC information will be differed by gender.
H6c	Rejected	The consumer purchasing intention after Type IIIof the company's E-ECSRC information will be differed by gender.
H7	Accepted	The consumer evaluation of a company will increase consumer purchasing intention after E-ECSRC.
H7a	Accepted	The consumer evaluation of a company will increase consumer purchasing intention after E-ECSRC of Type I.
H7b	Accepted	The consumer evaluation of a company will increase consumer purchasing intention after E-ECSRC of Type II.
H7c	Accepted	The consumer evaluation of a company will increase consumer purchasing intention after E-ECSRC of Type III.
H8	Accepted	The consumer perception of CSR will increase consumer purchasing intention after E-ECSRC.
H8a	Accepted	The consumer perception of CSR will increase consumer purchasing intention after E-ECSRC of Type I.
H8b	Accepted	The consumer perception of CSR will increase consumer purchasing intention after E-ECSRC of Type II.
H8c	Accepted	The consumer perception of CSR will increase consumer purchasing intention after E-ECSRC of Type III.

are helpful to improve consumer perception of CSR. Thirdly, the consumer purchasing intention will be differed from different type of a company's E-ECSRC information and the rank is: type III can significantly affect consumer purchasing much more than Type II; Type II can significantly affect consumer purchasing much more than Type I. Fourthly, gender is a moderating factor which can affect the relation between type of E-ECSRC and perception of CSR. We can conclude that, generally, female have stronger response to E-ECSRC information. At last, after all the three kinds of E-ECSR, consumer's evaluation of the company and consumer perception of CSR have significant positive relation with purchasing intention. By contrast, we found that after type I of E-ECSRC, consumer perception of CSR imposed the strongest impact to consumer purchasing intention, and after type II of E-ECSRC, consumer evaluation of the company have the strongest impact to consumer purchasing intention.

Based on this research, we can conclude that: it is a good choice to embed CSR information into the description in e-business. For Type Iand Type IIof the company's E-ECSRC information, the decision can be made by a company independently. For type III of the company's E-ECSRC information, an authoritative star-rated system is necessary and the rate should be dynamic by stakeholder. This conclusion enlightens the way of improvement of CSR conducted by government. For example, government could establish an authoritative star-rated system and embed it in the process of online retail.

The conclusion could provide a theoretical support to the marketing strategy for e-business and an operational tool for government regulation. Government could propose the online star-rated system for CSR just like the type III of the company's E-ECSRC information which is fictitious in this research. But the rater is the key to the success to the system. Future research could focus on the participation, rules and framework of the system.

ACKNOWLEDGMENT

This research is subsidized by the Young Fund of the MOE (Ministry of Education in China) Project of Humanities and Social Sciences (Project No. 16YJCZH139).

REFERENCES

Alvarado-Herrera, A., Bigne, E., & Aldas-Manzano, J. (2015). A Scale for Measuring Consumer Perceptions of Corporate Social Responsibility Following the Sustainable Development Paradigm. *Journal of Business Ethics*, *140*(2), 1–20.

Arsal, I., Woosnam, K. M., Baldwin, E. D., & Backman, S. J. (2010). Residents as travel destination information providers: An online community perspective. *Journal of Travel Research*, *49*(4), 400–413. doi:10.1177/0047287509346856

Berens, G., Riel, C. B. M. V., & Bruggen, G. H. V. (2005). Corporate Associations and Consumer Product Responses: The Moderating Role of Corporate Brand Dominance. *Journal of Marketing*, *69*(3), 35–48. doi:10.1509/jmkg.69.3.35.66357

Bhattacharya, C. B., & Sen, S. (2003). *Consumer-Company Identification: A Framework for Understanding Consume*. Academic Press.

Biehal, G. J., & Sheinin, D. A. (2007). The Influence of Corporate Messages on the Product Portfolio. *Journal of Marketing*, *71*(2), 12–25. doi:10.1509/jmkg.71.2.12

Bronn, P. S., & Vrioni, A. B. (2001). Corporate social responsibility and cause-related marketing: An overview. *International Journal of Advertising*, *20*(2), 207–223. doi:10.1080/02650487.2001.11104887

Brown, T. J., & Dacin, P. A. (1997). The company and the product: Corporate associations and consumer product responses. *Journal of Marketing*, *61*(1), 68–84. doi:10.2307/1252190

Chen, C. H., Bang, N., & Klaus, P. (2015). Exploring electronic word-of-mouth (eWOM) in the consumer purchase decision-making process: The case of online holidays - evidence from United Kingdom (UK) consumers. *Journal of Travel & Tourism Marketing*, *32*(8), 953–970. doi:10.1080/10548408.2014.956165

del Mar García de losSalmones, M. (2005). Influence of Corporate Social Responsibility on Loyalty and Valuation of Services. *Journal of Business Ethics*, *61*(4), 369–385. doi:10.1007/s10551-005-5841-2

Du, S., Bhattacharya, C. B., & Sen, S. (2011). Corporate Social Responsibility and Competitive Advantage: Overcoming the Trust Barrier. *Management Science*, *57*(9), 1528–1545. doi:10.1287/mnsc.1110.1403

Fatma, M., Rahman, Z., & Khan, I. (2016). Measuring consumer perception of CSR in tourism industry: Scale development and validation. *Journal of Hospitality and Tourism Management*, *27*, 39–48. doi:10.1016/j.jhtm.2016.03.002

Ferreira, A. I., & Ribeiro, I. (2017). Are you willing to pay the price? The impact of corporate social (ir)responsibility on consumer behavior towards national and foreign brands. *Journal of Consumer Behaviour*, 16.

Fishbein, M., & Azjen, I. (1975). *Beliefs, Attitude, Intention and Behavior: An introduction to theory and research*. Addison- Wesley.

Goldsmith, R. E., & Horowitz, D. (2006). Measuring Motivations for Online Opinion Seeking. *Journal of Interactive Advertising*, *6*(2), 2–14. doi:10.1080/15252019.2006.10722114

Gretzel, U., Yoo, K. H., & Purifoy, M. (2007). *Online Travel Review Study: Role & Impact of*. Online Travel Reviews.

Hennig-Thurau, T., Gwinner, K. P., Walsh, G., & Gremler, D. D. (2004). Electronic word-of-mouth via consumer-opinion platforms: What motivates consumers to articulate themselves on the Internet? *Journal of Interactive Marketing*, *18*(1), 38–52. doi:10.1002/dir.10073

Jin, L. Y. (2006). The effects of consumers' company identification on their evaluation of products and behavioral responses. *Nankai Business Review*, *9*(3), 16–21.

Kim, S. (2010). Transferring Effects of CSR Strategy on Consumer Responses: The Synergistic Model of Corporate Communication Strategy. *Journal of Public Relations Research*, *23*(2), 218–241. doi:10.1080/1062726X.2011.555647

Lafferty, B. A., & Goldsmith, R. E. (1999). Corporate Credibilitys Role in Consumers Attitudes and Purchase Intentions When a High versus a Low Credibility Endorser Is Used in the Ad. *Journal of Business Research*, *44*(2), 109–116. doi:10.1016/S0148-2963(98)00002-2

Lii, Y. S., & Lee, M. (2012). Doing Right Leads to Doing Well: When the Type of CSR and Reputation Interact to Affect Consumer Evaluations of the Firm. *Journal of Business Ethics*, *105*(1), 69–81. doi:10.1007/s10551-011-0948-0

Marin, L., Ruiz, S., & Rubio, A. (2009). The Role of Identity Salience in the Effects of Corporate Social Responsibility on Consumer Behavior. *Journal of Business Ethics*, *84*(1), 65–78. doi:10.1007/s10551-008-9673-8

Mohr, L. A., & Webb, D. J. (2005). The effects of corporate social responsibility and price on consumer responses. *The Journal of Consumer Affairs*, *39*(1), 121–147. doi:10.1111/j.1745-6606.2005.00006.x

Öberseder, M., Schlegelmilch, B. B., Murphy, P. E., & Gruber, V. (2014). Consumers Perceptions of Corporate Social Responsibility: Scale Development and Validation. *Journal of Business Ethics*, *124*(1), 101–115. doi:10.1007/s10551-013-1787-y

Sen, S., & Bhattacharya, C. B. (2001). Does Doing Good Always Lead to Doing Better? Consumer Reactions to Corporate Social Responsibility. *JMR, Journal of Marketing Research*, *38*(2), 225–243. doi:10.1509/jmkr.38.2.225.18838

Tao, W., & Wilson, C. (2016). The Impact of Corporate Communication on Company Evaluation: Examining the Message Effects of CSR. *Corporate Ability, and Hybrid Strategies*, *10*(5), 426–444.

Trudel, R., & Cotte, J. (2009). Does it pay to be good? *MIT Sloan Management Review*, *50*(2), 61–68.

Xie, H., Miao, L., & Kuo, P. J. (2011). Consumers responses to ambivalent online hotel reviews: The role of perceived source credibility and pre-decisional disposition. *International Journal of Hospitality Management*, *30*(1), 178–183. doi:10.1016/j.ijhm.2010.04.008

Xie, P. H., & Zhou, Z. C. (2009). Empirical study on the relationship between corporate Social esponsibility and consumer purchase intention in China. *Nankai Business Review*, *12*(1), 71–83.

KEY TERMS AND DEFINITIONS

Commodity Details: Means the descriptions about the commodity and consist of pictures and words (such as size, weight, raw material, advertorial etc.) under the E-business circumstance.

Company Evaluation: A variable to describe the evaluation of consumers to the corporation.

Consumer Experiment: A popular method to understand consumer behavior or consumer psychology by experiment and control some variables by using proper stimulants.

Corporate Social Responsibility (CSR, Also Called Corporate Conscience, Corporate Citizenship or Responsible Business): A form of corporate self-regulation integrated into a business model. CSR policy functions as a self-regulatory mechanism whereby a business monitors and ensures its active compliance with the spirit of the law, ethical standards and national or international norms.

CSR Communications(CSRC): Corporations try to communicate with mass about their CSR behavior in order to improve the public image or avoid social risk.

E-Business: A form of electronic commerce which allows consumers to directly buy goods or services from a seller over the Internet using a web browser. Consumers find a product of interest by visiting the website of the retailer directly or by searching among alternative vendors using a shopping search engine, which displays the same product's availability and pricing at different e-retailers. As of 2016, customers can shop online using a range of different computers and devices, including desktop computers, laptops, tablet computers and smartphones.

"Embedded CSRC in E-business" (E-ECSRC): It is found that lots of enterprises tried to lead a CSRC in the process of online retail. For example, some CSR information was included in a description of online products in the e-business, which we call it "Embedded CSRC in E-business" (E-ECSRC).

Perception of CSR: A variable to describe how consumer perceive the corporation's CSR.

Purchasing Intention: To describe the degree of the intention to pay for the commodity or service.

Star-Rated System for CSR: A popular way to show the online shop's credit score in E-business and yellow stars are used most. More stars mean better credit.

Chapter 8
Decision Models and Governance Mechanisms Considering Corporate Social Irresponsibility in Supply Chains

Bin Liu
Shanghai Maritime University, China

Pepei Ying
Shanghai Maritime University, China

Lin Peng
Shanghai Maritime University, China

ABSTRACT

This chapter describes corporate social irresponsibility (CSI) with a global variable, and supposes it has a negative influence on market demand to set up a decision model in supply chain for the joint effect of price and CSI on demand. Analyzing manufacturers' decisions on CSI under two supply chain structure settings including single-channel supply chain and competitive supply chain, the authors hope to design governance mechanisms to restrain the CSI performance of the manufacturers so as to provide a theoretic reference to the better fulfillment of sustainable development and social responsibility. Moreover, this chapter makes a simple extension on the asymmetry initial base demand, and the result is quite different from the symmetry initial base demand. It is found that an asymmetric CSI strategy can achieve equilibrium. A suggestion in this chapter is to intensify the competition in supply chain by supporting smaller enterprises and preventing complete monopoly.

DOI: 10.4018/978-1-5225-3537-9.ch008

INTRODUCTION

The problematic vaccine incident in 2016 revealed that the vaccine not stored in stringent cold chain was transported and sold to many cities and provinces, and vaccination might not achieve the expected protective effect; the melamine-tainted milk powder incident, the most impressive food additive issue in 2008, shocked the whole nation as soon as it was exposed, with almost all the domestic milk powder enterprises involved, among which Sanlu stood in the breach and brought about a crushing blow to the domestic milk powder industry. The disclosure of these negative events threw lots of enterprises into the crisis of survival and sustainable development, and some even went into liquidation. At the root of this was corporate social irresponsibility. In other words, these enterprises did not integrate social responsibility into their corporate culture and values. Enterprises' irresponsible behaviors indeed could reduce expenditure and bring immediate economic interests, but it would affect earning persistence (Wan & Liu, 2015). With listed companies as samples, Yang and Yang (2016) explored the relationship between corporate social responsibility (CSR) and financial performance, revealing that in the current Chinese context, the performance of CSR is not optimistic, since it not only doesn't have positive effect on sales, but also may lead to a cost increase and profit decrease.

However, in real life, similar product manufacturers compete against each other more and more intensely. Some manufacturers, to reduce production cost, bear no or little social responsibility. As a result, manufacturers have to weigh and balance between profit and social responsibility and that motivates us to study the following questions: Are manufacturers performance CSI always beneficial? On what condition can manufacturers do not perform CSI and undertake CSR?

This chapter takes manufacturers, a part of supply chain that goes wrong the most frequently and causes the most serious consequence, as a main research object to explore whether manufacturers really put immediate interests over corporate sustainable development and corporate social responsibility in supply chain competition. The authors explore manufacturers' CSI-based decision behavior in two supply chain structure settings: a single-channel supply chain and the competitive supply chain. In a single-channel supply chain, consumers are treated as a subject of decision-making to set up a three-stage Stackelberg game model. In a competitive supply chain, there is a Nash game in competition between manufacture 1 and 2 which hold the same initial base demand, while there is a Stackelberg game within each supply chain. Based on a reflection on this problem, this chapter hopes to get some suggestions to provide reference for corporate sustainable development.

BACKGROUND

In terms of corporate sustainable development (CSD), many scholars have discussed the relationship with CSR. Hu (2004) illuminated that CSR is an ethical basis for CSD and has practical significance to CSD. Xin and Liao (2013) argued that CSR has driving and restrictive effects on CSD. If an enterprise does not undertake the social responsibility, it can only gain temporary profit and achieve short-term development, but cannot develop sustainably. Hediger (2010) defined CSR as an action contributing to corporate profit maximization and social improvement, and pointed out that CSD requires corporate profit maximization, which shouldn't decrease with time. He also pointed out that if an enterprise undertakes the social responsibility, it could accumulate reputation and capital to enhance its values. Of course, some other scholars have confirmed that a certain positive correlation between CSR and corporate performance is conducive to its sustainable development. Sun and Li (2012) analyzed the 2011 report on CSR and IPO audit, concluding that for sustainable development, an enterprise should not only pay attention to economic benefits, but also actively undertake social responsibility. Wang and Hou (2009) confirmed that there is a positive correlation between performance of CSR and CSD. Carrasco and Buendía (2013) confirmed that if an enterprise integrates social responsibility into its culture, it could make self-innovation and internationalization a reality while implementing cultural innovation, so as to increase returns and promote social sustainable development. Wang and Han (2016) confirmed that CSR and internal control quality are significantly positively correlated with the level of CSD, and that a benign interaction between these two factors has a synergistic effect on CSD. All these researches directly reveal the positive correlation between CSR and CSD, while very few involve the influence of CSI on CSD.

In recent years, CSI has drawn increasing attention from scholars. As argued by Wan and Liu (2014), due to the fuzzy definition of CSI, many scholars equate CSI research to CSR research, or ignore CSI research by treating it as an appendage to CSR research. In the existing literatures, CSI researches are mainly focused on theoretical definition and empirical study. Armstrong (1977) argued that social responsibility is hard to define, while CSI is easy to realize since managers have the consciousness to avoid doing something. In his view, CSI is a suboptimal choice for a decision maker that does not consider the impact on various parties, but usually benefits one party at the cost of the overall interests. As pointed out by Jiang (2014), CSI refers to a behavior that an enterprise, for its part or another reason, does not undertake social responsibility by promoting the growth of social overall benefit as the public expects, and therefore brings an obvious negative impact to itself and causes damage to itself. Murphy and Schlegelmilch (2013) studied the historical and current trend of CSR and CSI, and divided it into several aspects: consumer, financial investment, marketing, concept theory and environment. Mishina, Dykes, Block and Pollock (2010) took 500 S&P manufacturers as research samples, discovering that the enterprises with high performance are more audacious about CSI. Armstrong and Green (2013) believed that pursuit of long-term profit could reduce the risk of decision on irresponsibility. Lin-Hi and Müller (2013) proposed to describe the frame of the relationship among CSR ("doing good" and "avoiding doing wrong"), CSI and perceived CSR, holding that "avoiding doing wrong" is the lowest standard for CSR. Sweetin, Knowles, Summey and Mcqueen (2013) found in the experiment that consumers tend to punish CSI. Wang and Mao (2015) turned their eyes to consumer attribution and expectation theory, discovering that both intentional act and unintentional act have a positive influence on consumers' perception of CSI.

The research of this aspect in supply chain is primarily focused on CSR, while rarely on CSI. Goering (2012) gave consideration to the two-part pricing problem concerning CSR in his research. Hsueh and Chang (2008) analyzed dispersion and concentration and considered the three supply chain network of CSR, finding that the coordination strategies distributed based on CSR are valid. Panda (2014) and Panda, Modak, Basu and Goyal (2015) analyzed CSR supply chain, indicating that contract negotiation could relieve channel conflict and coordinate channel members in distributing surplus profits, but manufacturers might suffer a loss. Ni, Li ang Tang (2015) studied CSR configuration in supply chain, concluding that when consumers make an active response to the performance of CSR in the overall supply chain, provided an enterprise undertakes its social responsibility, another enterprise would also undertake its social responsibility actively. Liang, Ni and Tang (2013) studied the effect of CSR performance-based product differentiation on a dual-channel supply chain, concluding that manufacturers would encourage retailers to bear CSR by offering them a low wholesale price, so as to increase the overall profitability of the supply chain. Modak, Panda, Sana and Basu (2015) described CSR with consumer surplus and researched pricing decision under both dispersion and concentration. Huang (2014) described CSR performance with a global variable and researched the impact of CSR on dual-channel supply chain operation under both dispersion and concentration.

As can be seen from the above analysis, the researches of CSI are focused on definition and practice, while there is no consideration given to the impact of CSI on pricing decision and manufacturers' strategy selection in supply chain. So, this paper analyzes CSI by modeling, depicts CSI level with a global variable by reference to Huang (2014)'s research, and supposes it has a negative influence on market demand. On this basis, we study the optimal pricing and manufacturers' decision of CSI strategies in supply chain, and carry out analyses from the perspective of CSI to provide reference for prevention and treatment of CSI, so as to promote CSD. Based on scholars' definition of CSI, here the authors hold that CSI is a lawful act by which an enterprise seldom or does not meet the public's expectation when it pursues profit maximization.

The rest of the paper is structured as follows. Section 3 develops the model including description and assumptions. Then, it is followed by two specific model structures: single-channel supply chain and competitive supply chain. Corporate governance mechanisms are designed to restrain the CSI performance of the manufacturers in Section 4. It makes an extension on the asymmetry initial base demand in Section 5. Finally, Section 6 contains the conclusion and future research.

THE MODEL

This chapter supposes manufacturers and retailers pursue profit maximization, while customers pursue utility maximization. Huang (2014) argued that since customers' identification with a responsible enterprise and its products is a mental perception, while distribution channel has a weak effect on demand, social responsibility has the same influence on demand in different channels. Based on his point of view, we assume that CSI also has the same influence on demand in different channels. In a single-channel supply chain, channel demand function has a linear relationship with price and CSI, and its expression is

$$D = a - bp - \lambda s \tag{1}$$

where $a > 0, 0 < \lambda < b$, and a represents initial base demand, b represents market's sensitivity to price, λ represents market's sensitivity to CSI, and $s(s > 0)$ represents CSI level. Here we define $\tau = \lambda s$ as CSI level of the marginal demand which reflects the demand changes by CSI level.

In a competitive supply chain, $D_i, p_i, w_i, s_i (i == 1,2)$ are used to represent the demand, retail price, wholesale price and CSI level in the ith channel respectively. According to consumer utility function structure, demand for the ith channel takes the following form where the initial base demand equals $a_i - \lambda s_i, i = 1,2$ when CSI is considered, which has precedent in works such as Zhang, Liu and Wang (2012):

$$D_i = \frac{a_i - \lambda s_i - \theta(a_{3-i} - \lambda s_{3-i}) - p_i + \theta p_{3-i}}{1-\theta^2}, i = 1,2 \tag{2}$$

where $\theta(0 \le \theta < 1)$ represents the channel substitutability. To simplify the calculation, we set $a_i = a_{3-i} = 1$, then $D_i = \frac{1-\lambda s_i - \theta(1-\lambda s_{3-i}) - p_i + \theta p_{3-i}}{1-\theta^2}, i = 1,2$ in the competitive supply chain.

Supposing manufacturers' unit production cost is $c(c > 0)$, and that $a - bc > 0$. Ni, Li and Tang (2010) argued that CSR performance can be measured by CSR investment, such as pollutant emission reduction, working condition improvement and charitable donation. So this paper holds that CSI level is measurable as well, and utilizing $\eta(\eta > 0)$ represents the influence coefficient of CSI on cost. Also, because CSI affects the continuity of corporate surplus, according to the principle of the CSR performance cost function (Ni, Li & Tang, 2010), the influence of CSI level on unit production cost is expressed as $2\eta\sqrt{s}$, so unit production cost is $c - 2\eta\sqrt{s} > 0$.

The Single-Channel Model

This chapter considers a single-channel supply chain comprising a manufacturer, a retailer and a potential consumer group. First, the manufacturer considers whether to perform CSI, and fixes the CSI level s and a wholesale price w; second, the retailer fixes a retail price p according to w; then, consumers decide whether to buy the products according to their attribute, and thus channel demand D is generated; finally, the retailer places an order to the manufacturer, and the order quantity equals demand D. As shown in Figure 1, the solid arrow points to the logistics direction in the supply chain, while the dotted arrow points to the information flow direction. As for CSI, the manufacturer has two choices: nonperformance of CSI (C0) and performance of CSI (C1). The superscript means the manufacturer's CSI choice, while subscript r and m mean the retailer and the manufacturer, respectively.

The calculation procedure is shown in Appendix 1. The combined influence of various parameters on different optimal CSI decision variables under single-channel model is shown in Table 1:

Figure 1. Single channel supply chain structure

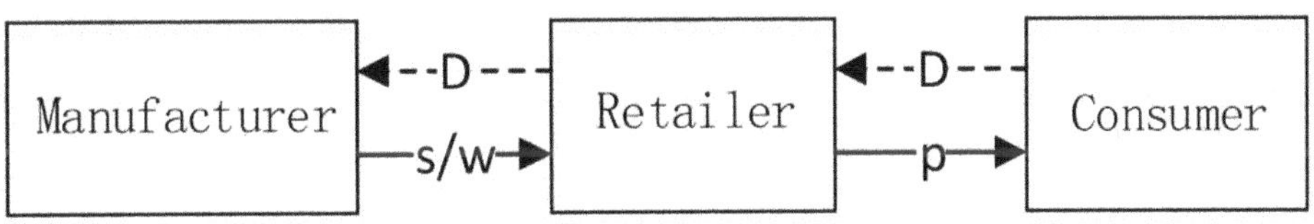

Table 1. With the increase of the parameters of the optimal decision variables

Parameters	C0		C1		
	w	p	s	w	p
a	↑	↑	—	↑	↑
b	↓	↓	↑	↓	↓
c	↑	↑	—	↑	↑
λ	—	—	↓	↑	↑
η	—	—	↑	↓	↓

A comparative analysis is carried out on the best prices, market demands and profits under the C0 and C1 strategies, aimed at proving manufacturers' CSI strategy can indeed improve the profitability. Comparing two strategies, we can conclude:

- **Theorem 1:** C0 is a dominant strategy if $\tau > \frac{a+bc}{3}$; otherwise C1 is a dominant strategy if $0 < \tau \le \frac{a+bc}{3}$.

Table 2 reports the equilibrium outcomes when $0 < \tau \le \frac{a+bc}{3}$. The default values of parameters are $a = 80, b = 10, c = 5, \eta = 0.8, \lambda = 6$. Here we fix the value of one parameter when others remain invariant. The result indicates that Theorem 1 is correct. In a perfect monopoly market, a decrease in unit production cost indeed can lower the price, since market's sensitivity to price is higher than its sensitivity to CSI, the overall market demand increases. Thus, the manufacturer can gain profit from the performance of CSI. Moreover, when the market is relatively sensitive to CSI, the manufacturers can merely gain low gains from the performance of CSI.

The Competitive Supply Chain Model

This section discusses the competitive supply chain model. Supposing there are only two competing supply chains on the market, the manufacturers on both sides supply homogeneous products for the target market. According to the normal conditions, supposing the manufacturers on both sides play a Nash game with each other and make a decision simultaneously under a complete information condition. The manufacturers in each supply chain play a Stackelberg game inside the supply chain and hold a dominant position, while the retailers serve as followers. The relational model researched in this section is more complicated than the single-channel model, so we suppose the production cost $c = 0$. Figure 2 describes the competition of the two supply chains. The manufacturers on both sides may form four combinations of CSI strategies: neither manufacturers performs CSI (C00), only manufacturer 1 performs CSI (C10), only manufacturer 2 performs CSI (C01), and both manufacturers perform CSI (C11).

The calculation procedure is shown in Appendix 2. The combined influence of various parameters on different optimal CSI decision variables under competitive supply chain model is shown in Table 3:

Table 2. Analysis of case in the single channel model

Parameter					C0				C1				
a	b	c	η	λ	w	p	D	Π_m	s	w	p	D	Π_m
80	10	5	0.8	6	6.5	7.25	7.5	11.25	1.78	4.9	5.92	10.12	20.67
80	10	5	0.8	8	6.5	7.25	7.5	11.25	1	5.3	6.25	9.5	18.05
80	10	5	1	6	6.5	7.25	7.5	11.25	2.78	4	5.16	11.67	27.22
80	12	5	0.8	6	5.83	6.25	5	4.16	2.56	3.91	4.65	8.84	13.02
80	10	6	0.8	6	7	7.5	5	5	1.78	5.4	6.17	7.67	11.76
100	10	5	0.8	6	7.5	8.75	12.5	31.25	1.78	5.9	7.42	15.17	46.01

Figure 2. Competitive supply chain structure

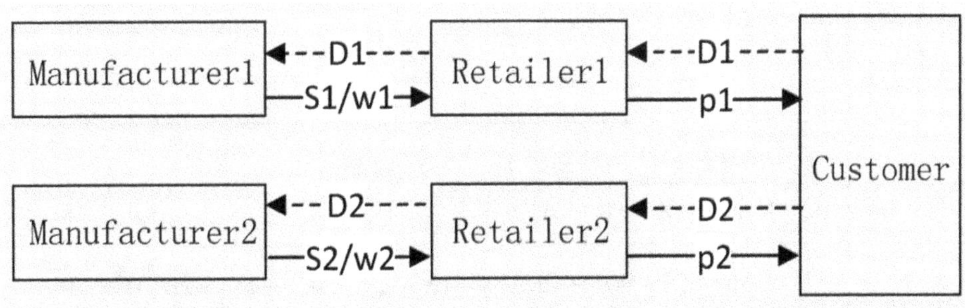

Table 3. With the increase of the parameters of the optimal decision variables

Parameters	C00		C10/C01					C11		
	w_i	p_i	s_i	w_i	w_{3-i}	p_i	p_{3-i}	s_i	w_i	p_i
θ	↓	↓	—	↓	↓	↓	↓	—	↓	↓
λ	—	—	↓	↑	↑	↑	↑	↓	↑	↑
η	—	—	↑	↓	↓	↓	↓	↑	↓	↓

According to the comparative analysis of the optimal decisions and corresponding optimal profits under C00, C10/C01 and C11, the optimal CSI level is the same under different strategies, which is only related to market's sensitivity to CSI λ and CSI's influence on cost η. Owing to the symmetry of the bilateral decisions in the competitive supply chain, an analysis is carried out here merely from the perspective of manufacturer 1. Since CSI helps manufacturers reduce the unit production cost, it seems that the manufacturers on both sides could gain more profits. The equilibrium profit of manufacturers in different CSI strategies is shown in Table 4.

From **Lemma 3-5**, we can know that

1. When $0 < \tau \le \dfrac{2-\theta-\theta^2}{6-\theta-3\theta^2}$, these four situations make sense. For manufacturer 1, C10 outperforms C11 and followed by C00 and C01;
2. When $\dfrac{2-\theta-\theta^2}{6-\theta-3\theta^2} < \tau \le \dfrac{8-2\theta-9\theta^2+\theta^3+2\theta^4}{24-25\theta^2+6\theta^4}$, only C11 does not make sense. For manufacturer 1, C10 outperforms C00 and followed by C01;
3. When $\tau > \dfrac{8-2\theta-9\theta^2+\theta^3+2\theta^4}{24-25\theta^2+6\theta^4}$, only C00 and C01 make sense. For manufacturer 1, C00 outperforms C01.

The corresponding results for manufacturer 2 can also be stated. And comparing the manufacturers' profits when they perform CSI and when they do not, we have the following Lemma 6.

Table 4. Equilibrium profit of manufacturers in different CSI strategies

	Π_m^*
i	
C00	$\frac{(2+\theta)(2-2\theta-\theta^2+\theta^3)}{(2-\theta)(1+\theta)(4-\theta-2\theta^2)^2}$
C10	$\frac{(2-\theta^2)(\eta^2(8-9\theta^2+2\theta^4)+\lambda(8-2\theta-9\theta^2+\theta^3+2\theta^4))^2}{(4-5\theta^2+\theta^4)(16-17\theta^2+4\theta^4)^2\lambda^2}$
C01	$\frac{(2-\theta^2)(\eta^2\theta(-2+\theta^2)+\lambda(8-2\theta-9\theta^2+\theta^3+2\theta^4))^2}{(4-5\theta^2+\theta^4)(16-17\theta^2+4\theta^4)^2\lambda^2}$
C11	$\frac{(2-\theta^2)(2-\theta-\theta^2)(\lambda+\eta^2)^2}{(2-\theta)(1+\theta)(4-\theta-2\theta^2)^2\lambda^2}$
3-i	
C00	$\frac{(2+\theta)(2-2\theta-\theta^2+\theta^3)}{(2-\theta)(1+\theta)(4-\theta-2\theta^2)^2}$
C10	$\frac{(2-\theta^2)(\eta^2\theta(-2+\theta^2)+\lambda(8-2\theta-9\theta^2+\theta^3+2\theta^4))^2}{(4-5\theta^2+\theta^4)(16-17\theta^2+4\theta^4)^2\lambda^2}$
C01	$\frac{(2-\theta^2)(\eta^2(8-9\theta^2+2\theta^4)+\lambda(8-2\theta-9\theta^2+\theta^3+2\theta^4))^2}{(4-5\theta^2+\theta^4)(16-17\theta^2+4\theta^4)^2\lambda^2}$
C11	$\frac{(2-\theta^2)(2-\theta-\theta^2)(\lambda+\eta^2)^2}{(2-\theta)(1+\theta)(4-\theta-2\theta^2)^2\lambda^2}$

Lemma 6

The manufacturer 1 benefits from its own performance of CSI when its rival does not perform CSI (from C00 to C10) if $0<\tau\le\frac{8-2\theta-9\theta^2+\theta^3+2\theta^4}{24-25\theta^2+6\theta^4}$, and when its rival performs CSI (from C01 to C11) if $0<\tau\le\frac{2-\theta-\theta^2}{6-\theta-3\theta^2}$. However, the manufacturer 1 is hurt by its rival's performance of CSI when it performs CSI (from C10 to C11) if $0<\tau\le\frac{2-\theta-\theta^2}{6-\theta-3\theta^2}$, and when it does not perform (from C00 to C01).

Lemma 6 shows that the performance of CSI indeed can help the manufacturers gain more profits and cause a loss in rivals' profits. Especially, if only one manufacturer performs CSI where $0<\tau\leq\frac{8-2\theta-9\theta^2+\theta^3+2\theta^4}{24-25\theta^2+6\theta^4}$, it will maximize its profits and do a damage to the rival. A primary cause for this phenomenon is that the performance of CSI can help the manufacturer lower its unit production cost and product price. As a result, the consumers transfer from the channel that the manufacturer does not perform CSI to the channel that the manufacturer does. Thus, for the rival that does not perform CSI, it has no choice but to perform CSI, so as to gain more profit and stop others from profit maximization. The game between the manufacturers on both sides is shown in Table 5.

- **Theorem 2:** C11 is an equilibrium if and only if $0<\tau\leq\frac{8-2\theta-9\theta^2+\theta^3+2\theta^4}{24-25\theta^2+6\theta^4}$; otherwise, C00 is an equilibrium if $\tau>\frac{8-2\theta-9\theta^2+\theta^3+2\theta^4}{24-25\theta^2+6\theta^4}$.

Theorem 2 indicates that there exist an equilibrium which depends on the extent of channel substitutability and the boundary values of τ, as shown in Figure 3. If $0<\tau\leq\frac{8-2\theta-9\theta^2+\theta^3+2\theta^4}{24-25\theta^2+6\theta^4}$, that is to say, CSI level of the marginal demand is small, thus both manufacturers are willing to perform CSI. As channel substitutability grows, products tend to perfect competition and the value of A reduces. The main reason of reduce might be that consumers pay more attention to the price than CSI. The manufacturers on both sides are definitely motivated to perform CSI, because if the manufacturer on one side does not perform CSI, it will get more kicks than halfpence. That is, it will have its profit lowered and meanwhile lead to an increase in its rival's profit. Therefore, it is a steady state that both sides perform CSI and neither wants to deviate from this practice. In this case, C11 is an equilibrium. If $\tau>\frac{8-2\theta-9\theta^2+\theta^3+2\theta^4}{24-25\theta^2+6\theta^4}$, demand change by CSI is large and both manufacturers will not perform CSI, because in this situation the manufacturers can benefit from their rivals' performance of CSI, thus neither is willing to perform CSI to make rival benefit and ultimately C00 is an equilibrium.

Table 6 reports the equilibrium outcomes when $0<\tau\leq\frac{8-2\theta-9\theta^2+\theta^3+2\theta^4}{24-25\theta^2+6\theta^4}$. The default values of parameters are $\theta=0.9,\eta=0.8,\lambda=12$. Here we fix the value of one parameter when others remain invariant. The result demonstrates that Theorem 2 is correct, namely the manufacturers on both sides achieve game equilibrium under C11 and gain more benefits. When only one side perform CSI (C10/

Table 5. Results of Manufacturers game

m_{3-i} \ m_i	Performance of CSI	Nonperformance of CSI
performance of CSI	$\Pi_{mi}^{C11*},\Pi_{m3-i}^{C11\ *}$	$\Pi_{mi}^{C01*},\Pi_{m3-i}^{C01\ *}$
nonperformance of CSI	$\Pi_{mi}^{C10*},\Pi_{m3-i}^{C10\ *}$	$\Pi_{mi}^{C00*},\Pi_{m3-i}^{C00\ *}$

Figure 3. Equilibrium result

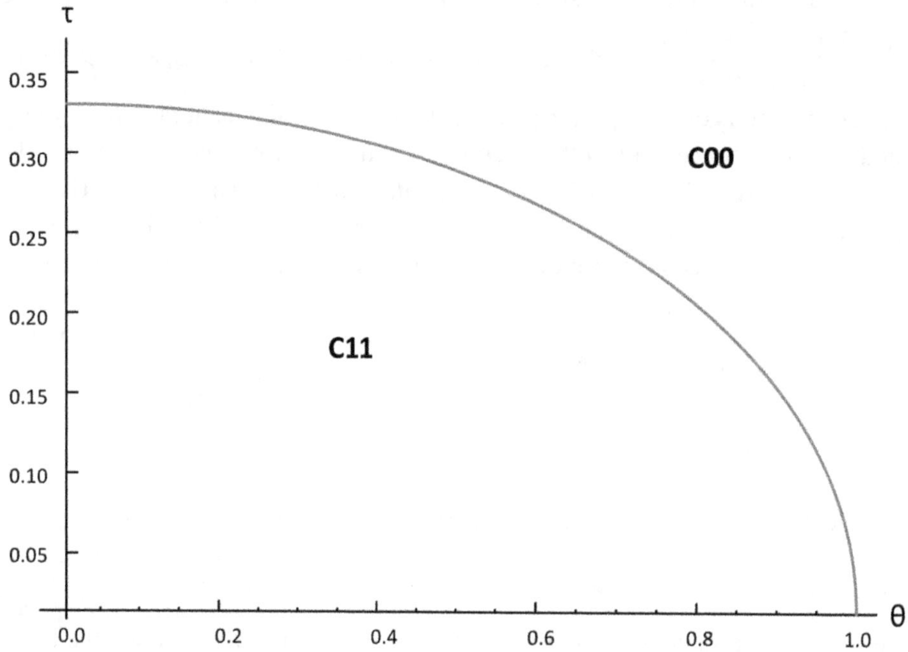

Table 6. Analysis in the competitive supply chain model with different values of (η,θ,λ)

Parameter	C00				C10/C01					C11				
	w_i	p_i	D_i	Π_{mi}	s_i	w_i/w_{3-i}	p_i/p_{3-i}	D_i/D_{3-i}	Π_{mi}/Π_{m3-i}	s_i	w_i	p_i	D_i	Π_{mi}
(0.8,0.2,12)	0.47	0.71	0.24	0.115	0.004	0.39/ 0.47	0.64/ 0.71	0.26/ 0.24	0.129/ 0.115	0.004	0.39	0.64	0.26	0.128
(0.8,0.2,10)	0.47	0.71	0.24	0.115	0.006	0.38/ 0.47	0.63/ 0.70	0.26/ 0.24	0.132/ 0.115	0.006	0.38	0.62	0.26	0.131
(0.8,0.9,12)	0.20	0.27	0.38	0.075	0.004	0.11/ 0.18	0.19/ 0.25	0.43/ 0.36	0.093/ 0.067	0.004	0.10	0.18	0.41	0.084
(0.8,0.9,10)	0.20	0.27	0.38	0.075	0.006	0.09/ 0.18	0.18/ 0.25	0.44/ 0.36	0.097/ 0.065	0.006	0.08	0.16	0.41	0.085
(1,0.2,12)	0.47	0.71	0.24	0.115	0.007	0.35/ 0.47	0.60/ 0.70	0.27/ 0.24	0.137/ 0.114	0.007	0.35	0.60	0.26	0.135
(1,0.2,10)	0.47	0.71	0.24	0.115	0.01	0.32/ 0.47	0.58/ 0.70	0.27/ 0.24	0.141/ 0.114	0.01	0.32	0.58	0.27	0.140
(1,0.9,12)	0.20	0.27	0.38	0.075	0.007	0.06/ 0.18	0.15/ 0.24	0.45/ 0.35	0.104/ 0.062	0.007	0.05	0.12	0.42	0.088
(1,0.9,10)	0.20	0.27	0.38	0.075	0.01	0.04/ 0.17	0.13/ 0.24	0.47/ 0.34	0.111/ 0.059	0.01	0.02	0.10	0.42	0.091

C01), it can gain more benefits, while the other side suffers losses. Besides, as can be seen, as the market becomes more sensitive to CSI, the manufacturers gain fewer benefits from the performance of CSI; as the substitutability of both channels rises, the manufacturers gain fewer benefits from the performance of CSI as well; but as CSI further affects cost, the manufacturers gain more benefits from the performance of CSI.

DESIGN OF CORPORATE GOVERNANCE MECHANISMS

Through the research of the CSI decisions of manufacturers in different scenarios, we find that nonperformance of CSI is a dominant strategy for manufacturers if and only if τ is large, otherwise performance of CSI is a dominant strategy if τ is small where τ is influenced by η and λ and in single channel also influenced by b. And the boundary value of τ in single channel is related to a,b and c, while in competitive supply chain is a function of θ. According to the characteristics of these parameters, we can design governance mechanisms from the enterprise internal and external. The parameters b and λ are related to consumers sensitivities which may be caused by asymmetry information between enterprises and customers, and this suggests us establish and perfect the corporate social responsibility disclosure mechanism in the enterprise internal, thus improving market sensitivity to CSI and reducing CSI level. The parameter θ is inversely proportion to the profit, which means strengthening market competition in supply chain can cause price war and reduce the manufacturer's profit. The parameter η is inversely proportion to the unit production cost and proportion to the CSI level, thus a symmetry reward-penalty mechanism in supply chain is essential to increase penalty cost. The symmetry reward-penalty mechanism (denoted with a superscript D) is the focus and is introduced detailedly in this section. Here the design method is the manufacturer 's output and the unit reward-penalty intensity (Wang & Deng, 2016). The manufacturer will be rewarded if it does not perform CSI; otherwise, it will be punished. Let $k(k>0)$ as an unit reward-penalty intensity, where the intensity of reward and penalty is the same.

In the Single Channel Model

With a symmetry reward-penalty mechanism, the manufacturer's profits in the single channel are $\Pi_m^{D0*}=\dfrac{(a-bc+bk)^2}{8b}$ and $\Pi_m^{D1*}=\dfrac{(b^2\eta^2+(a-bc+bk)\lambda)^2}{8b\lambda^2}$, respectively. Comparing these two strategies, we have

- **Theorem 3:** The symmetry reward-penalty mechanism can effectively constrain the manufacturer's CSI if $k\in[\dfrac{b\eta^2}{2\lambda},\dfrac{a+bc}{b}]$, and D0 will be a dominant strategy.

In the Competitive Supply Chain Model

With a symmetry reward-penalty mechanism, the manufacturer's profits in the competitive supply chain are $\Pi_{mi}^{D00*}=\dfrac{(1+k)^2(2+\theta)(2-2\theta-\theta^2+\theta^3)}{(2-\theta)(1+\theta)(4-\theta-2\theta^2)^2}$,

$$\Pi_{mi}^{D10*} = \frac{(\eta^2(8-9\theta^2+2\theta^4)+(8-2\theta-9\theta^2+\theta^3+2\theta^4+k(-8-2\theta+9\theta^2+\theta^3-2\theta^4))\lambda)^2}{8(4-\theta^2)(2-\theta^2)(1-\theta^2)(8-9\theta^2+2\theta^4)\lambda^2}$$

$$, \Pi_{mi}^{D01*} = \frac{(\eta^2\theta(8-9\theta^2+2\theta^4)+\begin{pmatrix}-32+8\theta+54\theta^2-9\theta^3-27\theta^4+2\theta^5+4\theta^6\\+k(-32-8\theta+54\theta^2+9\theta^3-27\theta^4-2\theta^5+4\theta^{66}))\lambda\end{pmatrix}^2}{16(8-9\theta^2+2\theta^4)^2(8-14\theta^2+7\theta^4-\theta^6)\lambda^2}$$

and $\Pi_{mi}^{D11*} = \dfrac{(2-\theta^2)(2-\theta-\theta^2)(\eta^2+\lambda(1-k))^2}{(2-\theta)(1+\theta)(4-\theta-2\theta^2)^2\lambda^2}$, respectively. Comparing these strategies, we have

- **Theorem 4:** The symmetry reward-penalty mechanism can effectively constrain the manufacturer's CSI where $k \in [\dfrac{\eta^2}{2\lambda}, \dfrac{2-\theta-\theta^2}{2-\theta^2}]$, and D00 will be an equilibrium strategy.

EXTENSION

We have considered two manufacturers holding almost the same initial base demand in the competitive supply chain environment, and setting $a_i = a_{3-i} = 1$. Here, in this section we consider two manufacturers holding the asymmetry initial base demand, and whether manufacturer 1's initial base demand is larger than manufacturer 2 or not, the results of the two cases are similar. Hence, we assume that the situation of manufacturer 1's initial base demand is larger and set $a_i = 3, a_{3-i} = 2$ (denoted with a superscript A in this situation). The equilibrium profit of manufacturers in the asymmetry initial base demand is shown in Table 7.

From the Table 7, we find that manufacturer 1's profit is larger than manufacturer 2's. And for manufacturer 1, it will perform CSI if $0 < \tau \le \dfrac{24-4\theta-27\theta^2+2\theta^3+6\theta^4}{24-25\theta^2+6\theta^4}$ and will not if $\tau > \dfrac{24-4\theta-27\theta^2+2\theta^3+6\theta^4}{24-25\theta^2+6\theta^4}$. But for manufacturer 2, the result is a little different. When channel substitutability is low $(0 \le \theta \le 0.944)$, it will perform CSI if $0 < \tau \le \dfrac{16-6\theta-18\theta^2+3\theta^3+4\theta^4}{24-25\theta^2+6\theta^4}$ and will not if $\tau > \dfrac{16-6\theta-18\theta^2+3\theta^3+4\theta^4}{24-25\theta^2+6\theta^4}$; otherwise when channel substitutability is sufficiently high $(0.944 < \theta < 1)$, Nash game works against manufacturer 2. Hence, unlike in the symmetry initial base demand, where C11 is an equilibrium if and only if $0 < \tau \le \dfrac{8-2\theta-9\theta^2+3\theta^3+2\theta^4}{24-25\theta^2+6\theta^4}$; otherwise, C00 is an equilibrium. Here we find that an asymmetric CSI strategy can also achieve equilibrium, as shown in Figure 4.

Table 7. Equilibrium profit of manufacturers in the asymmetry initial base demand

	Π_m^*
i	
A00	$\frac{(2-\theta^2)(24-4\theta-27\theta^2+2\theta^3+6\theta^4)^2}{(4-5\theta^2+\theta^4)(16-17\theta^2+4\theta^4)^2}$
A10	$\frac{(2-\theta^2)(\eta^2(8-9\theta^2+2\theta^4)+\lambda(24-4\theta-27\theta^2+2\theta^3+6\theta^4))^2}{(4-5\theta^2+\theta^4)(16-17\theta^2+4\theta^4)^2\lambda^2}$
A01	$\frac{(2-\theta^2)(\eta^2\theta(8-9\theta^2+2\theta^4)+\lambda(24-4\theta-27\theta^2+2\theta^3+6\theta^4))^2}{(4-5\theta^2+\theta^4)(16-17\theta^2+4\theta^4)^2\lambda^2}$
A11	$\frac{(2-\theta^2)(\eta^2(8-2\theta-9\theta^2+\theta^3+2\theta^4)+\lambda(24-4\theta-27\theta^2+2\theta^3+6\theta^4))^2}{(4-5\theta^2+\theta^4)(16-17\theta^2+4\theta^4)^2\lambda^2}$
3-i	
A00	$\frac{(2-\theta^2)(16-6\theta-18\theta^2+3\theta^3+4\theta^4)^2}{(4-5\theta^2+\theta^4)(16-17\theta^2+4\theta^4)^2}$
A10	$\frac{(2-\theta^2)(\eta^2\theta(-2+\theta^2)+\lambda(16-6\theta-18\theta^2+3\theta^3+4\theta^4))^2}{(4-5\theta^2+\theta^4)(16-17\theta^2+4\theta^4)^2\lambda^2}$
A01	$\frac{(2-\theta^2)(\eta^2(8-9\theta^2+2\theta^4)+\lambda(16-6\theta-18\theta^2+3\theta^3+4\theta^4))^2}{(4-5\theta^2+\theta^4)(16-17\theta^2+4\theta^4)^2\lambda^2}$
A11	$\frac{(2-\theta^2)(\eta^2(8-2\theta-9\theta^2+\theta^3+2\theta^4)+\lambda(16-6\theta-18\theta^2+3\theta^3+4\theta^4))^2}{(4-5\theta^2+\theta^4)(16-17\theta^2+4\theta^4)^2\lambda^2}$

- **Theorem 5:** In the asymmetry initial base demand, A11 is an equilibrium if $0<\tau\leq\frac{16-6\theta-18\theta^2+3\theta^3+4\theta^4}{24-25\theta^2+6\theta^4}$; A10 is an equilibrium if $\frac{16-6\theta-18\theta^2+3\theta^3+4\theta^4}{24-25\theta^2+6\theta^4}<\tau\leq\frac{24-4\theta-27\theta^2+2\theta^3+6\theta^4}{24-25\theta^2+6\theta^4}$; A00 is an equilibrium if $\tau>\frac{24-4\theta-27\theta^2+2\theta^3+6\theta^4}{24-25\theta^2+6\theta^4}$, where $0\leq\theta\leq0.944$. Otherwise, there is no Nash game where $0.944<\theta<1$.

Theorem 5 indicates that the existence of an asymmetric equilibrium depends on the extent of channel substitutability and CSI level of the marginal demand in the asymmetry initial base demand. If CSI level of the marginal demand and the channel substitutability are low ($0<\tau\le\frac{16-6\theta-18\theta^2+3\theta^3+4\theta^4}{24-25\theta^2+6\theta^4}$ and $0\le\theta\le0.944$), both manufacturers would benefit from performance of CSI. However, if CSI level of the marginal demand is a little bit high and the channel substitutability is low ($\frac{16-6\theta-18\theta^2+3\theta^3+4\theta^4}{24-25\theta^2+6\theta^4}<\tau\le\frac{24-4\theta-27\theta^2+2\theta^3+6\theta^4}{24-25\theta^2+6\theta^4}$ and $0\le\theta\le0.944$), then the manufacturer whose initial base demand is larger would still perform CSI, and the other manufacturer would rather stay out of the CSI game. Thereby, we find an asymmetric equilibrium. If CSI level of the marginal demand is sufficiently high and the channel substitutability is low ($\tau>\frac{24-4\theta-27\theta^2+2\theta^3+6\theta^4}{24-25\theta^2+6\theta^4}$ and $0\le\theta\le0.944$), both manufacturers would stay out of the CSI. If the channel substitutability is sufficiently high $(0.944<\theta<1)$, Nash game is counterproductive to the manufacturer whose initial base demand is smaller, and might take into account the Stackelberg game between manufacturer 1 and 2.

FUTURE RESEARCH DIRECTIONS

However, this paper meets with lots of problems needing further research when tentatively exploring how manufacturers should make pricing decisions and select strategies based on CSI in supply chain environment. This paper merely considers the Nash game between two manufacturers in competition, and it does not consider the Stackelberg game; secondly, this paper does not discuss the indicator system that could measure CSI level, but just makes a generalization by a global variable; finally, this paper develops a symmetrical reward-penalty mechanism and other governance mechanisms which suggest carrying out a further study with empirical research.

CONCLUSION

This paper depicts CSI level with a global variable, and takes manufacture as a main research object to explore how to make decisions on CSI in the supply chain, respectively analyzing the optimal pricing and CSI decision-making in both single-channel and competitive supply chain environments. The research result shows that manufacturers implement the optimal pricing strategy, and that in both single-channel and competitive supply chain environments, the performance of CSI can help manufacturers gain more benefits and lead to an increase in market demand if and only if CSI level of the marginal demand is small, otherwise no performance of CSI is a dominant strategy if CSI level of the marginal demand is large. In addition, as can be seen, whether manufacturers perform CSI depends on such two parameters as η and λ. As the market becomes more sensitive to CSI, some consumers that value product attribute give up buying CSI products. Hence, the manufacturers gain fewer benefits from the performance of CSI which limits the strategy selection on CSI of the manufacturers. As the effect of CSI on cost gets

Figure 4. Equilibrium result in the asymmetry initial base demand

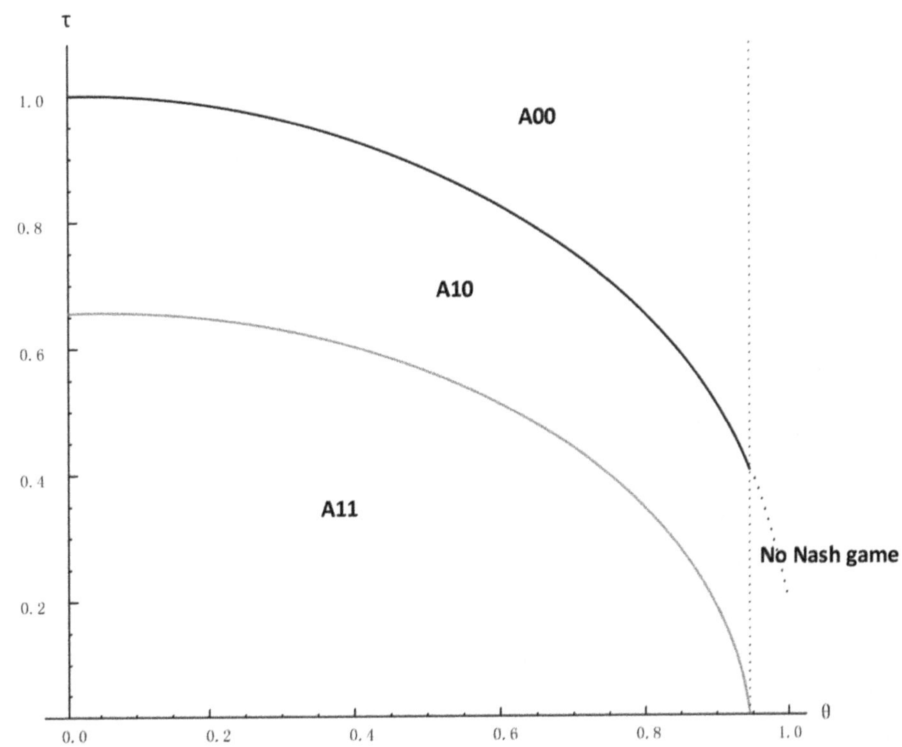

stronger, the performance of CSI can help manufacturers save the unit production cost more efficiently, which makes cost advantage more prominent, so as to expand the channel demand and let manufactures gain more benefits.

On this basis, a symmetry reward-penalty mechanism which takes a certain intensity of reward-penalty on the output is designed to restrain the CSI performance of the manufacturer and promote corporate sustainable development. And the unit reward-penalty intensity within a certain range can effectively restrain the CSI behavior of the manufacturer, which makes the profit of the manufacturer without CSI is bigger than performance CSI. Based on the above analysis, a key issue that governs the corporate social irresponsibility is not only to develop the symmetry reward-penalty mechanism, but also to enhance the national environmental protection awareness so as to resist the productions which are produced by companies with CSI. Moreover, the corporate social responsibility disclosure mechanism in the enterprise internal is essential to reduce the market price sensitivity and improve the sensitivity of CSI, so as to reduce information asymmetry between the customers and corporations. And strengthening market competition in supply chain is also effective for reducing the profit of the manufacturers.

Furthermore, this paper makes an extension on the asymmetry initial base demand and setting $a_i = 3, a_{3-i} = 2$. The result is quite different from that in the symmetry initial base demand. Firstly, when channel substitutability is sufficiently high $(0.944 < \theta < 1)$, Nash game is counterproductive to the manufacturer whose initial base demand is smaller. Secondly, we find that an asymmetric CSI strategy can be an equilibrium, if and only if $\frac{16-6\theta-18\theta^2+3\theta^3+4\theta^4}{24-25\theta^2+6\theta^4}<\tau\leq\frac{24-4\theta-27\theta^2+2\theta^3+6\theta^4}{24-25\theta^2+6\theta^4}$ where $0\leq\theta\leq 0.944$. Comparing Theorem 2 against Theorem 5, we find that if the manufacturers' respective initial base demand is comparable and almost the same, CSI level of the marginal demand is lower and can effectively restrain manufacturers' performance of CSI. So it is suggested to support smaller enterprises and prevent complete monopoly so as to intensify the competition in supply chain.

REFERENCES

Armstrong, J. S. (1977). Social Irresponsibility in Management. *Journal of Business Research*, *5*(3), 185–213.

Armstrong, J. S., & Green, K. C. (2013). Effects of corporate social responsibility and irresponsibility policies. *Journal of Business Research*, *66*(10), 1922–1927.

Carrasco, M. I., & Buendía, M. I. (2013). Corporate social responsibility: A crossroad between changing values, innovation and internationalisation. *European Journal of International Management*, *7*(3), 295–314.

Goering, G. E. (2012). Corporate social responsibility and marketing channel coordination. *Research in Economics*, *66*(2), 142–148.

Hediger, W. (2010). Welfare and capital-theoretic foundations of corporate social responsibility and corporate sustainability. *Journal of Socio-Economics*, *39*(4), 518–526.

Hsueh, C. F., & Chang, M. S. (2008). Equilibrium analysis and corporate social responsibility for supply chain integration. *European Journal of Operational Research*, *190*(1), 116–129.

Hu, X. (2004). Corporate Sustainable Development and Responsibility. Journal of Chongqing University of Posts and Telecomm unications. *Social Sciences*, *16*(2), 123–125.

Huang, S. (2014). Research on Pricing and Social Responsibility Decision Models in Dual-channel Supply Chain. *Commercial Research*, *56*(5), 153–161.

Jiang, L. (2014). A Review of Foreign Literature of Corporate Social Irresponsibility. *Foreign Economics & Management*, *36*(2), 13–23.

Liang, Y., Ni, D., & Tang, X. (2013). Corporate Social Responsibility Based Model on Dual-channel Competition in Supply Chains. *Chinese Journal of Management Science*, *21*(S2), 453–460.

Lin-Hi, N., & Müller, K. (2013). The CSR bottom line: Preventing corporate social irresponsibility. *Journal of Business Research*, *66*(10), 1928–1936.

Mishina, Y., Dykes, B. J., Block, E. S., & Pollock, T. G. (2010). Why good firms do bad things: The effects of high aspirations, high expectations, and prominence on the incidence of corporate illegality. *Academy of Management Journal*, *53*(4), 701–722. doi:10.5465/AMJ.2010.52814578

Modak, N. M., Panda, S., Sana, S. S., & Basu, M. (2014). Corporate social responsibility, coordination and profit distribution in a dual-channel supply chain. *Pacific Science Review, 16*(4), 235–249.

Murphy, P. E., & Schlegelmilch, B. B. (2013). Corporate social responsibility and corporate social irresponsibility: Introduction to a special topic section. *Journal of Business Research, 66*(10), 1807–1813.

Ni, D., Li, K. W., & Tang, X. (2010). Social responsibility allocation in two-echelon supply chains:Insights from wholesale price constracts. *European Journal of Operational Research, 207*(3), 1269–1279.

Ni, D., Li, X., & Tang, X. (2015). CSR Operations in Supply Chains: Mutual Incentive, CSR Allocation and Cooperation. *Chinese Journal of Management Science, 23*(9), 97–105.

Panda, S. (2014). Coordination of a socially responsible supply chain using revenue sharing contract. *Transportation Research Part E, Logistics and Transportation Review, 67*, 92–104.

Panda, S., Modak, N. M., Basu, M., & Goyal, S. K. (2015). Channel coordination and profit distribution in a social responsible three-layer supply chain. *International Journal of Production Economics, 168*, 224–233.

Sun, H., & Li, H. (2012). Chinese Corporate Social Responsibility and the Way of Corporate Sustainable Development. *Reformation and Strategy, 28*(6), 158-160, 185.

Sweetin, V. H., Knowles, L. L., Summey, J. H., & Mcqueen, K. S. (2013). Willingness-to-punish the corporate brand for corporate social irresponsibility. *Journal of Business Research, 66*(10), 1822–1830.

Wan, S., & Liu, F. (2014). Empirical Analysis on the Relationship between Corporate Social Responsibility and Corporate Social Irresponsibility. *Finance and Accounting Monthly, 22*, 3-8.

Wan, S., & Liu, F. (2015). Research on the Causes and Economic Consequences of Corporate Social Irresponsibility. *Friends of Accounting, 11*, 15–25.

Wang, H., & Han, B. (2016). Social Responsibility, Internal Control and Corporate Sustainable Development--Based on an Empirical Analysis of Chinese A-share Main Board Enterprises. *Journal of Beijing Technology and Business University, 31*(1), 75–84.

Wang, J., & Hou, T. (2009). Empirical Analysis of the Effects of Social Responsibility on Corporate Sustainable Development. *Science & Technology Progress and Policy, 26*(18), 94–96.

Wang, W., & Deng, W. (2016). Comparison between the Reward-penalty Mechanism with the Tax-subsidy Mechanism for Reverse Supply Chains. *Chinese Journal of Management Science, 24*(4), 102–110.

Wang, X., & Mao, W. (2015). Consumer Perceptions of Corporate Social Irresponsibility Initiatives: The Effects of Consumer Attribution and Expectation. *Journal of Beijing Institute of Technology (Social Sciences Edition), 17*(6), 74-80+112.

Xin, J., & Liao, X. (2013). Sustainable Development and Social Responsibility of Enterprises. *Journal of Central South University of Forestry and Technology, 7*(3), 77–80.

Yang, W., & Yang, S. (2016). An Empirical Study on the Relationship between Corporate Social Responsibility and Financial Performance under the Chinese Context——Based on Contrastive Analysis of Large, Small and Medium-size Listed Companies. *Chinese Journal of Management Science*, *24*(1), 143–150.

Zhang, R., Liu, B., & Wang, W. (2012). Pricing decisions in a dual channels system with different power structures. *Economic Modelling*, *29*(2), 523–533. doi:10.1016/j.econmod.2011.08.024

Chapter 9
Opportunities and Challenges for Sustainable Business Strategic Planning in Small and Medium Enterprises (SMEs)

Mauricio Quintero-Angel
Universidad Santiago de Cali, Colombia

Claudia C Peña-Montoya
Universidad Autónoma de Occidente, Colombia

Carlos Hernán Fajardo-Toro
Universidad EAN, Colombia

Andres Aguilera-Castillo
Universidad EAN, Colombia

ABSTRACT

Approximately 90% of companies are Micro, Small, and Medium Enterprises (MSMEs) and Small & Medium Enterprises (SMEs). Their importance in the economy lies in the fact that they account for almost 60% of the jobs in the world, number which could be greater in developing countries. All these companies have an environmental impact that is almost imperceptible when observed individually, but if observed and analyzed as a whole, this impact is actually relevant. It is therefore important for these companies to include the concept of sustainability within their processes, which is difficult given the informal way they have managed their activities and processes. On the other hand, there are two paradigms that rule the concept of sustainability, weak sustainability and strong sustainability. As it will be presented in this chapter, both concepts are opposed and companies are normally oriented towards weak sustainability, because they are more interested in economic growth than in caring for the environment.

DOI: 10.4018/978-1-5225-3537-9.ch009

INTRODUCTION

The different production and transformation methods and the technologies used throughout human history have had a direct impact on the environment that could not be easily absorbed by nature at the beginning of the XXI century (Rockström et al. 2009; Steffen et al. 2015). These changes in earth's systems meant the search for a new kind of relationship between mankind and nature, and the acknowledgement that both economy and society are sub-systems of a larger and finite global system: the biosphere.

In this context of significant transformation in the society-nature relationship balance, there is evidence of the need for important changes in the extractive and productive activities with an integrative and ecologically sustainable approach. Several countries have embraced the concept of sustainable development in their national legislation and have strengthened their environmental policies and standards, and many others have imposed non-tariff trade barriers by restricting the trade of products that use certain kinds of chemical substances and materials.

Several companies are including environmental aspects in their management in order to avoid the environmental restrictions on trade, and use green marketing elements to encourage the sale of products to consumers increasingly demanding environmental protection. These companies generally do not question the business model's sustainability, and environmental actions are only limited to including elements of cleaner production, product life cycle analysis, eco-labeling or being awarded with the ISO 14001 certification. These are interesting actions that can reduce environmental impact, but do not question the business model or the business' scale. This shows Jevons paradox, which according to Alcott (2005), shows that as technological improvement increases, so does the efficiency with which resources are used, therefore there is a high probability that their use and demand will increase.

In the case of SMEs, environmental initiatives such as cleaner production, product life cycle analysis, certification in an environmental management system or other Green Production Strategies for Sustainability, are usually not implemented due to the fact that they are associated to high costs and highly-trained professionals which are out of scope for SMEs. Being that SMEs represent over 90% of all companies (WIPO-World Intellectual Property Organization. n.d.) they are required to include environmental management aspects in order to reduce the environmental impact caused by the significant cumulative effect.

Being said, it is important for SMEs to get involved in environmental matters, starting with their strategic planning. Kotler & Armstrong (2003) suggest that strategic planning is essential for managers because it allows them to link past, present and future events, forcing companies to define better objectives and policies, resulting in improved synchronization of activities, control and performance.

The chapter analyzes two opposing paradigms of *sustainability* from a Latin American experience, and their economic and ecological implications for sustainable business strategic planning in the context of SMEs. On one hand, strong sustainability is based on technological skepticism and considers that it is not possible to replace many functions and ecosystem services with human-built elements, which means that the ecosystem base that supports socio-economic development should be maintained. On the other hand, weak sustainability corresponds to the contrary. Different strategies for both paradigms, as well as their uses, challenges and opportunities for being included in SMEs' strategic planning, will be analyzed.

BACKGROUND

SMEs and Sustainability Subjects

SMEs represent over 90% of the companies in the world (WIPO – World Intellectual Property Organization, n.d). According to the OECD and the World Bank, they represent 99% in Latin America and 92% in Europe (Portafolio, 2014). They account for approximately 60% of jobs globally jobs (World Bank, 2016; World SME Forum, n.d.) and 67% of jobs in Latin America according to the IDB, with 50% of those jobs related to the manufacturing sector (Inter-American Development Bank, 2013).

Based on the statistics mentioned above, the SMEs impact on the environment is clear and important. The necessity to include sustainability concepts inside the organizational culture and structure results vitally important, especially when reconciling innovation, sustainability and competitiveness which are a necessity for the survival of the company.

Klewitz & Hansen (2014) propose the relation between sustainability and innovation, and the importance to include the concept of eco-innovation within the SMEs strategic planning process. The eco-innovations would consider these three dimensions: i) Process Innovations, ii) Organizational innovations and iii) Product Innovations.

However, the successful achievement by SMEs to be sustainable is not an easy task, especially in developing countries where they have to deal with factors like difficulties finding qualified employees, lack of government control and support and financial difficulties (Günerergin et al. 2012).

The next section presents the importance of including the concept of sustainability in the companies' different processes, as well as a discussion on two different paradigms associated to sustainability: weak sustainability and strong sustainability, which will affect the organizations' strategic vision in different ways depending on the approach.

The Concept of Sustainability: Weak Sustainability and Strong Sustainability

The environmental crisis and the strong pressure of socioeconomic activities against fundamental biophysical processes that are necessary for the planet's balance gave rise in 1987 to the United Nation's concept of sustainable development, which has been widely diffused around the world and aims at orienting the nation's socioeconomic development towards ways of having a lower impact on nature. According to the United Nation's General Assembly:

Humanity has the ability to make development sustainable to ensure that it meets the needs of the present without compromising the ability of future generations to meet their own needs. The concept of sustainable development does imply limits - not absolute limits but limitations imposed by the present state of technology and social organization on environmental resources and by the ability of the biosphere to absorb the effects of human activities. But technology and social organization can be both managed and improved to make way for a new era of economic growth. (United Nations General Assembly, 1987: 43)

The United Nations General Assembly's guidelines on sustainable development have been accepted and diffused by the different governments, and practically every country has incorporated the concept as an important criteria for decision-making. Moreover, the terms "sustainability" and "sustainable development" are now part of the corporate vocabulary and strategic planning processes (Franceschi & Kahn 2003). These definitions have been widely debated in the past years because the "needs" are not clearly defined and because they are based on an optimistic vision of technology as the element that will allow sustained economic growth. Therefore, Munda (1997) defines that sustainability and sustainable development are different concepts, with different complexities, and subject to a different debate.

There are two variants or lines associated to sustainability: weak sustainability and strong sustainability. According to Neumayer (2003) each one is based on two different economic paradigms, where weak sustainability is based on the works of neoclassic economists Robert Slow and John Harteick and strong sustainability moves away from the neoclassic theories and is recognized as the paradigm of non-substitutability.

Munda (1997) states that weak sustainability justifies the use and damage of nature to obtain economic growth, while strong sustainability highlights the importance of harmony between nature and economic growth. Based on the work of Costanza & Daly (1992) and despite its linear nature, the mathematical expression developed by Guhl (2004) helps understand the concepts of weak and strong sustainability.

$K_t = K_n + K_m + K_h + \ldots + K_{ni}$, where,
K_t (Total Capital) a community's assets, services and products in a specific time t;
K_n (Natural Capital) natural resources and environmental services;
K_m (Manufactured Capital) Capital created by the community as a product of their activities;
K_h (Human Capital) The society's set of rules, norms and customs;
K_{ni} Other forms of capital;

According to the concept of weak sustainability, a process is considered sustainable when a society's total capital (K_t) increases for a period of time or remains constant, which means that natural capital (K_n) could be replaced by manufactured capital (K_m). On the other hand, natural capital (K_n) must be maintained or increased in strong sustainability.

Nevertheless, for Costanza & Daly (1992), weak sustainability tries to keep total capital (Kt) intact, without acknowledging the weight of each kind of capital that composes it, which means that the different capital types are more or less substitutable, at least within the limits of the economic activity levels and endowment of resources. Neumayer (2003) defines:

Loosely speaking, according to weak sustainability, it does not matter whether the current generation uses up non-renewable resources or dumps CO2 in the atmosphere as long as enough machineries, roads and ports are built in compensation. Because natural capital is regarded as being essentially substitutable in the production of consumption goods and as a direct provider of utility...(Neumayer 2003: 1).

In contrast, Pérez & Rojas (2010) argue that strong sustainability is based on technological skepticism and the principle of precaution, because it is impossible to replace many of the ecosystem functions and services by manmade elements. This implies that it is necessary to maintain the environment's capacity to endure development's patterns, that is, the eco systemic base that supports socioeconomic development (Pérez & Rojas 2010: 31).

For these authors, strong sustainability recognizes the existence of limits and laws of nature that have to be respected. They mention the limits that strong sustainability imposes for the suitable exploitation of natural resources and environmental services. These limits are defined by nature's capacity to reproduce and self-depurate, and to absorb and eliminate contamination. Under this approach, nature not only offers resources such as wood, water or land (which have an economic value), but also provides environmental services like forests, rivers and territories which are fundamental to support life. Therefore, economic growth has to be compatible with the growth dynamics of renewable resources, as well as the development of new alternative energy sources that allow sustainability of non-renewable resources (Pérez & Rojas 2010)

All the different authors support either one or both of the paradigms (Hediger 1999), but there is evidence of the pressure on nature to satisfy human needs and expectations (see Table 1). This pressure means that it is necessary to proceed based on the principle of precaution (Riechmann & Tickner 2002) and responsibility (De Sequeira 2009), which take an ethical position to face technological civilization, something that is close to the principles of strong sustainability.

Finally, these two opposite paradigms of sustainability represent a challenge to be included in business planning, especially because both paradigms have socio-ecological implications which are significantly different in the way of doing business.

Strategic Planning and its Importance for Sustainability

Importance of Strategic Planning in SMEs

Strategic planning is fundamental for SMEs in order to ensure growth and competitive advantage, because it allows organizing actions to reach the proposed objectives (Mora-Riapira et al. 2015). According to the study by Mora-Riapira et al. (2015), MSMEs and export and non-export SMEs in the commercial sector have poor levels of competitiveness in environmental management, related to the level of achievement in strategic planning. That is, companies that implement partial strategic plans tend to have lower

Table 1. Approaches on weak and strong sustainability (source: Peña-Montoya, 2016)

Reference	Approach	
	Strong	Weak
(Munda 1997)	Based on the assumption that certain kinds of natural capital are critical and not replaceable by human-built capitals.	Based on the assumption of the perfect replacement of the different types of capital
(Hediger 1999)	Keeping ecological capital intact over time	Involves different benefits from the environment with economic development
(Pérez & Rojas 2010)	Based on technological skepticism and the principle of precaution. Technological capacity will not prevent the restriction of natural resources from achieving economic growth, therefore this growth must be harmonious with the environment's ability to support it	Conceives nature as an economic resource necessary to sustain development represented in consumption, welfare or income per capita, ignoring the natural limits and complexity
(Nilsen 2010)	The economy and nature are considered complementary and both must be sustained	Possibility of substituting economy and nature to achieve the highest possible utility for human beings

performance in environmental aspects. The idea expressed above is related to the low competitiveness perceived in different factors (human resources, environmental management, etc.) by the partial implementation of strategic planning. The authors recommend applying strategic planning fully to raise the competitiveness levels and harmonize factors towards the achievement of business objectives.

SMEs report the absence of formality, strategic plans and a business strategy based on their economic uncertainty (Pastrana & Srirameshb 2014). On the other hand, the absence of strategic planning also influences the weakness in the creation of a corporate identity in which corporate social responsibility plays a leading role.

As part of the benefits derived from corporate social responsibility in Colombia, the improvement of organizational culture, company image and customer loyalty, as well as the engagement of the best employees, are reported. In addition, it is considered to reduce costs through the development of environmental projects (Pastrana & Srirameshb 2014), which suggests a reductionist view of the components of sustainability.

According to Klewitz & Hansen (2014), several strategic behaviors of SMEs are associated to sustainability. For example, these companies carry out actions associated to environmental and social aspects only in the normative scope. Likewise, other intermediate behaviors are carried out until the point of reaching innovation-based behavior, where proactive solutions are given to social and environmental aspects. This behavior, based on sustainability, generates a new business model with new principles on innovation and strong interactions with external stakeholders.

In spite of the importance of studying the challenges of SMEs in a globalized environment, some studies disregard the inclusion of sustainability in these challenges (Felício et al. 2016; Paul et al. 2017), focusing mainly on export-based success (Paul et al. 2017) and on the cultural context through the development of work and knowledge networks of international markets (Felício et al. 2016). Similarly, Martineau & Pastoriza (2016) propose a framework for the inclusion of SMEs in the international context where strategic planning and sustainability are not considered. This is due to the fact that international recognition is related to SMEs' profits from exports (Fernandez & Nieto 2006).

Based on the above, it is possible to suggest that SMEs have an important role in the international context, but these companies should also be prepared to face the responsibility they have with the sustainability approach and have an organizational structure ready to overcome that challenge, strengthening their business plans towards innovation and having an holistic vision associated to all strategic planning processes and activities.

Challenges and Opportunities When Weak and Strong Sustainability are Included in the Strategic Planning Process

According to the previous section, strategic planning is useful for any kind of company, large or small, consolidated or in the early stages of development. Kotler & Armstrong (2003) state that adequate strategic planning allows any company to anticipate changes in the environment and give quick response to them, as well as to be better prepared to face unpredictable situations.

At the beginning of the 21st century, giving a quick response to the sudden changes in the business environment worldwide has become relevant, as evidenced by Thomas & D'Aveni (2004) based on data from the US manufacturing sector between 1950 - 2002. They conclude: i) Industry leadership and competitive advantage are more likely to be lost and are more difficult to keep; ii) Production efficiency is no longer enough to maintain a competitive advantage or to survive; (iii) The only way of maintaining a long-term advantage is through innovation, since innovation destroys the position of established firms that do not adapt to the innovative pace, although innovation is rapidly imitated; (iv) Product life cycle is getting shorter; and (v) The barriers of entry of new competitors are difficult to maintain, so there is frequent entry and exit of companies from the market

In this context, SMEs in early stage of development have more possibilities of adapting to changes in the environment and to innovate in a way that improves their competitive position. This is basically because normally they do not have burocratic and rigid processes such as large companies do.

Because of the characteristics of large companies, some decisions are easily made; however, others are virtually impossible. Reengineering, cost accounting, continuous improvement, outsourcing, relocation of manufacturing processes in different territories, and all the changes that allow them to produce better, fast and in a less expensive way, will be well received and greatly implemented. However, asking large companies to be strategically agile, innovative or truly stimulating workplaces is virtually impossible (Hamel 2007).

The adaptation capacity of SMEs and companies in an early stage of development constitutes an opportunity that allows finding blue oceans that are defined as untapped market spaces and creating demand and opportunities for highly profitable growth. In the blue oceans, competition loses value because the rules of the game do not yet exist (Kim & Mauborgne 2005)

Therefore, SMEs have the opportunity that could be enhanced through innovation in management associated to their strategic planning processes, knowledge management, project management, capital budgeting, among others. For Hamel (2007), managerial innovation is anything that substantially modifies the way it is managed, or that ostensibly modifies the usual forms of organization and thereby promotes the company's purposes. In other words, managerial innovation modifies the way managers do what they do in order to improve organizational performance (Hamel 2007).

For SMEs, the inclusion of the concept of sustainability in their business strategy could be a challenge, a need, and an opportunity at the same time, and if the concept materializes, it could be considered managerial innovation. Especially for SMEs in developing countries, the inclusion of sustainability is a need and an opportunity basically because it allows complying with local laws and additionally enables the company to access new markets, gain recognition from its clients and the community, greater value in the market and even innovate compared to its competitors. However, managerial innovation through the inclusion of the concept of sustainability in planning is a challenge, and its degree of difficulty varies in regards to weak or strong sustainability.

It is more feasible to achieve weak sustainability, and there are numerous examples of companies that show concern on environmental issues but do not involve this dimension at the core of their strategy, but the issue appears tangentially without them making substantial changes in their business model. According to Peña-Montoya (2016), companies in general work under the weak sustainability outline, because they give more importance to economic growth than to the conservation of the ecosystem and the strengthening of social harmony with the environment.

In the case of the fashion industry, the leading companies of this sector follow a production strategy based on fast fashion. This approach is based on speed and consists of supplying and replenishing customers with different clothing according to the trend, and at the same time developing a new one. They try to link fashion with a low financial and psychological cost of consumption, easier to use and even easier to discard, because it has ultimately had low cost (Martínez 2007).

The fast fashion production model has facilitated the *democratic access to consumer goods*, which has forced the fashion industry in the world to an intensive textile production strategy which demands and increases the requirements for raw materials and produces a higher volume of waste. Politowicz (2013) says that fast fashion has also caused the textile industries at a global level to focus on massive consumption, with consequent large-scale contamination and waste generation.

While the fast fashion business model is not sustainable from the environmental standpoint, some companies with this business model include the ISO 14000 certification among their Green Production Strategies for Sustainability. Such is the case with ZARA and the other group brands (INDETEX 2017). Other groups do recycling campaigns such as the H&M group (H&M Group 2017) and try to reduce the pollutant emissions in their supply chain as is the case of GAP (Gap Inc 2017). Likewise, companies in the sector could adopt the principles of cleaner production, optimize the use of raw materials and replace polluting raw materials, optimize production and energy consumption in garments, use reverse logistics, implement lean manufacturing, among others.

For the Colombian case, the 2012 sustainability baseline for the fashion system (Ministerio de Comercio Industria y Turismo, 2012) that observed and analyzed 90 companies associated to this sector concludes that most of these companies do not have sustainability policies and corporate social responsibility associated to any strategy, action plans, and monitoring and control systems. Only 33% of the companies in the textile sector and 20% of the companies in the leather, footwear and leather goods industry met with the required standards. The sector's large companies have reports with the Global Report Initiative (GRI) methodology and social balances as a tool for transparency and sustainability management. However, most of the companies do not have sustainability reports; 93% of the companies in the leather, footwear and leather goods sector and 67% in the Textiles and Clothing sector do not produce any type of sustainability report.

Nevertheless, as evidenced by the fashion industry, companies implementing Green Production Strategies for Sustainability do not necessarily question the business model or scale. For example, the leading companies in the sector support their strategy in fast fashion, which is a case of business success (Ferdows et al. 2004), but it is not suitable from the environmental perspective because it encourages consumerism and this contributes to generating more waste and demand for raw materials, as has been mentioned above. They also support their strategy in ambitious expansion plans that lead them to have presence in over 80 countries worldwide, which means having a global supply chain. This is negative from an environmental point of view, due to the associated atmospheric emissions.

Thus, the inclusion of Green Production Strategies for Sustainability in SMEs is a great challenge because they try to survive in the market as they have limited personnel and resources to implement strong sustainability. Just as an example: optimizing reverse logistics for the recycling of waste requires highly trained personnel to estimate volumes, optimize routes, integrate managers, carry out process indicators, and more. However, besides it being a great challenge, considering the sustainability process and taking the time to include the concept in their strategic planning is an opportunity for companies to innovate, add value and reinvent themselves in the process.

The following is a hypothetical example, which explains the reinvention process and the opportunities that arise with managerial innovation by including the concept of strong sustainability in strategic planning. The company VPK is an SME from the forestry sector that has operations in Colombia, and it is dedicated to cutting and trading wood from several intertropical 20-hectare wild forests. These forests usually have 2000 trees, 600 with more than 60 cm in diameter at breast height (DBH) and the rest with smaller diameters. Trees over 60 cm are paid at 100 USD in the Colombian market, trees with diameters less than 60 cm are paid at 10 USD. To cut the wood they must use a crew of five loggers, which costs $15,000 per three months of work.

VPK could have an operating profit of $59,000 by cutting the forest and selling the wood. However, the company could replace these trees with a forest plantation and sustain that their practices are sustainable from the weak sustainability perspective, especially when the company hopes that the trees will grow to be able to re-cut the trees in a period of 15 years and with that be able to enter a new era of economic growth.

In general, the implementation of forest plantations corresponds to a trend identified by Toledo (2003), in which the tropical regions undergo trends of the agroindustrial modernization towards the specialization, that end up transforming the whole variety of landscapes and their corresponding wealth of flora and fauna to give way to agricultural monocultures, livestock (pastures) and/or forestry (plantations), generating inadequate levels of quality of life for its inhabitants.

The weak sustainability perspective is limited, being that logging has an irreversible environmental impact by affecting the structure and ecology of that wild forest and the dynamics of wildlife populations in the forest, replacing a wild community with a population and eliminating trees that have grown for more than 30 years.

VPK could, however, enter into a reinvention process and in its strategic planning consider the concept of strong sustainability. In this case, the interest of strong sustainability would be to keep the forests with 2000 trees, since they not only provide wood but also oxygen, shelter for some species and other ecosystem services. Therefore, logging practices and the establishment of forest plantations are incompatible with strong sustainability, so the company should reinvent itself. It should enter into a strategic planning and reinvention process. For most traditional entrepreneurs, this would seem to be an unfeasible process, because how could they generate money while maintaining the forest and even generating more employment? Therefore, the need to promote creativity and training in environmental issues for entrepreneurs and professionals is evident.

VPK company should follow the normal planning process and analyze the ease of entry of competitors, the possibilities of appearance of substitute products, all the factors that may affect the negotiation with buyers and the situation of suppliers, identify the competitors and all the companies in the sector, the substitute products, among others, according to a methodology that prioritizes its governing body.

In the process the company should identify global trends, development and prospects in various sectors. For example, they could identify that in the case of toys and games, according to Euromonitor International (2016), in the short term in emerging markets it is most likely that financial difficulties will continue affecting the toy expenditure, but in the long run growth continues to be positive, as there is a favorable environment for the reduction of the size of the households and the increase of the child population. In addition, it could be identified that in emerging markets the increase in the use of mobile internet has contributed to the increase of online sales and especially of traditional toys and games (Euromonitor International 2016).

From this point of view, the company VPK could begin generating ideas on the adjustments to its business model in the planning process. Brainstorming does not have to be complex, and there are helpful tools such as the user matrix, costumer journey canvas, empathy maps, ERAF, SWOT analysis, the double diamond system, prototyping, the Jobs to be Done, among others (Kumar 2012; Clayton M. et al. 2016). An initial exercise might be to ask the sectors of interest, what the demands and/or needs of their clients are. Which products could be manufactured? Could they be sold in the market? Would they have a differential price?

Once the product or group of products and the value offer have been defined, it is desirable to make a business model canvas (Osterwalder & Pigneur 2010). This is a tool to document current business models and evaluate the creation of new ones, through a visual graphic template that allows describing the company's product or value proposition, the infrastructure, customers and finance. Once the business model has been validated, the company must adjust its internal learning and innovation processes and for their implementation.

According to all the aforementioned concepts and following the example, the company VPK should make the strategic decision to maintain the forests in order to meet with strong sustainability. It could then maintain the forest and not cut down any trees if it decides to take advantage of the growing use of mobile internet in emerging markets and the growth of online sales, and sell certificates of adoption/ lease of a tree for a monthly fee for their protection. If the company charges a monthly fee of 3 USD for each tree, a forest with 20 hectares and 2000 trees would have an income of 72,000 USD and an operating profit slightly higher than that associated with the sale of timber.

This way the company VPK would enter into a very different kind of business, in a blue ocean, that requires low investment and does not require the cutting of trees, since it demands a web page with Search Engine Optimization-SEO to have a better position in the Internet search engines, Search Engine Marketing- SEM to put advertising on the Internet search engines, and establishing a digital marketing strategy that attracts people from all over the world and companies to sponsor trees and donate resources for their conservation.

Likewise, the company VPK could eventually cut down some trees but maintain the forest, cutting them at a rate below the forest growth rate. This way it could cut a tree and make board games similar to Jenga for example, which is a game created by Leslie Scott in the early 1970s, and currently marketed by Hasbro (Scott 2010), consisting of a tower of 54 blocks of wood, in which one block is removed at a time and it is located at the top to create a higher tower.

In the case of the company VPK, it would be a game made out of fine tropical wood and the product would gain an added value, because it would he contributing to conservation of a wild forest and it could be marketed in zoos and natural parks in the United States and Europe. One thousand games could be obtained out of a 60-cm-diameter and 5-m-tall tree. When selling them at 15 dollars to retailers, this would generate an income of 15000 USD, so cutting 5 trees in the first year would generate an income of 75000 usd. This is higher than the income from the sale of all the raw timber. In fact, the sale of wood for the company VPK is a business that, besides affecting the forest, is not profitable, being that the company would cut the wood and then have to plant a forest and wait 15 years before being able to re-cut the wood.

As the aforementioned example shows, strong sustainability implies constraints on business development that affect the natural base. Also, it implies a restriction on the business scale, since under the strong sustainability-oriented strategic planning, in the case of the games, business growth through the sale of games should be subordinated to the growth rate of the forest and other ecological criteria so as not to affect the natural structure and dynamics of the forest. However, the company could add more value to its products and obtain higher prices to increase its profitability. The company could, for example, engrave the figures with animals or plants, among other things.

Some critics of the strong sustainability paradigm might argue that this is utopic and that it cannot be implemented or that it would affect the essence of businesses by not being able to generate profit. However, the trees sponsorship example would show that it does not necessarily have to be so, because the same tree could be sponsored by several people and could generate a steady growth in income.

The new challenge arises of defining a role or profile of the entrepreneur under strong sustainability. It is clear that a company's purpose is to generate profit, but a company should also offer quality employment, protect nature and give back to society. It is therefore required for business schools to promote training in ecology and environmental management.

Finally, although the chapter has emphasized on the viability and importance of strong sustainability, it is necessary to invoke a principle of reality, which indicates that weak sustainability is in many cases desirable and necessary, compared to the idea of not doing anything to protect nature. The principle of reality demands the existence of weak sustainability, as in the case of a corporation established with many jobs and cash flows that cannot be removed or easily redesigned and needs to optimize the use and waste of materials, products, energy or fuel.

Finally, it is important to emphasize that for González & Guzmán (2006) there is no equilibrium of nature but several equilibria, so that ecosystems vary from one equilibrium to another. The rejection of the existence of an original state of equilibrium of the ecosystems forces to deny the existence of a single and objective state that is sustainable. Therefore sustainability, more than a normative definition, is a goal which involves changes in the economic, political and even ethical systems, and leads to change in the way human beings relate to the environment.

FUTURE RESEARCH DIRECTIONS

A future research approach will be the implementation of environmental policies in MSMEs and SMEs in developing countries, especially given the shortage of economic resources they have and the low level of professionalization. This causes a strategic lack of vision that affects their competitiveness, and leads to a minimization of the use of renewable and non-renewable natural resources.

CONCLUSION

There is a growing demand worldwide in the business sector for the establishment of Green Production Strategies for Sustainability, since in recent years global environmental policies and regulations have become more stringent and constitute non-tariff barriers to trade, and because consumers are increasingly interested in knowing the environmental impact of products in the buying process.

This chapter emphasized on SMEs from a Latin American perspective, considering that they represent approximately 90% of companies that although individually have a low impact, their cumulative impact can be significant. SMEs generally have the environmental impact associated with the nature of their business, especially those related to the production and transformation of goods. Usually this impact is reflected on waste and emissions where the cumulative impact is high, such as the fashion and leather tanning industries in Colombia.

Faced with these environmental impacts, some companies implement actions to reduce their effects on nature, although these actions are not usually the focus of their business strategies. However, due to increased pressure on environmental issues, many companies begin to make adjustments to their business processes and models that are enhanced by green marketing strategies, but most only make very limited changes to satisfactorily protect the ecological base that supports socio-economic activities.

In relation to sustainable practices, there are two paradigms that guide the companies' environmental management, including weak and strong sustainability. These are two opposing paradigms, where weak sustainability justifies the use and deterioration of nature to achieve economic growth, while strong sustainability claims the opposite.

The challenge for companies, and especially for SMEs, is to include environmental aspects in their strategic planning to achieve significant environmental changes. Strategic planning is a process that looks towards a differentiation that gives the company a competitive advantage and that establishes a promise of value for consumers. An important part of this promise of value has to include environmental aspects, but unfortunately both the environmental aspects and the elaboration of an adequate strategic plan are generally elements that SMEs barely address.

Weak sustainability is easy to achieve because it does not necessarily imply changes in direction in the business strategy. It is enough for an SME to include actions of product life cycle analysis, cleaner production, reverse logistics, etc., looking towards the reduction of the negative environmental impact that does not necessarily challenge the model and scale of the business.

Strong sustainability implies a great challenge and at the same time an opportunity, since it allows the company to reinvent itself and innovate to protect the natural base that sustains the ecosystems. It can be said that this type of sustainability means more efficient processes such as the need to merge together goods and services in order to generate a low impact and arrive at a production process that is subject to the recovery rate of natural resources.

Achieving all this means a series of innovations in both administrative processes and operations associated with the organization, and going beyond quality and other existing certifications. It is a question of organizations acquiring and applying best practices due to the fact that they have great flexibility and capacity for adaptation, but these innovation processes will be difficult or nonviable for managers and professionals not concerned with environmental issues. The concept of strong sustainability should guide the business' strategic planning, because this vision conserves our natural ecosystem and the productive activity.

REFERENCES

Alcott, B. (2005). Jevons paradox. *Ecological Economics*, *54*(1), 9–21. doi:10.1016/j.ecolecon.2005.03.020

Clayton, M. C. (2016). Know Your Customers "Jobs to Be Done". *Harvard Business Review*, *94*(9), 54–62.

Costanza, R., & Daly, H. E. (1992). Natural Capital and Sustainable Development. *Conservation Biology*, *6*(1), 37–46. doi:10.1046/j.1523-1739.1992.610037.x

De Sequeira, J. E. (2009). El principio de responsabilidad de Hans Jonas. *Bioethikos*, *3*(2), 171–193.

Euromonitor International. (2016). *Toys and Games: Global Trends, Developments and Prospects*. Available at: http://www.euromonitor.com/toys-and-games-global-trends-developments-and-prospects/report

Felício, J. A., Ieva Meidutė, I., & Kyvik, Ø. (2016). Global mindset, cultural context, and the internationalization of SMEs. *Journal of Business Research*, *69*(11), 4924–4932. doi:10.1016/j.jbusres.2016.04.054

Ferdows, K., Lewis, M. A., & Machuca, J. A. (2004). Rapid-fire fulfillment. *Harvard Business Review, 82*(11), 104–117.

Fernandez, Z., & Nieto, M. J. (2006). Impact of ownership on the international involvement of SMEs. *Journal of International Business Studies, 37*(3), 340–351. doi:10.1057/palgrave.jibs.8400196

Franceschi, D. & Kahn, J.R., 2003. Beyond strong sustainability. *International Journal of Sustainable Development & World Ecology, 10*(3), 211–220. Available at:.10.1080/13504500309469799

Gap Inc. (2017). *Addressing the Impacts of Our Supply Chain*. Retrieved February 13, 2017, from: http://www.gapincsustainability.com/environment/protecting-our-shared-environment/addressing-impacts-our-supply-chain

González, M., & Guzmán, G. I. (2006). *Tras los pasos de la insustentabilidad. Agricultura y medio ambiente en perspectiva histórica*. Barcelona: Icaria Editorial.

Guhl, A. (2004). *Coffee and landscape change in the colombian countryside 1970-2002*. Miami, FL: University of Florida.

Hamel, G. (2007). *The future of management*. Boston: Harvard Business School Press.

Hediger, W. (1999). Reconciling "weak" and "strong" sustainability. *International Journal of Social Economics, 26*(7/8/9), 1120–1144.

H&M Group. (2017). *Sustainability look good, do good, feel good*. Retrieved March 13, 2017, from: http://about.hm.com/en/sustainability.html

INDETEX. (2017). *Sustainability*. Retrieved March 13, 2017: https://www.inditex.com/es/sustainability

Kim, W. C., & Mauborgne, R. (2005). *Blue ocean strategy: How to create uncontested market space and make the competition irrelevant*. Boston: Harvard Business School Press.

Klewitz, J., & Hansen, E. (2014). Sustainability-oriented innovation of SMEs: A systematic review. *Journal of Cleaner Production, 65*, 57-75.

Kotler, P., & Armstrong, G. (2003). *Fundamentos del Marketing*. Pearson Education.

Kumar, V. (2012). *101 Design Methods: A Structured Approach for Driving Innovation in Your Organization*. Hoboken, NJ: John Wiley & Sons, Inc.

Martineau, C., & Pastoriza, D. (2016). International involvement of established SMEs: A systematic review of antecedents, outcomes and moderators. *International Business Review, 25*(2), 458–470. doi:10.1016/j.ibusrev.2015.07.005

Martínez, A. M. Á. (2007). Un modelo de empresa innovadora y flexible: El caso Zara. *Revista de Investigaciones Políticas y Sociológicas, 6*(1), 69–80.

Ministerio de Comercio Industria y Turismo. (2012). *Informe Sostenibilidad 2012, Sector Sistema Moda*. Retrieved March 20, 2017 from: http://static.globalreporting.org/report-pdfs/2013/ef86ec29136d077c-71c790ab7324d762.pdf

Mora-Riapira, E. H., Vera-Colinab, M., & Melgarejo-Molina, Z. A. (2015). Planificación estratégica y niveles de competitividad de las Mipymes del sector comercio en Bogotá. *Estudios Gerenciales*, *31*(134), 79–87. doi:10.1016/j.estger.2014.08.001

Munda, G. (1997). Environmental economics, ecological economics, and the concept of sustainable development. *Environmental Values*, *6*(2), 213–233. doi:10.3197/096327197776679158

Neumayer, E. (2003). *Weak versus strong sustainability: exploring the limits of two opposing paradigms*. Northampton, UK: Edward Elgar.

Nilsen, H. R. (2010). The joint discourse reflexive sustainable development—From weak towards strong sustainable development. *Ecological Economics*, *69*(3), 495–501. doi:10.1016/j.ecolecon.2009.11.011

Osterwalder, A., & Pigneur, Y. (2010). *Business model generation: a handbook for visionaries, game changers, and challengers*. John Wiley & Sons.

Pastrana, N., & Srirameshb, K. (2014). Corporate Social Responsibility: Perceptions and practices among SMEs in Colombia. *Public Relations Review*, *40*(1), 14–24. doi:10.1016/j.pubrev.2013.10.002

Paul, J., Parthasarathyc, S., & Gupta, P. (2017). Exporting challenges of SMEs: A review and future research agenda. *Journal of World Business*, *52*(3), 327–342. doi:10.1016/j.jwb.2017.01.003

Peña-Montoya, C. C. (2016). *Gestión sostenible de residuos sólidos industriales mediante la integración de estrategias asociadas con la lógistica de reversa* (Doctoral Thesis). Universidad del Valle, Colombia.

Pérez, M.A., & Rojas, J.H. (2010). *Desarrollo sostenible. Principios, aplicaciones y lineamientos de la política en Colombia*. Programa editorial Universidad del Valle.

Politowicz, K. (2013). *Design Lifetimes: A Manifesto for Strategic Change*. London: University of the Arts London.

Riechmann, J., & Tickner, J. (2002). Un principio para reorientar las relaciones de la humanidad con la Biosfera. In J. Riechmann & J. Tickner (Eds.), El Principio de Precaución en medio ambiente y salud pública: de las definiciones a la práctica. Barcelona: Icaria.

Rockström, J. (2009). A safe operating space for humanity. *Nature*, *461*, 472–475. 10.1038/461472a

Scott, L. (2010). *About Jenga: The Remarkable Business of Creating a Game That Became a Household Name*. Austin, TX: Greenleaf Book Group LLC.

Steffen, W., Richardson, K., Rockstrom, J., Cornell, S. E., Fetzer, I., Bennett, E. M., & Sorlin, S. et al. (2015). Planetary boundaries: Guiding human development on a changing planet. *Science*, *347*(6223), 1259855. doi:10.1126/science.1259855 PMID:25592418

Thomas, L. G., & D'Aveni, R. A. (2004). *The Rise of Hypercompetition in the US Manufacturing Sector 1950-2002*. Tuck School of Business at Dartmouth.

Toledo, V. M. (2003). Hacia un modelo de conservacion bio-regional en las regiones tropicales de mexico: biodiversidad, sustentabilidad y pueblos indigenas. In *Hacia una Evaluación de las Áreas Naturales Protegidas del Trópico*. Xalapa, Veracruz, México: Universidad Veracruzana. Retrieved March 13, 2017 from: http://www.era-mx.org/biblio/politica/Toledo2003.pdf

United Nations General Assembly. (1987). Report of the World Commission on Environment and Development : Our Common Future. Oslo, Norway: Author.

WIPO-World Intelectual Property Organization. (n.d.). *About the SMEs Section*. Retrieved March 22, 2017 from: http://www.wipo.int/sme/en/about_sme.html

Chapter 10
Green Innovation in the Absence of Social Trust:
A Case Study of MengNiu Dairy

Li-Chung Chang
University of Electronic Science and Technology of China (Zhongshan Institute), China

Wei Li
University of Electronic Science and Technology of China (Zhongshan Institute), China

Yu Wu
Changsha University of Science and Technology, China

Sang-Bing Tsai
University of Electronic Science and Technology of China (Zhongshan Institute), China

Chao-Feng Hsu
University of Electronic Science and Technology of China (Zhongshan Institute), China

Cheng-Kuang Wang
University of Electronic Science and Technology of China (Zhongshan Institute), China

ABSTRACT

In recent years, food safety has become an urgent problem to be solved, especially in China's dairy industry. After the 2008 Chinese milk scandal, how to rise again by green innovation is a very important issue. As a representative of China's dairy enterprises and with the enterprise positioning of "a milk manufacturing service provider that commits itself to human health", MengNiu Dairy has achieved "MengNiu speed" and "MengNiu miracle" attracting worldwide attention within ten years. Therefore, the green innovation strategy implemented by MengNiu Dairy is worthy of research. In this paper, MengNiu Dairy, the leading dairy enterprise in China, was taken as the research object, and the author mainly applied narrative research method to analyze the green innovation strategy implemented by MengNiu Dairy from all aspects, and summarized that through the combination of charity marketing and green innovation, MengNiu Dairy developed the system and model of sustainable strategy, which had far-reaching significance in its development and adapting to the current environment at home and abroad.

DOI: 10.4018/978-1-5225-3537-9.ch010

INTRODUCTION

With the development of China's economy and the establishment of sustainable development in China, people's environmental protection awareness has been continuously improved, and some new green concepts have been introduced by all countries, like mushrooms after a rain (Li, 2013). In the current society that advocates the concept of sustainable development, human's pattern of consumption is becoming increasingly civilized and green innovation is gradually rising, and how should enterprises meet the green innovation demand, develop green products, and implement green marketing has become a key issue to be solved; besides, sustainable development will also become the necessity of historical development (Gao, 2011).

In recent years, the continuous emergence of China's food problems brought enterprises many problems in selling products and people continue to question about the product safety; therefore, how to solve the mistrust between enterprises and consumers becomes a difficulty of enterprise development.

Green innovation has two meanings. One is that enterprises unify their own benefits, consumer benefits and environmental protection benefits together in the production and operation process; the other is that enterprises take environmental protection as management guiding ideology, green culture as value, and green consumption of consumers as the center and the starting point (Ma and Deng, 2014). The key point is that enterprises take green and health as leading factors in each link, including all aspects of production and sales, and will obtain the proof from relevant departments, which has relatively high credibility. For the choice of production technology, product design, the use of raw materials, establishment of production routine, establishment of packaging methods, disposal of waste, and other aspects, enterprises pay attention to the protection of the environment and take responsibility for people's health, so that the enterprises can balance the consumer demand, enterprises benefits, and environmental protection, thus better shaping the enterprise image.

Through the implementation of green investment, enterprises reduced the consumption of material resources per unit of products and improved the utilization ratio of resources, which can not only save material costs, but also reduce pollution and pollution control costs, thereby reducing production costs; in this way, greater economic and social benefits can be achieved with less green investment, and it has played a crucial role in both enterprises profits and environmental protection, achieving the purpose of a win-win business (Li, 2015). Such an innovative model is suitable for long-term development of enterprises, as well as sustainable requirements, and a series of such marketing strategies can be summarized as sustainable strategies. Only by implementing the sustainable strategy can enterprises achieve better long-term stable development and continue to grow.

In years of development, MengNiu Dairy has been committed to "strive to be a consumer-centric century-old nutrient and healthy food enterprise that leads innovation" and taken "focus on nutrition and health and bring a little bit of happiness for more people in every single moment" as its own responsibility. How did MengNiu Dairy change its enterprise image, strengthen people's impression of MengNiu Dairy's green health, and become an everlasting enterprise of the dairy industry is worthy of research. Therefore, this paper took MengNiu Dairy as an example and analyzed that how could it carry out green innovation through charity marketingeting under the background of losing the social trust.

LITERATURE REVIEW

Creativity, Innovation, and Originality

Creativity is to create a new solution in the use of a new idea and to change a field for a long time. The model of choosing a field at first, and then entering a discipline, in which a problem is solved with a new solution, thus changing or creating a new field can be called as creativity. According to the researches of scholars, Yu (1988) concluded that creativity refers to the process of using a variety of known information to create a unique object with social meaning and personal value by people and the comprehensive performance of ability and the non-ability factors of the people. In psychology, the meaning of creativity has always been controversial. McKinnon (1962) emphasized the adaptability and originality of creativity when explaining creativity. He believed that the real creativity should meet three conditions: the first one is novel and unusual; the second one is the adaptability or reality; the third one is unique insight. In the research on the content of originality, Ding, Liang, and Hui (2010) proposed that the static definition of originality is valuable and feasible new ideas. At the same time, they also proposed that the dynamic definition of originality is creative thinking, namely, the thinking that can produce valuable new ideas. In economics, the origin of concept of innovation was proposed by the American economist Joseph Alois Schumpeter. He believed that innovation is to introduce "a new combination" of a new production factor and production method into the production system, which includes the introduction of a new product, the introduction of a new production method, the opening up of a new market, the obtainment of a new source of supply of raw materials or semi-finished products, and a new organization form. Zhang (2006) argued that innovation refers to that the matter is unprecedented or has substantial progress, can be implemented or can be confirmed, can produce beneficial social effects, and is conducive to the coordinated development of human and ecology.

Green Innovation

Green innovation is the management process of using a profitable and environmentally acceptable way to identify, anticipate, and meet consumer and social needs. Ma and Deng (2014) believed that green innovation is a new idea, which conforms to the requirements of current sustainable development strategy, pays attention to the protection of ecological environment, and is conducive to promoting the coordinated development of economy and ecological environment.

Charity Marketing

Through the induction of charity marketing literature, Chen (2010) put forward that there are three main components of charity marketing, including enterprises, non-profit organizations that enterprises cooperated with, and consumers. charity marketing is a marketing activity that combines profit-making enterprises with non-profit organizations, so charity marketing can also be called collaborating marketing or social event marketing. He believed that charity marketing is a new marketing model, which combines enterprises' business activities and non-profit organizations or public welfare activities, thereby improving consumers' purchasing intention, brand loyalty, and enterprise image recognition, and it is a kind of "win-win" or even "triple win" marketing strategy. Chen Xiufen (2010) believed that the charity marketing is a kind of marketing behavior of enterprises, which adhere to the people-oriented thought, take social progress as the starting point, care for public welfare undertakings, communicate with consumers through public welfare activities, influence consumers through the establishment a good enterprise image, so that the consumers will develop preference and give preference to the enterprises' products when making purchase decision.

Theoretical Gap (Research Content)

Based on the above, the author found that the model of combination of green innovation and charity marketing is currently a theoretical gap. Such a sustainable green innovation model includes sustainability of enterprise products, sustainability of prices, sustainability of channels, sustainability of promotions, and other aspects. Sustainability is a way of combining enterprise benefits, social environment, and consumer demand, which requires that what the enterprises have done shall be conducive to the maintenance of consumers' health and social environment. It not only provides guarantee for the immediate interests, but also provides the direction of future development; therefore, to implement sustainable marketing model is a new model for enterprises to become everlasting enterprises.

CASE ANALYSIS

MengNiu Dairy Profile

Headquartered in the Inner Mongolia, PRC, MengNiu Dairy is one of the leading production enterprises of milk, yogurt, and dairy products. Inner Mongolia MengNiu Dairy Co., Ltd. (MengNiu Dairy for short) was established in August 1999 with the total assets of more than 10 billion Yuan, nearly 30,000 employees, and annual production capacity of six million tons of dairy products.

In 2005, MengNiu Dairy's output of liquid milk and ice cream was No.1 in China and its sales of dairy products ranked No.2 in China. MengNiu Dairy's main business is manufacturing liquid milk, ice cream, and other dairy products. So far, it has established more than 20 production bases in 16 provinces, autonomous, and municipalities throughout the country, including Horinger Base, which can produce more than 400 items of five series products of liquid milk, yogurt, ice cream, dairy products, and cheese; with the excellent quality, MengNiu Dairy's products cover home market and are exported to the United States, Canada, Mongolia, Southeast Asia and other countries and regions.

The positioning of MengNiu Dairy is "a milk manufacturing service provider that commits itself to human health", which has achieved "MengNiu speed" and "MengNiu miracle" attracting worldwide attention within ten years. By the end of 2008, MengNiu Dairy's main business income reached 23.9 billion Yuan with the average annual increase of 104%, which is the first dairy enterprise with income over 20 billion Yuan in China. Its market share of the main products is over 35%; its UHT milk sales ranks No.1 in the world; its liquid milk, ice cream, and yogurt sales ranks No.1 in China; its export volume of dairy products and the number of export countries and regions also rank No.1 in China.

With the intensifying competition, in order to meet the needs of modern large-scale production and high-quality dairy products, MengNiu Dairy constantly strengthen and guide the intensive and standardized construction of milk source. Until now, by cooperating with Modern Farming in the form of participation, MengNiu Dairy has built 13 modern pastures with ten thousand cows each throughout the country. In order to better guarantee the quality of milk source, with the support of local governments and under the guidance of MengNiu Dairy, the construction of MengNiu milk source base achieved the intensification and standardization. Moreover, together with Chinese Academy of Inspection and Quarantine, MengNiu Dairy established a "dairy joint laboratory" and actively devoted itself into scientific research of dairy products and innovative products, which created a unique self-dependent innovation way and won the honor of "The Most Creative China Enterprise".

MengNiu Dairy takes "focus on nutrition and health and bring a little bit of happiness for more people in every single moment" as its own responsibility, takes "to be a consumer-centric century-old nutrient and healthy food enterprise that leads innovation" as the root, and take health as the goal of producing dairy products all the time. In order to appeal to consumers to join environmental protection undertaking, MengNiu donated the first one million Yuan to China Environmental Protection Foundation (CEPF for short), established the "Blue Sky & Green Land" Fund together with CEPF, launched the first project focusing on wetland protection, and launched public benefit activity of "Joint Haze Prevention" on Weibo at the same time, inviting internet users to join the environmental protection, concern about and improve the atmospheric environment, and contribute to the environmental protection.

Green Innovation Strategy Implemented by MengNiu Dairy

Product Strategy

The 2008 Chinese milk scandal was a heavy blow to the dairy industry in China. Although MengNiu Dairy was a leading enterprise of dairy products in China, it was inevitably affected. The distrust of consumers made MengNiu Dairy realize the importance of green products R&D, therefore, MengNiu Dairy increased investment in green products R&D.

1. **Establishment of Green Milk Source Base and Green Eco-Industrial Chain:** MengNiu Dairy put forward that the green milk source strategy should be implemented based on the construction of ecological grassland, which should be taken as the guarantee of green milk source, and the high-quality grass should be introduced. MengNiu Dairy firmly believed that high-quality milk source only comes from high quality grassland and MengNiu Dairy cannot survive without high-quality milk source. Therefore, in early 2014, MengNiu Dairy put forward the win-win pasture management project named "W-W project"; at the same time, it continued to merge enterprises with high-quality milk source and introduced 14,000 cows from Australia and New Zealand; in May 2013, MengNiu Dairy increased its shareholding proportion of Modern Farming, which is the largest cow breeding enterprise in China; in November 2013, MengNiu invested 365.7 million Yuan in the original ecological animal husbandry. Through the continuous investment and mergers and acquisitions, the proportion of MengNiu Dairy's intensive and large-scale pastures was nearly 100%. Through the improvement of infrastructure and standardization of operation management of all cooperative pastures, the mechanized milking was totally achieved progressively and safe feed feeding was also totally achieved. Moreover, MengNiu Dairy has invested more than 400 million Yuan in environmental protection, including the establishment of large-scale sewage treatment plants for recycling the sewage.

Table 1. Progress of nutrition and health performance of MengNiu Dairy

	2013	2014	2015
Outgoing product batches for quality and safety monitoring (%)	100	100	100
Rate of milk source from large-scale and intensive pastures (%)	94	94	100
Recycling rate due to safety and health factors during product transportation and sales (%)	0	0	0
Total amount of funds invested in technology or R&D (10,000 Yuan)	5700.7	6697.5	8190.0
Number of scientific research personnel	2515	2583	2588
Number of newly increased patent licenses	123	94	150

Source: 2014-2015 Sustainability Report of China MengNiu Dairy Co., Ltd.

2. **Recognition of Green Products and Correct Market Positioning:** According to the characteristics of the consumer needs, MengNiu Dairy carried out market segments and accurate positioning of products, developed different green products, and produced different products according to the characteristics and demands of different populations. Classification of MengNiu Dairy's products based on age: milk for baby, such as the serious of future star, milk for young students, and high-calcium low-fat milk for the elderly; classification of MengNiu Dairy's products based on taste: strawberry flavor, orange flavor, grape flavor, peach flavor, etc.; classification of MengNiu Dairy's products based on drinking time: milk for breakfast, milk for dinner, milk for leisure, etc.; classification of MengNiu Dairy's products based on income level of consumers: pollution-free products, green products, and organic products, especially the newly launched high-quality yogurt—yogurt containing bifidobacterium BB-12, developed for the white-collar workers, which had two new flavors of yellow peach with oat and walnut with oats, providing consumers with higher nutritional value. In recent years, MengNiu Dairy has increased the lactoprotein content of pure milk to 3.2g/100ml, which was 6.9% higher than the national minimum standard.
3. **Green Packaging:** Food safety and environmental pollution are closely related to the materials used in food packaging, and the consumption of packaging materials in China is increasing every year. In recent years, green packaging has been advocated internationally and various enterprises have also applied new packaging into products successively, of which MengNiu Dairy was the first to use packaging materials from the sustainable development forests, thus achieving the green packaging of milk. MengNiu Dairy not only used the green packaging, but also paid attention to the recycling of packaging. Consumers who send ten or twenty empty milk boxes to MengNiu Dairy's fixed recycling stations can receive a box of milk, and such an environmental protection way can better reflect the social responsibility of the enterprise. The packaging of "Green Mood" of MengNiu Dairy has successfully attracted people's attention, the new packaging of which was designed with grassland and green characters in order to better reflect the characteristics of fresh, green, and original ecology of green products.

Price Strategy

In startup stage, under the background of no market, no factories, no milk source, and lacking of funds, MengNiu Dairy implemented the strategy of "market first, factory second", namely pursuing the maximization of market share first; therefore it set relatively low prices in order to rapidly achieve large market share in a short period of time. In the normal operation period, MengNiu Dairy began to divide low-profit products and high-profit products to meet the needs of different markets. In terms of product pricing, MengNiu Dairy has not only regular milk for the public, but also high-grade milk for high-end crowd. However, it rarely drifted into price change, which won the trust of consumers.

The cost of green products is often relatively high, but the enterprises shall develop the concept of "paid use of the environment" when setting the prices of green products, and the pricing of green products shall face the mass consumers and balance it with the prices of similar products. The establishment of sub-brand Milk Deluxe of MengNiu has taken advantage of psychological pricing of consumers. Milk Deluxe mainly grasped the consumer psychology of pursuing health products, so it created a logo of green transformation of the dairy industry and shaped the enterprise's green and healthy image. Green products are more easily to be accepted by consumers who pursue nutrient, healthy, and environment friendly products, and such consumers always have a strong purchasing power and are willing to pay a higher price. Therefore, MengNiu Dairy applied market-skimming pricing into Milk Deluxe-dominated green products, and that consumers are more likely to purchase ecological, scarce, and high-quality products from prairies.

Place Strategy

MengNiu Dairy is unique in place. First, determine place partners. The place relationship of MengNiu Dairy is loose-type in contract form mainly, that is, to cooperate based on contract; the other place relationship is close-type in ownership form, that is, to become close partner by transferring share to customers.

1. **Initial Place Strategy of MengNiu Dairy:** The initial place strategy of MengNiu Dairy was "weakening the first level, maintaining the second level, and winning the third level". In developing period, MengNiu Dairy transformed "weakening" into "stabilizing", but it was operated according to "weakening the first level" actually. MengNiu Dairy's place strategy was also reflected in the three marketing principles, namely, customer value, competitive advantage, and centralized advantage. MengNiu Dairy was transformed from 4P, the traditional marketing tool to 4C and applied 4C for adaption, namely, transforming from product to customer; from price to cost; from place to convenience; from promotion to communication.
2. **Logistics Supply Model:** Currently, MengNiu Dairy has three models of supply. Traditional model: "company + farmer"; mature model: "company + scale pasture"; innovation model: "company + OEM supplier".
    a. **Traditional Model:** "company + farmer": MengNiu Dairy's "company + farmers" model adopts the process of "decentralized raising+ concentrated milking + unified processing". "Decentralized feeding" of "farmers" makes every cow be carefully taken care of; "concentrated milking" of "milk station" makes the past "collecting milk" changed into "collecting cows". The company indirectly controls the entire milk source through controlling the milk station.
    b. **Mature Model:** "company + scale ranch": This model can fully monitor milk source strictly to ensure the perfect quality of milk. MengNiu Dairy established a dozen major production bases in MaAnShan, Anhui; Shangzhi, Heilongjiang; Zhangjiakou, etc. to operate 10,000-scale modern dairy farming pastures, which have become the main milk base for MengNiu Dairy

Table 2. Distribution Structure Comparison of MengNiu Dairy

Traditional distribution structure	Existing distribution structure
Dairy processing company	Dairy processing company
Regional agent	Branch office/dealer
Province-level agent	Wholesaler, retailer, and consumer
City-level agent	
County-level agent	
Wholesaler, retailer, and consumer	

Data source: collated and analyzed by this research

Figure 1. Existing distribution structure of MengNiu Dairy
Data source: collated and analyzed by this research

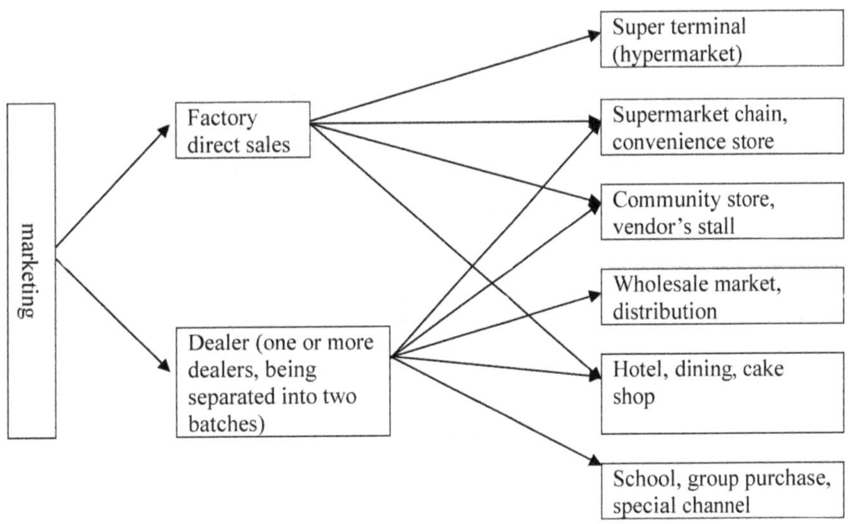

to supply the surrounding areas, so as to avoid "long-distance raid" of milk. In that way, MengNiu Dairy shortens the logistics distance and ensures the freshness of products.

c. **Innovation Model:** "company + OEM supplier": With uneven distribution of dairy resources and fierce market competition, one of the knacks of MengNiu Dairy is to wantonly merger and purchase local enterprises to carry out OEM production. OEM suppliers located around are only responsible for production under quality supervision of MengNiu Dairy. The expansion model of "production workshop" makes MengNiu Dairy establish dozens of branch factories within eight years only. Now, countrywide MengNiu products still rely on OEM model.

3. **Logistics Distribution Mode:**
 a. **E-Commerce-Based Direct Sales Model:** With the rapid growth of MengNiu Dairy, in addition to traditional wholesale and retail, home delivery service comes into being. This model is mainly suitable for first-tier cities, which have high requirements for the freshness of milk and pursuit convenience, such as Beijing, Shanghai, Guangdong, etc.; therefore, telephone

ordering, online ordering and home delivery service have become a competitive delivery mode.

b. **Specialty-Store-Based Terminal Sales Model:** MengNiu Dairy controls terminal network through developing special stores, thus further improving the network to meet the challenges of competitors. Special stores are built in economically developed cities. Vertical management of the chain monopoly system greatly enhances control strength of market and place, and can extend in-depth distribution system directly to the consumer level; moreover, it can supplement market gap of place, thus increasing market coverage.

4. **Green Delivery Model:** Currently, MengNiu products have spread to the whole country, including Hong Kong, and Macao, and even been exported to Southeast Asia. Some products have relatively short shelf life, so there is a high demand for temperature during transport. This poses a higher demand for time control and temperature control of transportation. In order to send the milk to supermarkets in the shortest time and effective storage conditions with the lowest cost, MengNiu Dairy takes the following measures:

First, shorten the transport radius. Low-temperature products such as yogurt have short shelf life, coupled with high demand for freshness of consumers, so in order to ensure timely delivery of products, MengNiu Dairy minimizes the transport radius. When MengNiu Dairy expands the product line to the yogurt, its production layout extends from Inner Mongolia to the Yellow River line and Yangtze River line gradually, aiming to make milk production base be close to the market and thus ensure low-temperature products to be sent to stores and supermarkets quickly. In the retail terminal, some large supermarkets have established a long-term cooperation with MengNiu Dairy to get access to freezers directly via special transportation tools, thus avoiding deterioration in process of transportation and significant losses of supermarkets. MengNiu Dairy has established a scientific and immobilized cold chain logistics management and operation system by integrating cold chain resources of manufacturers, social resources and its own resources continuously.

Second, a reasonable choice of transport mode. At present, MengNiu products are mainly transported by automobiles and train containers. For long-distance transport of low-temperature products, in order to ensure the high quality and speed of products, MengNiu Dairy often uses more costly automobile transport. For example, low-temperature products transported from Beijing to Guangzhou are all transported by automobiles, which, although expensive than rail transport, is timelier. In order to better understand the situation of automobile transport, MengNiu Dairy installed GPS system for some transport vehicles, and the GPS system can track vehicles, including whether they are normally running, their location, speed, temperature inside truck beds, etc. The installation of GPS avoids deterioration of products due to long-time parking of drivers. For products with longer shelf life, they are usually transported by train containers, because they are cost-saving compared with automobile transport.

Third, whole-journey cold chain protection. Low-temperature dairy products must be in 2°C-6°C in the whole journey to ensure quality. MengNiu products adopt low-temperature and closed transport mode in the operation sequence of "cow-milk station-milk tanker-factory". No matter where MengNiu is, it can send the squeezed raw milk to production workshop in six hours by its refrigerated transport system to ensure fresh taste and rich nutrition of milk. When dairy products leave the factory, refrigerated trucks are prepared to ensure low-temperature transport. In the retail terminal, the dairy products are put into the freezer to ensure quality.

The logistics supply and distribution model of MengNiu Dairy all pursuit flat and efficient marketing channels to ensure the quality of products, save the cost of logistics, set green image and promote sustainable development of MengNiu Dairy.

MengNiu Dairy has also strengthened certification management of ISO14000 and green logo, and certification application contributes to establishment of internal environmental management system, thus guiding MengNiu Dairy to improve product design, production technique and productive process (Sustainable development report of China MengNiu Dairy Co., Ltd., 2014-2015).

Promotion Strategy

Promotion strategy is an indispensable part of the enterprise marketing strategy. MengNiu Dairy's promotion strategies are advertising and public relations mainly.

1. **Advertising Strategy:** Advertisement means "publicizing widely", whose purpose is to let consumers understand the product information and generate purchasing desire. Successful advertisement can seize the psychology of consumers and is in line with the needs of the masses to achieve product promotion. There are many slogans for MengNiu Dairy, whose theme is "prairie, blue sky, and health"; for example, the familiar ones, "Lingering fragrance from the prairie", "Deep love for prairie, creamy milk by prairie", "Green, your mood, my pursuit ", "Welcome to MengNiu prairie", etc. Those slogans all present us with blue sky, white milk, and green prairie, and confirming consumers' belief of nutrition, health and nature of MengNiu milk.
2. **Public Relations:** Public relations mean the relationship between the enterprise and the public, which need to be properly handled in marketing activities to establish positive image of the enterprise and the brand, and thereby promoting sales. MengNiu Dairy is successful and distinctive in public relations. In 1999, MengNiu Dairy began to develop dairy industry, committing to revitalize Inner Mongolia initially. MengNiu Dairy had no factory, no brand, and no market. Under that premise, MengNiu Dairy invested a third of start-up capital equal to three million Yuan for public service advertising. In 1999, Hohhot advertised in more than 500 pieces of outdoor billboards. "Developing dairy industry, revitalizing Inner Mongolia economy", "Boundless prairie revitalizes Yili Group, Xingfa Group, MengNiu Dairy…we cheer for Inner Mongolia, we take Inner Mongolia off" etc. were sprung up in Inner Mongolia. Although advertising expense was high, MengNiu Dairy didn't immediately promote their products, but care about city orientation of Inner Mongolia and develop public welfare. As still waters running deep, MengNiu Dairy made a small bundle to make it different. MengNiu Dairy successfully seized consumers through advertising. It was this bold advertising awareness that bought MengNiu Dairy sales revenue equal to 44 million Yuan in six months since its establishment.

In spring of 2003, "SARS" raged in China, causing people's extreme panic. Since dairy products can strengthen human immunity and promote calcium absorption, they became scarce goods. At that time, it is undoubtedly a business opportunity for dairy enterprises to enhance dairy price. MengNiu Dairy seized this opportunity, but surprisingly, it prohibited dealers enhancing prices. MengNiu Dairy played hard to get for long-term interests, winning identification of consumers through "SARS" disaster relief. In "SARS" period, MengNiu Dairy increased the amount of public service advertising to enhance everyone's health awareness, and took the lead in donating one million Yuan to the Ministry of Health,

Table 3. Public welfare activities MengNiu Dairy participated in

Time	Projects	Value (Yuan)
1999	Revitalize the economy of Inner Mongolia	4 million
2003	SARS disaster relief	12 million
2004	Donate milk to teachers nationally	over 30 million
2005	Set up milk risk fund	10 million
2006	Milk donation for the first batch of 500 poor primary schools	110 million
2008	Disaster relief of Southern snow	10 million
2008	Earthquake Relief in Sichuan	12 million

Data source: collated and analyzed by this research

becoming the first enterprise to fight "SARS" by donation. Afterwards, MengNiu Dairy continued donating money and milk to health workers and consumers of 30 cities nationwide. Through a series of public welfare behavior, MengNiu Dairy set up noble social responsibility and quickly cultivated a large number of loyal consumers.

In 2006, Premier Wen Jiabao said that "I hope every Chinese, especially children can drink 500g milk every day" during the inspection (Big river newspaper, 2006). MengNiu Dairy answered the call quickly, and held a massive milk donation project, in which MengNiu Dairy sent fresh dairy products valuing hundreds of millions Yuan to poverty campus with the slogan of "500g milk a day, strengthening Chinese people". The charity activities held by MengNiu Dairy firmly grasped the psychology of consumers, for which MengNiu Dairy has been developing so fast in deep-root.

Since 2013, in order to help more children to receive education, MengNiu Dairy held a public welfare activity named "MengNiu Dairy Helps You Attend School", which contributed to complete their education for children of many poor employees. In the same year, with the support of China Environmental Protection Foundation, MengNiu Dairy set "Blue Sky & Green Land Special Fund", which belongs to association marketing with public welfare undertaking. In the market promotion activities held by MengNiu Dairy, consumers only need to join certain link of public welfare activities, MengNiu Dairy will donate RMB one Yuan for the fund. Over 1,500,000 people participated in the activity for support, and all funds are used to prevent and control air pollution, protect natural habitat and biodiversity projects. In addition, MengNiu Dairy also launched "Joint Haze Prevention", which is also an important part of "Blue Sky & Green Land Special Fund" activity. MengNiu Dairy has never stopped its pace to public welfare undertaking. Supported by Tencent Charity Foundation and Sina micro public welfare, MengNiu Dairy promoted "Exchange Calories" together with 58 partners, in which certain calories consumers consumed were exchanged to a pair gloves to give warm to rural children in winter.

In 2014, MengNiu Dairy launched educational public welfare activity—"I will Go to School in My Hometown", and actively invited social celebrities to give lessons in rural schools, through which meeting rural children's desire for knowledge; among them, "The Most Beautiful Lesson" micro-film advertising was awarded as "Best Public Welfare Drama Film of the First Session of China Public Welfare Micro Film Festival in 2015".

In 2015, in the 21st "World Day to Combat Desertification and Drought", No. 53 well launching ceremony of large-scale livelihood project—"Charity Well Project" was held in Ulanqab city, Inner Mongolia, taking practical action to help the local rural herdsmen to get out of droughts.

From the above events it can be seen that MengNiu Dairy correctly understands the connotation of charity marketingeting and links public welfare undertaking and corporate interests well, which makes MengNiu Dairy get more sales opportunities and take on more social responsibilities. In addition, MengNiu Dairy is not blind to do all public welfare activities; it selects public welfare undertaking suitable for its corporate culture concept for investment and carries out activities focusing on health, green, environmental protection, etc. Moreover, MengNiu Dairy grasps the feelings of consumer group, injects the concept of green, healthy and credible products into them and attracts more consumers to participate in public welfare activities held by MengNiu Dairy, which shortens the distance between MengNiu Dairy and consumers and makes them be more and more optimistic about the future of MengNiu Dairy. MengNiu Dairy also takes the opportunity to actively respond to government calls to help the government solve difficulties and meanwhile promote MengNiu products properly. As the saying goes, harvest comes after perseverance. MengNiu Dairy adheres to this principle, and it has been commits to public welfare undertaking, which enhances the image of MengNiu Dairy significantly, and makes MengNiu Dairy embark on sustainable development.

MengNiu Dairy injects "green, environmentally friendly, natural and sustainable" awareness in a series of marketing measures based on the general marketing strategy; what's more, public welfare undertaking make it stand out from dairy products, and develop stably in long term; therefore, we name the marketing strategy of MengNiu Dairy "sustainable marketing strategy".

RESEARCH FINDINGS

In the past, there was little research analyzing and summarizing green innovation and charity marketing. This research, through analyzing the case of MengNiu Dairy, summed up a sustainable strategy model which integrates charity marketing and green innovation.

Green innovation emphasizes the environmental protection of marketing, highlighting the unity of consumers' health, corporate interests, and the ecological environment. charity marketing combines the enterprise and public welfare undertaking to improve the enterprise image, bear more social responsibilities and achieve sustainable development of the enterprise. These two marketing methods are in line with the needs of consumers in current era, and enable enterprises develop stably, fast, and far. Green marketing and public marketing complement each other: green marketing focuses on development of enterprises in all aspects, maximizing the quality from production to sales; charity marketing can help enterprises to establish high-quality and committed image. Figure 2 shows the sustainable strategic model combining green innovation and charity marketing.

CONCLUSION

MengNiu Dairy applies sustainable marketing strategy with the natural, green, environmentally friendly and healthy concept to operation and management, strategy development and implementation of the enterprise, striving to unifying ecological environment, corporate interests and consumers' health and achieving win-win situation.

Figure 2. Sustainable strategy of green innovation
Data source: collated and analyzed by this research

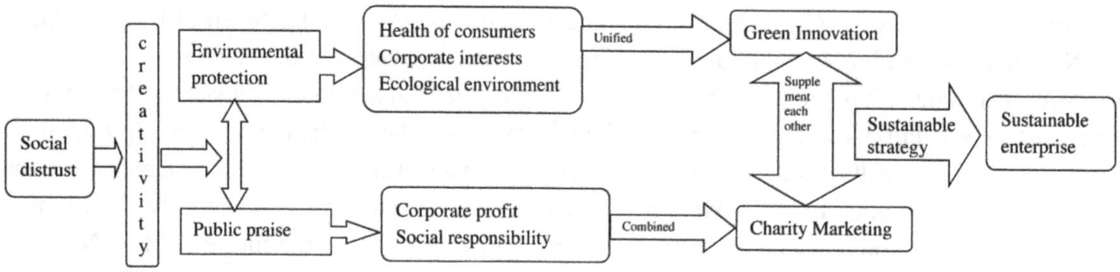

Figure 3. Industry chain of MengNiu Dairy
Data source: collated and analyzed by this research

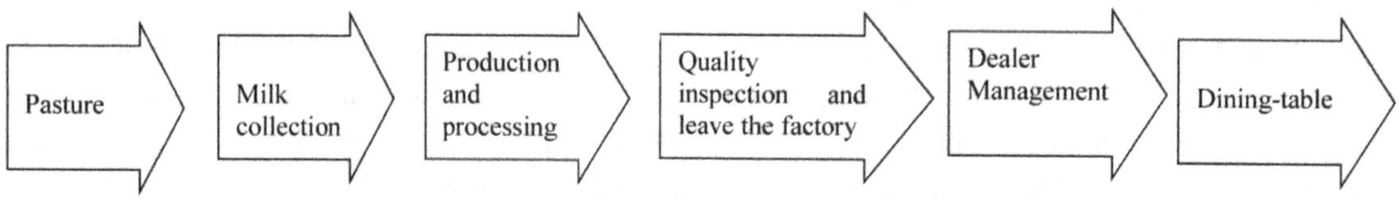

Taking ecological grassland construction as the basis, MengNiu Dairy established a "pasture-table" natural nutritional quality production line. Firstly, milk source is taken as a key link in the sustainable industry chain. In order to improve the quality of milk, MengNiu Dairy actively promotes the modernization of pastures, strictly performing access, grading and daily management, and constantly transforming from supervision-oriented milk management to technical service-oriented milk management, so as to ensure milk quality from source. Moreover, MengNiu Dairy introduced the most stringent food safety standards of international pastures to be out to improve the quality of milk in all aspects. In production management and control, MengNiu Dairy takes 4Q, namely, quality planning, quality control, quality assurance, quality support as the core to guard a pass of product development and procurement, production and stock removal of raw materials. In channel, MengNiu Dairy established 32 cold chain quality model markets, strengthening the construction and routine maintenance of cold chain, ensuring the product to be delivered at 2-6 degrees, and sparing no effort to create a sustainable green industry chain. In this year, in order to strengthen communication with consumers, MengNiu Dairy promoted an activity named "transparent factory", a new platform for consumers interaction, in which consumers understood the process of production and processing in MengNiu factory, and felt "a little happiness matters" created by MengNiu Dairy from visual, sensory, and taste bud aspects.

Sustainable innovation is a new sustainable strategy aimed at eco-environmental protection and centered on green and environmental protection to meet the health needs of consumers and seek long-term development. In addition to realize corporate interests and meet requirements of consumers, it pursuits harmony between men and nature. Take MengNiu Dairy as an example, it applies green innovation and charity marketing to enterprise management, ignoring immediate interests to go farther, higher and more stably. MengNiu Dairy resolutely adheres to green and sustainable strategy, from control of milk source, production process to transportation channel, which makes us understand the implementation of sustainable development is not shown in a certain stage, but need to be persisted in throughout. MengNiu Dairy committed itself to public welfare undertaking, which teach us that bearing social responsibility is a wise move for the development of enterprises, and the public welfare undertaking investment helps to establish the brand image; however, not all public welfare undertaking need to be participated in, the enterprise need to choose proper public welfare activities which are consistent with the enterprise, and keep it going, making public welfare undertaking be a long-term strategic behavior.

However, green products are costly in both production process and sales link, so it is difficult to invest high technology in all links to produce green products. Cost decreasing and benefit increasing are difficulties for enterprises to implement sustainable marketing strategy.

REFERENCES

Boynd, H. William, & Massy. (2002). Marketing Management. New York: Harcourt Brace Jovanvich.

Bi, S. (2011). *Research on Green Marketing Strategy of Enterprises in the Background of Ecological Civilization*. Ocean University of China.

Big River Newspaper. (2006). Retrieved from http://news.sohu.com/20060425/n242984539.shtml

Chen, Z. (2010). Charity marketing: Concept, Theory and Literature Review. *Market Modernization, 2010*(20), 34-35.

Chen, X. (2010). *Exploration of Chinese Charity Marketing Model*. Academic Press.

Martin, D., & Schouten, J. (2014). *Sustainable Marketing*. Shanghai Renmin Press.

Ding, G. (2010). Research on Creative Connotation. *Journal of Chongqing University of Technology: Social Science*, *24*(11), 77–80.

Gao, Y. (2011). *Research on Marketing Channel Innovation of MengNiu Dairy Company*. Jilin University.

Hua, H. (2000). Discussion on 4P Theory and 4C Theory. *Journal of Beijing Institute of Finance and Commerce, 2000*(4), 36-37.

Huang, W. (2008). *Marketing Strategy Analysis of MengNiu Dairy*. Xiamen University.

Li, J. Y. (2013). *Study on Green Marketing Strategy of Hua Peng Food*. Central South University of Forestry and Technology.

Li, X. (2015). The Research and Prospect of Green Innovation Research. *Research and Development Management*, *27*(2), 1–11.

Ma & Yao. (2014). Discussion on Green Marketing. *Management Supervisor, 2014*(5X), 275-275.

MacKinnon, D. W. (1962). The Personality Correlates of Creativity: A Study of American Architects. In G. S. Neilsen (Ed), *Proceeding of the XIV International Congress of Applied Psychology*, (vol. 2, pp. 11-39). Copenhagen: Munksgarrd.

Qi, Y. (2010). Practice Low Carbon: MengNiu Dairy Build Green Industry Chain. *Advertiser Market Observer, 2010*(8), 50-50.

Sonya, S., Jordan, J., & Mazar, N. (2015). *Green Consumerism: Moral Motivations to a Sustainable Future*. Current Opinion in Psychology.

Wang, L. (2014). Research on Interaction and Strategic Innovation of Green Logistics and Green Marketing. *Logistics Technology, 2014*(5), 25-26.

Wei, X. (2007). Charity marketing—Super Strategy of MengNiu Dairy. *Modern Communication, 2007*(1).

Wu, Y. (2005). *Marketing*. Wuhan University of Technology Press.

Yang, J. (2010). Discussion on Green Marketing. *Management & Technology of SME Every Ten Days, 2010*(6), 110-110.

Yin & Xin. (2015). A Comparative Study of 4P and 4C Marketing Theory. *Journal of Huaihai Institute of Technology (Humanities & Social Sciences Edition), 2015*(2), 88-90.

Yu, X. (1988). The definition, structure and hierarchy of creativity. *Contemporary Youth Research, 1988*(8), 25-28.

Yan, Z. (2006). *Research on Characteristics of Distribution Channels of MengNiu Dairy*. University of International Business and Economics.

Zhang, R. (2011). International Green Marketing Strategy of China. *China Business & Trade, 2011*(24), 20-21.

Zheng, P. (2015). Sustainable Development and Green Marketing. *China Economist, 2015*(7), 271-271.

Section 3
Sustainability and Management

Chapter 11
Green Entrepreneurship in Transitional Economies:
Breaking Through the Constraints of Legitimacy

Yang Gao
Dalian University of Technology, China

Sang-Bing Tsai
Foshan University, China & Dalian University of Technology, China

Shuo Zhang
Dalian University of Technology, China

Guodong Li
Civil Aviation University of China, China

ABSTRACT

With environmental pollution, climate change and resource scarcity being serious global issues, green entrepreneurship is increasingly seen as an approach to simultaneously address economic performance, environmental impact and social responsibility. As green entrepreneurship needs to consider both venture performance and social responsibility, it will be subject to legitimacy constraints at the institutional level. Especially for transition economies in different countries, institutional uncertainty is a key factor. Better green performance can be easily achieved if green startups have a higher level of legitimacy. However, against the background of transitional economies, the increase in institutional uncertainty will damage the promotion of political legitimacy and make the enterprises subject to political legitimacy constraints lose their green performance. Overall startups in different industries need to develop green actions under the pressure of institutional constraints, and governments also need to improve policies and regulations quickly.

DOI: 10.4018/978-1-5225-3537-9.ch011

INTRODUCTION

With the deterioration of global environment and the worsening energy contradiction, balancing the conflicts between economic development and environmental protection has caused great difficulties for governments, academia and industries all around the world. Multiple events, including the oil spill of British Petroleum in the Gulf of Mexico and serious PM2.5 in China, have made the governments and enterprises realize the inevitability and urgency of a green transformation. As the subject of production, enterprises play a significant role in the development of national economic and product market. However, when enterprises provide consumers with products and service bringing convenience for them, many production processes will destroy the sustainable development of ecological environment. Therefore, an important component of addressing sustainable development issues and environmental degradation will hinge on making up for and improving the lack of social responsibility of corporations (Rockström, J., Lannerstad, M., & Falkenmark, M., 2007; Lin, C.S., Chang, R.Y., & Dang, V.T., 2015). As the society's sensitivity to environmental problems, energy consumption and pollution continues to rise, these elements become the important factors threatening enterprise legitimacy. So, there is an increasing voice for corporations to develop green businesses that takes both ecological benefits and economic benefits into account, that is, GE. It is recognized by governments, ventures, customers and other stakeholders, as well as integrates business entrepreneurship and sustainable development, simultaneously considering the sustainable development of environment, economy and society (Allen & Malin, 2008; Isaak, 1997, pp.80-90). Therefore, whether enterprises with GPO can really obtain green performance and how legitimacy constraints influence the achievement of green performance in green companies have drawn the attention of business and academia.

GE refers to entrepreneurial behavior in an enterprise's green innovation of products, services and market development to generate profit while considering environmental protection (Cohen & Winn, 2007; Dean & McMullen, 2007). Green entrepreneurship theory gradually developed with the integration of entrepreneurship theory and social enterprise theory. Under the dual functions of eco-orientation and market-orientation, enterprises are not likely to be burdened by GE as it will provide vast resources for enterprise development (Hartman & Stafford, 1997; Soto-Acosta, P., Cismaru, D.M., Vătămănescu, E.M., & Ciochină, R.S., 2016). According to the definition of EO by Miller and Friesen (1983), this study believes that green entrepreneurial corporations need to own green entrepreneurial orientation (GEO), which contains green innovation, green risk taking and green proactiveness. As green entrepreneurial corporations need to undertake social responsibility, it is particularly vital for them to gain legitimacy. Suchman (1995) proposes that legitimacy refers to the overall evaluation of external stakeholders about appropriateness of corporate behavior. The content involves both social responsibility and other standards such as competitive advantage and enterprise scale. Breaking through legitimacy constraints and achieving green performance have been the main purposes of enterprises which launch green entrepreneurship activities.

Because proactive green-oriented businesses not only concern about business performance but also consider social responsibility, the relationship between GEO and green performance will be restricted by institutional environment. As an institutionally embedded binding mechanism, the study of enterprise legitimacy has attracted attention (Shirokova, G., Bogatyreva, K., Beliaeva, T., & Puffer, S., 2016). However, present research encompasses two views. The first takes strategic theory as a starting point and treats legitimacy as a strategic resource which ventures are prone to lose, as well as claims that the ultimate purpose of seeking legitimacy is to obtain mobilized resources and securing approval and legitimacy from external stakeholders will be beneficial for start-ups to realize opportunity and value; this 'legitimacy—resources—business growth' model of causal relationships is widely accepted (Zimmerman & Zeitz, 2002; Lounsbury & Glynn, 2001; Navis & Glynn, 2011). Nevertheless, a large amount of scholars questioned it and indicated that legitimacy couldn't be regarded as the key resource for start-ups to overcome the disadvantage of starting anew because experienced and rational entrepreneurs are more likely to pursue more easily obtained resources in order to build their business models (Aldrich & Fiol, 1994; Zimmerman & Zeitz, 2002; Delmar & Shane, 2004). Combining the above arguments, this chapter tends towards the latter view and believes that legitimacy should exist as a limiting mechanism embedded in the institution and green-oriented entrepreneurship should not deem legitimacy as a necessary resource but take appropriate strategic action according to different contexts.

Because the causal model of 'GEO– legitimacy – resources - business growth' does not reflect the essence of legitimacy in GE, a comparative exploration about the effects on varying degrees of legitimacy on the impacts between GEO and green performance appears to be more valuable, meanwhile, this kind studies are rare. As a result, this chapter will explore the relationship between GPO and green performance and reveal the moderating effect of legitimacy. More importantly, the premise upon which it is possible to integrate legitimacy theory and green entrepreneurship theory is the existence of stable and effective systemic rules on which legitimacy evaluations depend. However, this isn't consistent with lots of transitional economies, including China. Due to the substantial institutional uncertainty created by economies, judging legitimacy guidelines produces higher heterogeneity in transition economies (Scherer, A.G., Palazzo, G. & Seidl, D., 2013). Thus, this chapter combines the situation of transition economy in China and explains the relationship among GPO, legitimacy and green performance under institutional uncertainty.

Apart from the introduction, the paper is structured along eight sections. Sections 2-4 provide a description of GE and relevant content, legitimacy and transitional economy. Section 5 is devoted on the methodologies and data for the model. Section 6 provides the results. And Section 7 provides the approaches to achieve green performance. Finally, the last section just summarizes the key issues that have arisen in this paper.

GEEEN ENTREPRENEURSHIP AND RELEVANT CONTENT

Green Entrepreneurship and Green Entrepreneurial Orientation

Green entrepreneurship (GE), also known as environmental, ecological or sustainable entrepreneurship, primarily represents the process of identifying, evaluating and utilizing economic opportunities which arise during market failure and facilitates sustainable developments of enterprises (Shane & Venkataraman, 2000; Gibbs & O'Neill, 2014). The core of GE is to discover the future market opportunity in order to develop a green market through innovation and commit to corporate social and ecological responsibility (Shepherd & Patzelt, 2011; Cohen & Winn, 2007). Because of GE's longer payback period coupled with its social responsibilities, GE often requires policies that encourage, support and are widely recognized by society, and the enterprises also play a double role in ecological construction and entrepreneurship (Ndubisi, N.O. & Nair, S.R., 2009). In addition, enterprises dedicated to GE must proactively achieve the twin objectives of marketing and ecology, making both green participation orientation and entrepreneurial orientation important (Nelson, O.N. & Sumesh, R.N., 2009; Al-Saleh, Y. & Mahroum, S.,

2015). Thus, in the study of GE, the GEO and the recognition of legitimacy under institutional theory of enterprises are significant factors in determining the success of GE enterprises.

GEO is derived from the concept of entrepreneurial orientation (EO). The proposition of EO is mainly to show the distinctive characteristics between enterprises and other conservative enterprises, that is, innovation, risk taking and proactiveness (Miller, D., 1983, 2011; Wales, W.J., 2016, pp.3-15). This not only illustrates unique entrepreneurial strategic position of enterprises but also represents the inherent entrepreneurial orientation of start-ups (Lumpkin, G.T.; Dess, G.G., 1996; Hall, J.K., Daneke, G.A, & Lenox M J, 2010). EO indicates the overall competitive orientation and tendency of organizational behavior in entrepreneurial enterprises which is applicable to GE. GEO embodies not only the market competition orientation of GE enterprises but also their green participation orientation (Kuckertz, A. & Wagner, M, 2010). GEO contains green innovation, green risk-taking and green proactiveness, respectively representing the tendency of GE enterprises to keep ahead of the competitors by implementing green technological innovation, make risky investment decisions in the face of uncertainty and lead the industry through upfront actions. GEO proposed in this chapter is a concept at the organizational level.

At present, although there are many studies about EO and entrepreneurial performance, the results vary widely. Some scholars indicate that EO is conducive to enhance entrepreneurial performance, and this impact is particularly obvious in the initial start-up phase and growth phase (Kreiser, P.M., 2011, pp.1025-1050; Eshima, Y. & Anderson, B.S., 2016). Others states that EO may have a negative effect or no effect on entrepreneurial performance (Stam, W, 2008, pp.97-111). The difference between EO and entrepreneurial performance stems from the entrepreneurial activity itself, which is dynamic and highly uncertain (Cope, J., 2005, pp.373-397; Schepers, J., Voordeckers, W., Steijvers, T. & Laveren,E, 2014). Whether enterprise performance is enhanced relies largely on specific entrepreneurial behaviors under different circumstances (Bruton, G., Susanna, K., Donald, S., & Wright, M., 2015). Therefore, there are few studies about whether enterprises with high GEO can realize high green performance. Meanwhile, compared with financial performance, green performance is only reflected in product, production process, customer safety and other aspects. So, the relationship between EO and green performance should be different.

Green Proactiveness Orientation and Green Performance

As an important dimension of GEO, green proactiveness orientation (GPO) represents the tendency of enterprises to be more proactive in adopting GE behaviors compared to its competitors (Miller, D. & Friesen, P.H., 1983). Green proactiveness-oriented enterprises tend to be industry leaders, examine industry trends, seize opportunities in advance, thus taking early actions to cope with change (Deng, Q. & Ji, S., 2015). GPO is a combination of initiative and behavior orientation and on behalf of the enterprises that take the initiative in order to have an effect on environment, rather than passive acceptance. It reflects a forward-looking tendency of enterprises, including accurate prediction of industry trends, avoidable production of low green-potential products, and guidance of customer acceptance about green products and consumption patterns, etc. (Cohen, B. & Winn, M.I., 2007). Thus, taking green actions before competitors, such as pushing out new green products, new service patterns, new green management modes and green technologies, along with having the foresight to examine development of the industry, can reflect the meaning and nature of GPO (Boulding, W. & Christen, M.,2001; Morris, M.H., Kuratko, D.F., & Covin, J.G., 2010).

Green performance represents a reduction in negative impact of an enterprise on the environment while achieving a balance between business and environment interests (Wagner, M. & Schaltegger, S., 2004; Huang, M. & Wang, B.). Therefore, green performance measures an enterprise in which business and environmental needs interact (Mavragani, A., Nikolaou, I.E., & Tsagarakis, K.P., 2016; Olsthoorn, X., Tyteca, D., & Wehrmeyer, W., 2001). Achieving green performance requires ventures not only to decline pollution, waste and energy consumption, and improve customer safety but also to survive and develop (Zhu, Q. & Sarkis, J., 2004; Yang, C.S., Lu, C.S., & Haider, J.J., 2013). Based on the decreasing negative impact on the environment, green performance includes multiple measurement indicators, each representing a variable (Wagner, M. & Schaltegger, S., 2004). These variables contains reduced use of water, solid waste, energy, soil contamination, toxic inputs, non-renewable resources, wastewater emissions, and emissions into the noise, air, smell/odor emissions, risk of severe accidents and landscape damage.

The main objective of proactiveness-oriented enterprises is to gain proactive advantage and a demand premium (Boulding, W. et al., 1988; Alonso-Almeida, M.D.M., Bremser, K., Llach, J., 2015). Advantages contain deep and lasting customer impressions, strong brand recognition and high cost of customers switching to other start-ups. Meanwhile, early entrants of industry will also gain a low-cost advantage from having a lead in technology, thus downshifting the learning curve (Makadok, R., 1998, pp.683-696; Lopez, L.E. & Roberts, E.B., 2002). Enterprises who take proactiveness orientation can gain significant performance advantages. Similarly, GPO inherits the nature characteristics of proactiveness orientation, reflected in taking GE actions before competitors. Choosing advance actions of green orientation as the objective will achieve the green proactive advantage and the lead in green performance.

In the study of EO, which doesn't consider green performance as a prerequisite, an enterprise that takes proactive orientation will not necessarily achieve entrepreneurial performance (Cope, J., 2005, pp.373-397). The proactive advantage of pioneers is likely to become a proactive disadvantage eventually, and the relationship between proactiveness orientation and entrepreneurial performance is not a simple linear one. However, there is a linear relationship between GPO and green performance, which won't be influenced by the behaviors or the number of the followers. Therefore, enterprises having higher levels of GPO will have better green performance improvements.

LEGITIMACY AND ITS MODERATING EFFECT

Legitimacy

Several scholars have defined legitimacy (Parsons, T., 1960; Meyer, J.W. & Scott, W.R., 1983; Scott, W.R., 1995). Suchman (1995) conducted a comprehensive review of previous and put forward the concept of legitimacy which is widely recognized in academia: legitimacy is the general perception or assumption that the activities of an entity are appropriate, proper or desirable within the socially constructed system of norms, values, beliefs, and definitions. Enterprises are confronted with the constraints of legitimacy, incorporating the restraints of sociopolitical legitimacy and cognitive legitimacy (Becker-Blease, J.R. & Sohl, J.E., 2015). Sociopolitical legitimacy refers to the level of acceptance of key stakeholders, the public and government officials about the behavior and products, and low social legitimacy will result in registration difficulties, public rejections of product standards, financing problems, etc (Palmås, K., Andersson Schwarz, J., & Larsson, S., 2014). Cognitive legitimacy refers to the level of social understanding and acceptance of start-ups, which is largely determined by the degree of external recognition of new ventures and their products and services. Low cognitive performance is reflected in a low level of society-wide knowledge or difficult acceptance of products or services by the ventures, or in a low customer evaluation of products or services (Delmar, F. & Shane, S., 2004; Li, X., Wang, C., Xu, X., 2014).

Cognitive legitimacy pertains to the category of external legitimacy which is capable of being obtained by enterprises. External legitimacy is from outside an organization, in line with the support of suppliers, partners and customers for companies and the evaluations of companies by government, public media and regulators. Most existing studies emphasize legitimacy from an external perspective (Kostova, T. & Zaheer, S., 1999; Rottig, D. & ReusT, H., 2009). According to this, the study measures external legitimacy from two aspects: PL (government and regulatory agencies) and CL (suppliers, customers and partners).

The Moderating Effect of Legitimacy

In the field of entrepreneurship, it is common to discover the integration of legitimacy and opportunities. For realizing the business value of opportunities, experienced entrepreneurs tend to view commercial legitimacy (CL) significantly, regarding low CL as a key factor that leads to the difficult realization of opportunity value. This situation makes experienced entrepreneurs concentrate on CL because a higher level of legitimacy ensures a more intense spirit of innovation, and the more likely it is that a start-up will obtain the competitive advantage (Li, X. et al., 2014; Shane, S., 2000, pp.448-469; Czinkota, M., Kaufmann, H.R., & Basile, G., 2014). And in green research, compared to the value of ordinary entrepreneurship opportunities, the value of GE is more difficult to realize, because green enterprises are faced with the dual constraints of sociopolitical legitimacy and CL (Fraser, S. & Greene, F.J., 2006). Therefore, particularly for experienced entrepreneurs, constant learning behavior makes them less optimistic and focus on the evaluation of green opportunities (Kuckertz, A. & Wagner, M., 2010; Westhead, P., Ucbasaran, D., & Wright, M., 2005). As a result, when an enterprise selects GE to be ahead of its competitors, entrepreneurs should carefully think of the degree of recognition by external stakeholders because to a large extent, the level of political and commercial legitimacy will affect the realization of green opportunity value and green performance. In other words, for green entrepreneurs, various stakeholders rely on different standards and specifications to judge the legitimacy of GE, so the pressure of legitimacy will push enterprises to engage in GE more actively. Furthermore, when an enterprise perceives the level of political and commercial legitimacy goes higher, it will receive more attention and more transparent information. Thus, enterprises with higher political and commercial legitimacy own much more pressure to improve their green performance, which subsequently helps proactive green enterprises obtain green performance.

TRANSITIONAL ECONOMY AND INSTITUTIONAL UNCERTAINTY IN CHINA

Transitional Economy

'Transitional economy' refers to the transitional process from a planned economy to a market economy. Unlike developed western countries, a lot of developing countries coexists planned and market economies, government intervention and market mechanism, as well as traditional and modern economic development modes (Walder, A.G., 1995, pp.263-301; Li, Y., Peng, M. W., & Macaulay, C. D., 2013). In the situation of transitional economy, enterprises also have characteristics of ambidexterity, such as continuously swaying and balancing between commercial economy and environmental protection, diversification and specialization, entrepreneurship and keeping (Oi, J.C., 1995, pp.1132-1149; Tan J., 2001, pp.359-376). The transitional economy makes firms in developing countries, including China, to show the unique features in the selection of entrepreneurial behavior, and mature western theories are no longer suitable for transitional economies. Undoubtedly, a systematic study in entrepreneurship behavior and GE behavior of a corporation in transitional economy will offer a vital theoretical contribution for entrepreneurship management and green management, and a realistic guidance for getting rid of economic crisis.

Institutional Uncertainty

Institutional uncertainty is an important feature of transitional economies (Wu, K.J., Liao, C.J., Tseng, M.L., & Chou, P.J., 2015). It is a broad concept, encompassing various forms of uncertainties within the political environment (Kelly, B., Pástor, Ľ., & Veronesi, P., 2016). Because such uncertainties are hard to measure, which means only fewer proxies can be used in systematic cross-country investigation (Brunetti, A. & Weder, B., 1998; Simón-Moya, V., Revuelto-Taboada, L., & Guerrero, R.F., 2014). From the analysis of existing empirical results, we know different scholars used a variety of indicators to measure, containing political violence, governmental instability and policy uncertainty (Barro, R. J., 1991, pp.715-721; Mauro, P., 1991, pp.681-712; Brunetti, A., Kisunko, G., Weder, B., 1997). Although stability and few political conflicts are unique advantages of China, policy instability is a severe disadvantage caused by transitional economy. Such disadvantages include existing policy gaps in some areas, unclear government definition of key issues in the policy implementation process and poorly detailed observations and recommendations of many policies (Puffer, S.M., McCarthy, D.J., & Boisot, M., 2010; Sheng, S., Zhou, K.Z., & Li, J.J., 2011). Due to scarce innovation, stability and effectiveness in some policies, it is hard to turn them into valid entrepreneurship behavior. Therefore, to some extent, policy uncertainty is representative of institutional uncertainty under the transitional economy in China.

The Moderating Effect of Institutional Uncertainty

For enterprises carrying out green entrepreneurship, transitional economy has a substantial impact. Because institutional transition breaks through the restrictions of a planned economy, adding a private economy greatly stimulates the vitality of market competition. In the fierce competition, ventures tend to pursue short-term performance goals but long-term ones, and corporation behavior tends to be conducive to business performance rather than environmental performance (Orlitzky, M., Siegel, D.S., & Waldman, D.A., 2011). When institutional uncertainty is high, policy shackles for enterprises become less strict, and companies will continue to seek for and drill loopholes and gaps of policies. Particularly when the industry standards are not clear, institutional pressure on green enterprises declines along with the power to obtain political legitimacy (PL), resulting in damaged green performance. Nevertheless, in the study of institutional theory, formal and informal institutions show a mutual substitution effect on the process of guiding economic behaviors. When formal institutions are incomplete, informal norms and values will play a stronger role (Li, Y., Peng, M.W., & Macaulay, C.D., 2013; Peng, M.W. & Heath, P.S., 1996). Thus, when institutional uncertainty is high, it's more significant for enterprises to obtain recognition from the suppliers, partners and the public (Huang, T., Wu, F.; Yu, J., & Zhang, B., 2015).

In this context, influences of informal institutions are more vital and the facilitating roles of CL become more obvious between GPO and green performance.

The Moderating Effect of Institutional Uncertainty in Transitional Economy

Connecting the institutional theory and green entrepreneurship theory under transitional economy, this study finds out that the moderating effect of PL on the relationship between GPO and green performance becomes more negative as institutional uncertainty increases, such that the three-way interaction of GPO, PL, and institutional uncertainty is negatively related to green performance. However, the moderating effect of CL becomes more positive. That is, as a positive indicator, legitimacy is able to promote the relationship between GPO and green performance, but this conclusion will alter under the background of transitional economics. Under a high level of institutional uncertainty, CL will become a stimulus, while the promotion of PL will be restricted. In addition, this chapter analyzes the influences of institutional uncertainty on the moderating effect of legitimacy between GPO and green performance in a transitional economy. On one hand, this chapter finds that institutional uncertainty strengthens the moderating effect of CL. On the other hand, this uncertainty damages the moderating effect of PL. Considering this issue in the context of transitional economy in China, where there are few political conflicts and a stable government, the instability of policies can have a serious impact on business. Policies in many sectors are not stable due to gaps in policies. And it is difficult to implement certain standards because green start-ups must pay attention to both business performance and social responsibility, and institutional uncertainty has an even greater impact on them. In an institution environment which is highly uncertain, because of the gaps in policies and regulations, and incompletely specific green standards for implementation, pioneering GE start-ups will continue to drill policy 'loopholes' without being punished. As a result, the institutional pressure on those companies declines a lot and the impetus to obtain PL will reduce, suppressing the catalytic effects of PL between GPO and green performance. On the contrary, with the increase of institutional uncertainty, enterprises need to obtain recognition from stakeholders to achieve higher business performance. In this circumstance, the drive to gain CL will aggrandize and the power of CL exerts greater pressure on those ventures, thus making CL play a more prominent role in elevating the relationship between GPO and green performance. To a certain extent, this interprets the reason why in transitional economy in China, the enterprises find it so difficult to implement green standards, drill policy'loopholes', prefer to seek'Guanxi', and excessively focus on business performance, thus leading to the profit pursuit, malicious competition and serious product homogeneity.

DATA AND METHOD

Development of Survey Instrument

According to Murphy (1996), using appropriate control variables in the study of entrepreneurial performance can avoid performance measurement prejudice. Thus, this chapter chooses enterprise size, business scale, sector, establishment time and stage of development as control variables.

Figure 1. The model

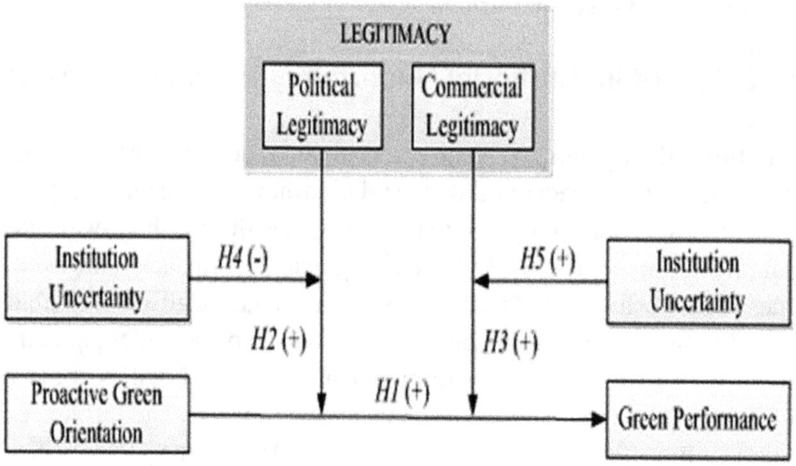

On account of the extensive theoretical background in previous section, this chapter identified the following demanded measurements. All items were measured on a Likert five scale, where an evaluation of one exhibits that the statement doesn't fit the reality of the start-up to which the respondent belongs and an evaluation of five shown a high fit. Besides, the measures of EO have been mentioned in previous study. This chapter adopts the scale of proactiveness orientation from Covin and Slevin (1989) and uses three statements to measure GPO. Table 1 presents the specific statements and factor loadings of GPO, and the extraction method of Table 1 to Table 5 is Principal Component Analysis.

PL and CL belong to external legitimacy of enterprises, including the recognition of governments, suppliers, partners, regulators and customers. We select four statements from the discussion about legitimacy of Suchman (1995) and Rottig (2008) to measure PL and CL. Tables 2 and 3 present these statements and corresponding loading factors, indicating a high significance affirmative of the statements.

Institutional uncertainty and market uncertainty are main characteristics of transitional economies. Puffer et al. (2010) and Sheng et al. (2011) have carried out the thorough researches on institutional uncertainty from the aspects of policy completeness and operability. On the basis of their results, we select five questions to measure institutional uncertainty (Table 4). Moreover, Zhu and Sarkis (2004) and Yang et al. (2013) put forward four items to evaluate green performance, which contains six questions expressed in Table 5.

Table 1. Rotated factor matrix for factor analysis of GPO

Factors		Cronbach's Alpha
1. We usually launch green initiatives which peers followed or reacted to.	0.848	0.806
2. We tend to be the industry 'leader', often offers new green products, new technologies or green management models before peers.	0.855	
3. Corporate executives look at future trends of industry frequently, understand opportunities for green development and take early actions to manage changes.	0.842	

Table 2. Rotated factor matrix for factor analysis of political legitimacy

Factors		Cronbach's Alpha
1. The corporation was highly praised by government departments.	0.855	0.884
2. The corporation was recognized by regulators.	0.873	
3. The government recommended us as a model enterprise frequently.	0.874	
4. The government visited our enterprise regularly.	0.842	

Table 3. Rotated factor matrix for factor analysis of commercial legitimacy

Factors		Cronbach's Alpha
1. The corporation is widely recognized by customers.	0.847	0.864
2. The corporation is widely recognized by suppliers.	0.824	
3. The products of enterprise have been highly evaluated by peers.	0.882	
4. The products of enterprise have been highly evaluated by customers.	0.831	

Table 4. Rotated factor matrix for factor analysis of institutional uncertainty

Factors		Cronbach's Alpha
1. Key policies related to business operation are incomplete.	0.786	0.854
2. Relevant industry standards are unclear.	0.766	
3. Relevant industry policies are fuzzy.	0.821	
4. There are gaps in related institutions and policies.	0.832	
5. Relevant policies are short of specific guidance and operational suggestion for implementation.	0.776	

Table 5. Rotated factor matrix for factor analysis of green performance

Factors		Cronbach's Alpha
1. Compared to peers, we release less air pollution.	0.820	0.897
2. Compared to peers, we cause less sewage.	0.797	
3. Compared to peers, we have less solid waste.	0.823	
4. Compared to peers, our cost of dealing with poisonous, harmful and hazardous materials is lower.	0.724	
5. Compared to peers, our cost of energy consumption is lower.	0.865	
6. Compared to peers, we have a lower rate of environmental accidents.	0.849	

Data Gathering

This chapter conducted a questionnaire survey. And our sample concentrated on data from start-ups within the last eight years due to Li and Atuahene-Gima's (2001) definition of start-ups as enterprises that have been established for no more than eight years. The survey distribution lasted six months, from February to July 2016, and the research areas contain Beijing, Shanghai, Shenyang, Hangzhou Changchun and Xiamen. These areas have various levels of development (i.e., central developed cities, coastal cities in South, and economically less active cities in North) to avoid the impacts of different degrees of regional economy development. Enterprises involved in this research include green construction and materials, low-carbon energy sources, eco-friendly materials, green furnishing, green process transformation, waste disposal services and training, and energy-saving equipment.

The data collection process was divided into three procedures. In the first procedure, the survey questionnaires were distributed through e-mails, and 120 responses were collected. In the second procedure, questionnaires were given to MBA students and 110 responded. In the third procedure, questionnaires were obtained onsite from mid-level and senior managers of solicited firms and from researchers completing the questionnaire face to face, with 63 responses gathered. In a word, 297 responses were collected, and 62 of which were eliminated due to incomplete or insufficient responses. A total of 235 samples were valid and utilized in the study of 22 variables, explaining the valid combination according to the requirements stated by Bentler and Chou (1987) that the sample number must be at least ten times of observed variables (Bentler, P. M. & Chou, C.P., 1987). A t-test was also conducted on enterprise size, sector, year established, ownership and business size of the three groups and no striking difference was found. As a result, they can be used in combination.

General Descriptive Analysis

After the gather of large samples, descriptive statistical analysis was carried out. On one hand, the results of a general descriptive analysis could reflect the fundamental features of sampled corporations. On the other hand, these results could also confirm the versatility and reproducibility of other findings. The results of a general descriptive analysis would help the study by providing salient characteristics of samples, thus replenishing the assessment of the research objects.

Examination of Validity

The study measured both content validity and construct validity. Content validity indicates whether the content of the scale can accurately cover the range or extent of the concept or meaning to be measured, and this can usually be measured subjectively(Zikmund, W. G., Babin, B. J., Carr, J. C., & Griffin, M., 2013) . Construct validity encompass both convergent validity and discriminant validity. Convergent validity, which can be explained by composite reliability (hereinafter referred to as CR) and average variance extracted from latent variables (hereinafter referred to as AVE). As for discriminant validity, this study referenced Fornell and Larcker (1981) and carried out tests by comparing the average root mean square values of AVE with different latent variables and the correlation coefficients of these variables.

Examination of Reliability

The reliability test refers to the degree of consistency in measurements by different measurers taken with the same instruments, reflecting the level of approximation of repeated measurements under the same conditions, that is, whether a series of projects measure the same concept or not. Generally, this test can be implemented by observing the internal consistency of the measurement instruments (Dawson, J.F. & Richter, A.W., 2006). In the Likert-scale method, the most commonly used method is the L.J. Cronbach α coefficient. Although an α coefficient value of 0.70 is on the lower boundary of the scale, it is still acceptable.

Correlation Analysis

Correlation analysis concentrates on the correlation among dependent variables, independent variables and control variables. The correlation coefficient is a commonly used measurement and the method is able to be the Pearson's product moment correlation method and the two-sided test. If the correlation coefficient among independent variables is greater than 0.75, the variables are considered too similar and collinearity might be present.

Harman's single-factor test was applied to analyze the entire questionnaire. In the absence of rotation, the explained variance of the first indicator extracted exceeded 50%. So, it could be believed that there was a severe bias in the common measurement method.

Regression Analysis

In this chapter, regression analysis was used to verify the relationship among GPO and business performance and the moderating effect of legitimacy and institutional uncertainty. The study used three factors for regression analysis. First, the determination coefficient of the model reflects the explanatory power of the regression equation for the dependent variables. This chapter used R2 to reflect the fitness of regression equation in order to avoid the influence from the number of independent variables on R2. Second, the regression model is said to be valid when the regression coefficient is not 0 at significance value of less than 0.05 or 0.01. Third, the variance inflation factor (VIF index) reflects the collinearity between independent variables. Collinearity is exhibited by a large VIF index. When $0 < VIF < 10$, multicollinearity generally does not exist.

EMPIRICAL RESULTS

Results of General Descriptive Analysis

Table 6 presents the features of the samples gathered. In summary, most enterprises have small and medium scale; the year number of enterprises were founded clustered around 3 to 8 years; the sector distribution was relatively dispersed; the proportion of companies with turnover under 5 million yuan was

Table 6. Characteristics of survey samples

Enterprise Size		Sectors			
1-50 people	17.9	Real estate			18.5
50-100 people	43.4	Agriculture, forestry and livestock			3.0
100-300 people	17.4	Energy, environmental protection			20.2
300-500 people	20.0	Environmental transformation and training			3.8
More than 500 people	1.3	Materials			14.5
		Accommodation, catering			3.8
		Manufacturing			28.5
		Construction			7.7
Business Size		**Ownership**		**Year Established**	
1 million or less	34.0	State-owned	11.5	1 year or less	7.6
1 million 3 million	19.6	Foreign investment	40.4	1-3 years	26.8
3 million to 5 million	15.3	Private	31.1	3-5 years	24.7
5 million to 10 million	18.5	Joint venture	17.0	5-8 years	40.9
10 million to 30 million	11.5				
More than 30 million	1.1				

about 70%; enterprise ownership was primarily foreign funded; and domestic ownership was privately funded. On the premise of involving green businesses, the samples satisfied the demand of diversity and the overall descriptive statistics reflected basic features of the objects under observation, laying the foundation for further analysis. Numbers present in Table 6 is percentage (%).

Results of Validity Examination

To verify content validity, this chapter conducted a literature review. Two expert representatives were also invited to examine the research questions and determine whether they were scientific and reasonable, and each of them is from academe, business and government sectors. According to suggestions from the expert panel, this chapter adjusted some indicators to enhance the clarity, technicality and reliability of the survey tools. The questionnaire designed by this study was in line with scientific and rational demands, that is, it met the demands of content validity.

The questionnaire passed the demands of convergent validity test, discriminant validity and test factor analysis. Average values gained in the convergent validity test for all five factors were greater than 0.5 (Tables 1 to 5), which meets the research standards. The load factors obtained in the factor analysis are all greater than 0.5 and the proportion of the cumulative of variance explained is greater than 50% (Tables 1 to 5), successfully satisfying the research requirements. The discriminant validity test result exhibited good discriminant validity among these variables. As shown in Table 7, the diagonal values are representative of the square root of AVE, which is greater than the coefficients in the corresponding row and column.

Table 7. Descriptive statistics and correlation matrix

	Mean	Std. Deviation	Enterprise Size	Year Established	Sector	Business Size	Ownership	GP	GPO	PL	CL	IU
Enterprise size	3.03	1.417	N/A									
Year established	2.99	0.993	0.544**	N/A								
Sector	6.66	3.325	-0.375**	-0.194**	N/A							
Business size	2.86	1.912	0.683**	0.568**	-0.295**	N/A						
Ownership	2.54	0.907	0.292**	0.539**	-0.067	0.306**	N/A					
Green Performance	3.1887	0.96600	-0.038	-0.090	-0.039	0.014	-0.203**	0.773				
GPO	3.1319	1.10042	0.041	-0.143*	-0.085	0.091	-.153*	0.424**	0.758			
PL	2.538298	1.0035308	0.055	0.034	-0.009	0.020	.185**	-0.171**	-.305**	0.81		
CL	2.520213	.9492271	0.090	0.094	0.029	0.050	.164*	-0.112	-.327**	0.587**	0.789	
Institutional uncertainty	3.934468	0.8377595	-0.043	-0.056	-0.086	-0.043	-0.173**	0.140*	0.255**	-0.224**	-0.212**	0.737

**. Correlation is significant at the 0.01 level (2-tailed), *. Correlation is significant at the 0.05 level (2-tailed).
N=235, N/A indicates analysis not applicable; the diagonal line lists root square values of average.

Results of Reliability Examination

The measured alpha coefficients of Cronbach are as follows: the GPO Cronbach's alpha is 0.806, the PL Cronbach's alpha is 0.884, the CL Cronbach's alpha is 0.864 and the institutional uncertainty Cronbach's alpha is 0.854. The alpha coefficient values are between 0.80 and 0.90, indicating a good confidence level in the measured factors.

Results of Correlation Analysis

Table 7 shows the descriptive statistical terms and correlation analysis. Under a significance degree of 0.01, the four main variables in this chapter - GPO, PL, CL and green performance are obviously related to institutional uncertainty. The three major constructs of PL, institutional uncertainty and GPO exhibit significant correlation. Among the control variables, business size and number of years established and other research constructs have a general correlation.

Correlation analysis exhibits that the correlation coefficients among variables are all less than 0.65, according to the standards put forward by Tabachnick and Fidel (Tabachnick, B.G., Fidell, L.S., Osterlind, S.J., 2001). The possibility of the results that may be threatened by multi-collinearity is small. Single-factor method of Harman is to test common method bias. In the absence of rotation, the first extracted factor illustrates only 28.75% of the variance, which is less than the 50% common standard. A single indicator doesn't appear, and the independent and dependent variables are loaded on different indicators, suggesting there is no single indicator to illustrate multiple variances. Thus, the issue of common method bias isn't discovered.

Regression Analysis Results

When consider the control variables, this chapter uses regression analysis to judge causal relationships among dependent variables, independent variable and control variables. Before carrying out the regression analysis, threats influencing the accuracy of the results from multi-collinearity must be removed. The VIF of various regression coefficients is calculated in the process of regression analysis. The results display that the maximum VIF is 1.59, far below the critical value of 10, thus reducing the possibility of any threat from multi-collinearity.

The moderating effect of GPO on green performance and that of political and commercial legitimacy on the relationship between GPO and green performance are calculated through multivariate regression models. This chapter has established regression model 1:

$$Y=\alpha+\alpha_1X+\alpha_2X_1+\alpha_3X_2+\alpha_4X_3+\alpha_5XX_1+\alpha_6XX_2+\alpha_7XX_3+\alpha_8XX_2X_3+\alpha_9XX_1X_3+\mu_m$$

where Y represents green performance, X represents GPO, X_1 represents PL, X_2 represents CL, X_3 represents institutional uncertainties, μ_m represents error, α represents a constant, α_i represents the regression

Table 8. Logistics regression analysis summary

Variable	Step 1	Step 2	Step 3	Step 4
(Constant)	0.676	0.403	0.250	0.263
Enterprise size	-0.066	-0.081	-0.009	-0.052
Year established	-0.014	0.079	0.060	0.074
Sector	-0.046	-0.024	0.013	-0.013
Business size	0.119	0.028	-0.023	-0.015
Ownership	-0.216**	-0.170**	-0.108	-0.078
GPO		0.416***	0.355***	0.294***
Political Legitimacy		-0.061	-0.058	-0.004
Commercial Legitimacy		0.089	0.068	0.052
Institutional Uncertainty		0.010	-0.001	-0.024
GPO*PL			0.224**	0.321***
GPO* CL			0.184**	0.201**
Institutional Uncertainty* GPO				0.107
Institutional Uncertainty* PL				0.030
Institutional Uncertainty* CL				-0.112
Institutional Uncertainty* CL*GPO				0.211*
Institutional Uncertainty* PL* GPO				-0.394**
Adjusted R Square	0.030	0.178	0.311	0.337
R Square Change	0.051	0.159	0.133	0.040
F Change	2.459**	11.323***	22.594***	2.794**

*p <0.05, **p < 0.01 ***p < 0.001.

coefficient and XX_n represents the interaction of two constructs (e.g., XX_1 represents the interaction between GPO and PL).

In inspecting the moderating effects, this chapter refers to the method proposed by Baron and Kenny (1986), utilizing multiple variable regression to verify our views.

In the regression model, the adopted means contains four steps:

Step 1: Only consider the effects from control variables.
Step 2: Consider the primary impacts of independent and control variables.
Step 3: Add the interaction between PL and CL, and think over the moderating effect.
Step 4: Investigate the sub-mediating effect between institutional uncertainty and legitimacy in institutional economy.

In step 2, the significant coefficient is 0.416, $p < 0.001$, indicating that the green proactive effect has an obviously positive impact on green performance. In step 3, the significant coefficient of the interaction between GPO and PL on green performance is 0.224, $p=0.007$, and the significant coefficient of the interaction between GPO and CL on performance is 0.184, $p=0.024$, displaying that PL and CL have significant promotional effects on GPO and green performance, as shown in Figures 2 and 3. Furthermore, after the two interactions introduced in step 3, the influence of GPO on green performance is still significant, with the significant coefficient value being 0.355 at $p < 0.001$.

In step 4, Table 7 exhibits that the significant coefficient GPO × PL × institutional uncertainty has a value of -0.394, $p=0.001$, while the significant coefficient GPO × PL has a value of 0.321, $p < 0.001$, indicating that institutional uncertainty inhibits the moderating effect of PL between GPO and green performance (Figure 4). Moreover, the significant coefficient GPO × CL × institutional uncertainty has a value of 0.211, $p=0.066$, while the significant coefficient GPO × CL has a value of 0.201, $p=0.020$, indicating that institutional uncertainty contributes to the moderating effect of CL between GPO and green performance (Figure 5).

HOW TO ACHIEVE GREEN PERFORMANCE

Because proactive green-oriented enterprises have taken advantages of achieving green performance, this conclusion encourages both business managers and entrepreneurs in various industries to take GE actions ahead of their competitors, encompassing the early launch of new and environmental products, technology and green management models. Senior managers ought to usually look at the future industry trends, know green development opportunities and act early in order to respond to changes. Compared to their peers, these actions are beneficial for enterprises to obtain the lead in emission, energy efficiency and safety.

As for green ventures in various industries, proactiveness is correct, however, the constraints of external legitimacy shouldn't be ignored. Pioneering green enterprises must not only manage their perception of governments and regulators but also attach great importance to the evaluations made by their suppliers, partners and customers. The two legitimacies are both institutional restricts and positive drives to proactive green enterprises in obtaining green performance. Nevertheless, for the governments, the lower the legitimacy of green start-ups, the more likely they will carry out non-green business activities. So, this kind of enterprises should gain the focus of green supervision and inspection.

Figure 2. The moderating effect of political legitimacy

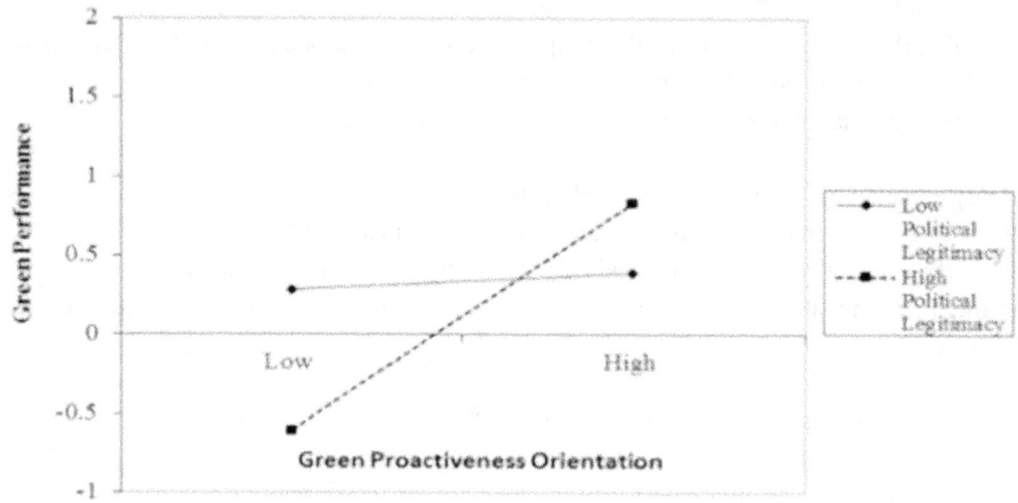

Figure 3. The moderating effect of commercial legitimacy

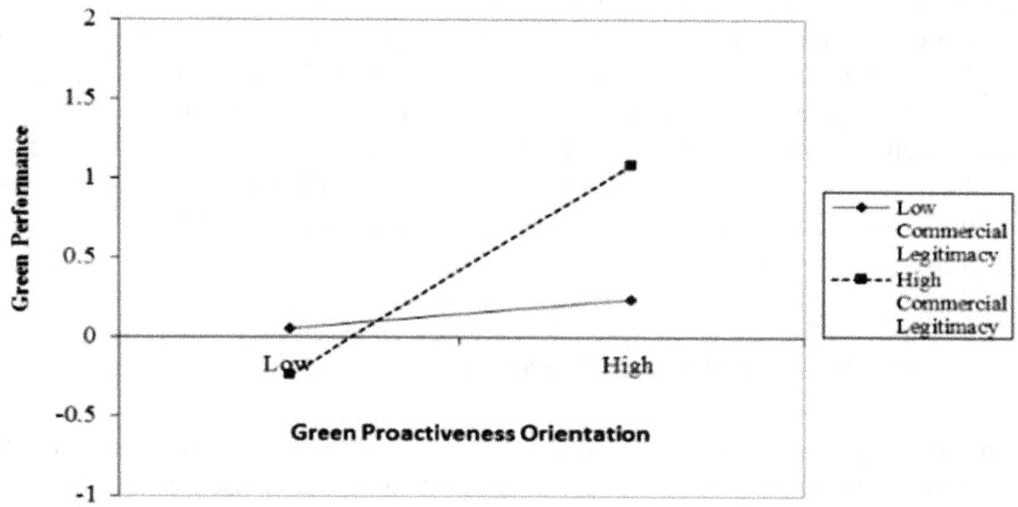

For some western developed countries, the institutional context is relatively stable. As a result, enterprises that develop GE activities must focus on the constraints of government and market simultaneously, which is useful for green ventures to seek for achieving green performance. In contrast, in a mount of developing countries, the high level of institutional uncertainty lead to poor PL restricts. Numerous companies will excessively pursue CL, resulting in vicious competition and homogeneous products. Therefore, this provides a warning for some governments: Apart from promoting institutional standards as soon as possible, increasing the strengths of moral propaganda, the government also ought to accelerate the embedment of green standards into society through extensive publicity, so that the social norms of informal institution can participate in promoting operations of green enterprises.

Figure 4. Sub-mediating effect between institutional uncertainty and political legitimacy

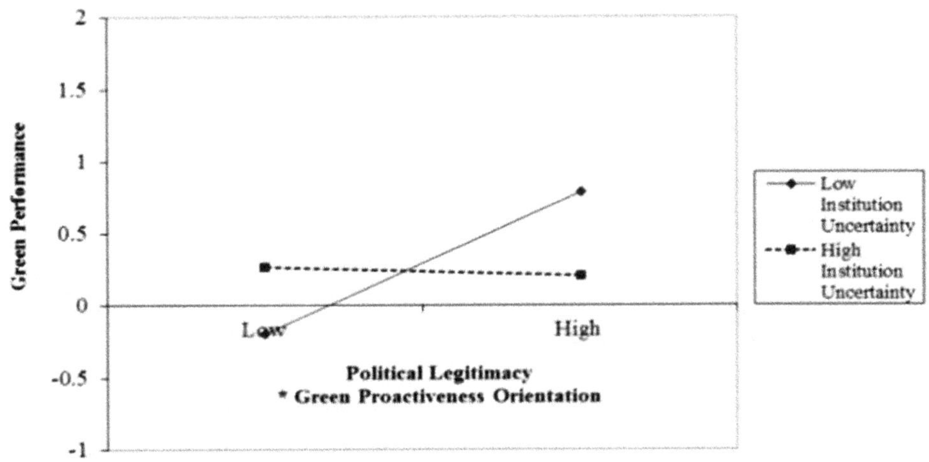

Figure 5. Sub-mediating effect between institutional uncertainty and commercial legitimacy

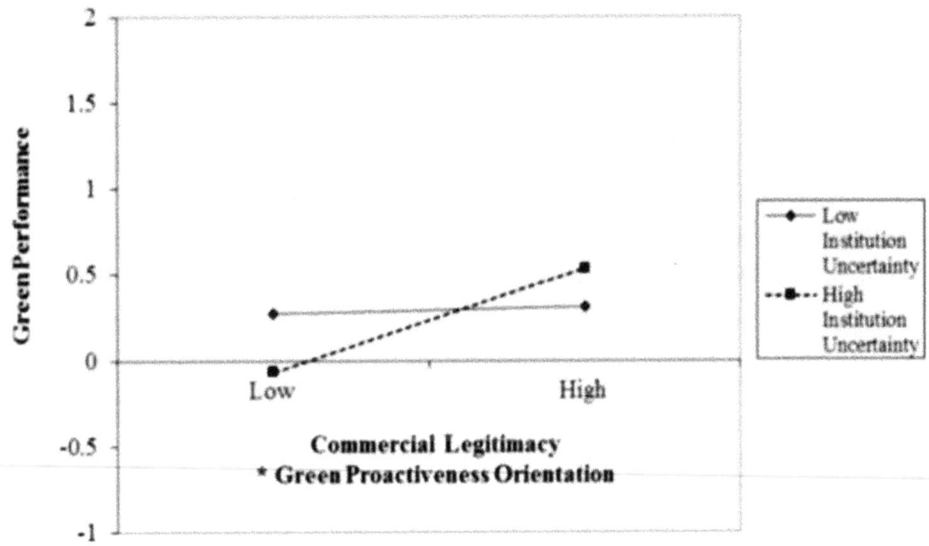

CONCLUSION

This chapter reviewed entrepreneurship theory, green entrepreneurship theory, legitimacy theory and transitional economy theory. And this chapter also verified the relationship between GPO and green performance empirically, the moderating effect of institutional legitimacy between the two and the influences of institutional uncertainty on the moderating effect of legitimacy. Our conclusion provides another perspective on GE, not only in systematically integrating institution theory and entrepreneurship theory but also enriching the entrepreneurial scenes, which contribute to green entrepreneurship theory, situation theory and legitimacy theory.

In addition, this chapter finds out that there is a positive correlation between GPO and green performance. It also confirms that an enterprise engaging in GE actions ahead of its competitors is able to gain the advantages in green performance, encompassing the gains in pollution emissions, energy costs and unexpected incidents. Besides, this chapter concludes that both political and commercial legitimacy play a promoting role in achieving green performance of proactive green-oriented enterprises. That is to say, the pressure exerted by legitimacy restrict is also a powerful force, and institutional restrict is a positive variable: green ventures need to not only gain recognition from government and regulators but also concentrate on the evaluations from suppliers, partners and customers. These two legitimacies are both institutional restrict and positive powers for proactive green enterprises to obtain green performance. Considering the transitional economy in China, institutional uncertainty would strength the promoting effects of CL between GPO and green performance while impairing that of PL. For GE needs to balance the relationship between business performance and social responsibility, it will be subject to the influences of institutional legitimacy. And the institutional uncertainty existing in transitional context can have an effect on the impacts of institutional legitimacy on green enterprises. This chapter sufficiently proves that when institutional uncertainty is high, Chinese green companies will tend to seek for'Guanxi', drill policy 'loopholes' and pursue profit and malicious competition. As a result, institutional uncertainty under the context of transitional economics will inhibit the relationship between GPO and green performance, even aggravating market uncertainty and making green strategies become no longer suitable for developing countries. Therefore, the findings will be vital warnings for the governments.

Conclusions of us enrich the entrepreneurship theory. Unlike the findings about the relationship between proactiveness orientation and entrepreneurial performance, a unique perspective emerged from the study that supports a particular conclusion - GPO and green performance have a significant positive correlation. Different from the causal relationship of 'legitimacy—resources—business growth', this chapter treated legitimacy as an institutionally embedded binding mechanism and explored how different levels of legitimacy affect the relationship between GPO and green performance. This chapter also added the factor of a transitional economy and revealed the influence of institutional uncertainty on the moderating effect of legitimacy on the relationship between GPO and green performance. This conclusion enriched entrepreneurship research scenarios, expanding and improving the universality of green entrepreneurship theory and institution theory.

REFERENCES

Al-Saleh, Y., & Mahroum, S. (2015). A critical review of the interplay between policy instruments and business models: Greening the built environment a case in point. *Journal of Cleaner Production, 109*, 260–270. doi:10.1016/j.jclepro.2014.08.042

Aldrich, H. E., & Fiol, C. M. (1994). Fools rush in? The institutional context of industry creation. *Academy of Management Review, 19*(4), 645–670.

Allen, J. C., & Malin, S. (2008). Green entrepreneurship: A method for managing natural resources? *Society & Natural Resources, 21*(9), 828–844. doi:10.1080/08941920701612917

Alonso-Almeida, M. D. M., Bremser, K., & Llach, J. (2015). Proactive and reactive strategies deployed by restaurants in times of crisis: Effects on capabilities, organization and competitive advantage. *International Journal of Contemporary Hospitality Management*, *27*(7), 1641–1661. doi:10.1108/IJCHM-03-2014-0117

Anderson, B. S., Kreiser, P. M., Kuratko, D. F., Hornsby, J. S., & Eshima, Y. (2015). Reconceptualizing entrepreneurial orientation. *Strategic Management Journal*, *36*(10), 1579–1596. doi:10.1002/smj.2298

Baron, R. M., & Kenny, D. A. (1986). The moderator–mediator variable distinction in social psychological research: Conceptual, strategic, and statistical considerations. *Journal of Personality and Social Psychology*, *51*(6), 1173–1182. doi:10.1037/0022-3514.51.6.1173 PMID:3806354

Barro, R. J. (1991). A cross-country study of growth, saving, and government. In *National saving and economic performance* (pp. 271–304). University of Chicago Press.

Becker-Blease, J. R., & Sohl, J. E. (2015). New venture legitimacy: The conditions for angel investors. *Small Business Economics*, *45*(4), 735–749. doi:10.1007/s11187-015-9668-7

Bentler, P. M., & Chou, C. P. (1987). Practical issues in structural modeling. *Sociological Methods & Research*, *16*(1), 78–117. doi:10.1177/0049124187016001004

Boulding, W., & Christen, M. (2001). First-mover disadvantage. *Harvard Business Review*, *79*(9), 20–21.

Brunetti, A., Kisunko, G., & Weder, B. (1997). *Institutions in transition: reliability of rules and economic performance in former socialist countries* (Vol. 9708). World Bank Publications.

Brunetti, A., & Weder, B. (1998). Investment and institutional uncertainty: A comparative study of different uncertainty measures. *Weltwirtschaftliches Archiv*, *134*(3), 513–533. doi:10.1007/BF02707928

Bruton, G., Khavul, S., Siegel, D., & Wright, M. (2015). New financial alternatives in seeding entrepreneurship: Microfinance, crowdfunding, and peer-to-peer innovations. *Entrepreneurship Theory and Practice*, *39*(1), 9–26. doi:10.1111/etap.12143

Cohen, B., & Winn, M. I. (2007). Market imperfections, opportunity and sustainable entrepreneurship. *Journal of Business Venturing*, *22*(1), 29–49. doi:10.1016/j.jbusvent.2004.12.001

Cope, J. (2005). Toward a dynamic learning perspective of entrepreneurship. *Entrepreneurship Theory and Practice, 29*(4), 373-397.

Cottrell, T., & Sick, G. (2002). Real options and follower strategies: The loss of real option value to first-mover advantage. *The Engineering Economist*, *47*(3), 232–263. doi:10.1080/00137910208965035

Covin, J. G., & Slevin, D. P. (1989). Strategic management of small firms in hostile and benign environments. *Strategic Management Journal*, *10*(1), 75–87. doi:10.1002/smj.4250100107

Czinkota, M., Kaufmann, H. R., & Basile, G. (2014). The relationship between legitimacy, reputation, sustainability and branding for companies and their supply chains. *Industrial Marketing Management*, *43*(1), 91–101. doi:10.1016/j.indmarman.2013.10.005

Dawson, J. F., & Richter, A. W. (2006). Probing three-way interactions in moderated multiple regression: Development and application of a slope difference test. *The Journal of Applied Psychology*, *91*(4), 917–926. doi:10.1037/0021-9010.91.4.917 PMID:16834514

Dean, T. J., & McMullen, J. S. (2007). Toward a theory of sustainable entrepreneurship: Reducing environmental degradation through entrepreneurial action. *Journal of Business Venturing*, *22*(1), 50–76. doi:10.1016/j.jbusvent.2005.09.003

Delmar, F., & Shane, S. (2004). Legitimating first: Organizing activities and the survival of new ventures. *Journal of Business Venturing*, *19*(3), 385–410. doi:10.1016/S0883-9026(03)00037-5

Deng, Q., & Ji, S. (2015). Organizational Green IT Adoption: Concept and Evidence. *Sustainability*, *7*(12), 16737–16755. doi:10.3390/su71215843

Doganova, L., & Karnøe, P. (2015). Clean and profitable: entangling valuations in environmental entrepreneurship. *Moments of Valuation: Exploring Sites of Dissonance*, 229-48.

Dowling, J., & Pfeffer, J. (1975). Organizational legitimacy: Social values and organizational behavior. *Pacific Sociological Review*, *18*(1), 122–136. doi:10.2307/1388226

Eshima, Y., & Anderson, B. S. (2016). Firm growth, adaptive capability, and entrepreneurial orientation. *Strategic Management Journal.*

Fisher, G., Kotha, S., & Lahiri, A. (2016). Changing with the times: An integrated view of identity, legitimacy, and new venture life cycles. *Academy of Management Review*, *41*(3), 383–409. doi:10.5465/amr.2013.0496

Fornell, C., & Larcker, D. F. (1981). Evaluating structural equation models with unobservable variables and measurement error. *JMR, Journal of Marketing Research*, *18*(1), 39–50. doi:10.2307/3151312

Fraser, S., & Greene, F. J. (2006). The effects of experience on entrepreneurial optimism and uncertainty. *Economica*, *73*(290), 169–192.

George, G., Robley Wood Jr, D., & Khan, R. (2001). Networking strategy of boards: Implications for small and medium-sized enterprises. *Entrepreneurship & Regional Development*, *13*(3), 269–285. doi:10.1080/08985620110058115

Gibbs, D., & ONeill, K. (2014). Rethinking sociotechnical transitions and green entrepreneurship: The potential for transformative change in the green building sector. *Environment & Planning A*, *46*(5), 1088–1107. doi:10.1068/a46259

Hall, J. K., Daneke, G. A., & Lenox, M. J. (2010). Sustainable development and entrepreneurship: Past contributions and future directions. *Journal of Business Venturing*, *25*(5), 439–448. doi:10.1016/j.jbusvent.2010.01.002

Hartman, C. L., & Stafford, E. R. (1997). Green alliances: Building new business with environmental groups. *Long Range Planning*, *30*(2), 184–149. doi:10.1016/S0024-6301(96)00111-2

Huang, M., & Wang, B. (2014). Evaluating green performance of building products based on gray relational analysis and analytic hierarchy process. *Environmental Progress & Sustainable Energy*, *33*(4), 1389–1395.

Huang, T., Wu, F., Yu, J., & Zhang, B. (2015). Political risk and dividend policy: Evidence from international political crises. *Journal of International Business Studies*, *46*(5), 574–595. doi:10.1057/jibs.2015.2

Isaak, R. (1997). Globalisation and green entrepreneurship. *Greener Management International*, 80-90.

Kelly, B., Pástor, Ľ., & Veronesi, P. (2016). The price of political uncertainty: Theory and evidence from the option market. *The Journal of Finance*, *71*(5), 2417–2480. doi:10.1111/jofi.12406

Kostova, T., & Zaheer, S. (1999). Organizational legitimacy under conditions of complexity: The case of the multinational enterprise. *Academy of Management Review*, *24*(1), 64–81.

Kreiser, P. M. (2011). Entrepreneurial orientation and organizational learning: The impact of network range and network closure. *Entrepreneurship Theory and Practice*, *35*(5), 1025–1050. doi:10.1111/j.1540-6520.2011.00449.x

Kuckertz, A., & Wagner, M. (2010). The influence of sustainability orientation on entrepreneurial intentions—Investigating the role of business experience. *Journal of Business Venturing*, *25*(5), 524–539. doi:10.1016/j.jbusvent.2009.09.001

Lankoski, L. (2000). *Determinants of environmental profit: An analysis of the firm-level relationship between environmental performance and economic performance*. Helsinki University of Technology.

Li, H., & Atuahene-Gima, K. (2001). Product innovation strategy and the performance of new technology ventures in China. *Academy of Management Journal*, *44*(6), 1123–1134. doi:10.2307/3069392

Li, X., Wang, C., & Xu, X. (2014). Rethinking new ventures cognitive legitimacy: An experimental study. *Systems Research and Behavioral Science*, *31*(3), 437–446. doi:10.1002/sres.2281

Li, Y., Peng, M. W., & Macaulay, C. D. (2013). *Market–political ambidexterity during institutional transitions*. Academic Press.

Lieberman, M. B., & Montgomery, D. B. (1988). First-mover advantages. *Strategic Management Journal*, *9*(S1), 41–58. doi:10.1002/smj.4250090706

Lin, C. S., Chang, R. Y., & Dang, V. T. (2015). An integrated model to explain how corporate social responsibility affects corporate financial performance. *Sustainability*, *7*(7), 8292–8311. doi:10.3390/su7078292

Lopez, L. E., & Roberts, E. B. (2002). First-mover advantages in regimes of weak appropriability: The case of financial services innovations. *Journal of Business Research*, *55*(12), 997–1005. doi:10.1016/S0148-2963(01)00200-4

Lounsbury, M., & Glynn, M. A. (2001). Cultural entrepreneurship: Stories, legitimacy, and the acquisition of resources. *Strategic Management Journal*, *22*(6-7), 545–564. doi:10.1002/smj.188

Lumpkin, G. T., & Dess, G. G. (1996). Clarifying the entrepreneurial orientation construct and linking it to performance. *Academy of Management Review*, *21*(1), 135–172.

Makadok, R. (1998). Can first-mover and early-mover advantages be sustained in an industry with low barriers to entry/imitation? *Strategic Management Journal*, *19*(7), 683–696. doi:10.1002/(SICI)1097-0266(199807)19:7<683::AID-SMJ965>3.0.CO;2-T

Maurer, J. G. (1971). *Readings in organization theory: Open-system approaches*. Random House.

Mauro, P. (1995). Corruption and growth. *The Quarterly Journal of Economics, 110*(3), 681–712. doi:10.2307/2946696

Mavragani, A., Nikolaou, I. E., & Tsagarakis, K. P. (2016). Open Economy, Institutional Quality, and Environmental Performance: A Macroeconomic Approach. *Sustainability, 8*(7), 601. doi:10.3390/su8070601

Meyer, J. W., & Scott, W. R. (1983). Centralization and the legitimacy problems of local government. *Organizational Environments: Ritual and Rationality, 199*, 215.

Miller, D. (2011). A reflection on EO research and some suggestions for the future. *Entrepreneurship Theory and Practice, 35*(5), 873–894. doi:10.1111/j.1540-6520.2011.00457.x

Miller, D., & Friesen, P. H. (1983). Strategy-making and environment: The third link. *Strategic Management Journal, 4*(3), 221–235. doi:10.1002/smj.4250040304

Morris, M. H., Kuratko, D. F., & Covin, J. G. (2010). *Corporate entrepreneurship & innovation*. Cengage Learning.

Murphy, G. B., Trailer, J. W., & Hill, R. C. (1996). Measuring performance in entrepreneurship research. *Journal of Business Research, 36*(1), 15–23. doi:10.1016/0148-2963(95)00159-X

Navis, C., & Glynn, M. A. (2011). Legitimate distinctiveness and the entrepreneurial identity: Influence on investor judgments of new venture plausibility. *Academy of Management Review, 36*(3), 479–499.

Ndubisi, N. O., & Nair, S. R. (2009). Green entrepreneurship (GE) and green value added (GVA): A conceptual framework. *International Journal of Entrepreneurship, 13*, 21.

Nelson, O. N., & Sumesh, R. N. (2009). Green entrepreneurship and green value added. *J. Int. J. Entrep, 13*, 21–31.

Oi, J. C. (1995). The role of the local state in Chinas transitional economy. *The China Quarterly, 144*, 1132–1149. doi:10.1017/S0305741000004768

Olsthoorn, X., Tyteca, D., Wehrmeyer, W., & Wagner, M. (2001). Environmental indicators for business: A review of the literature and standardisation methods. *Journal of Cleaner Production, 9*(5), 453–463. doi:10.1016/S0959-6526(01)00005-1

Orlitzky, M., Siegel, D. S., & Waldman, D. A. (2011). Strategic corporate social responsibility and environmental sustainability. *Business & Society, 50*(1), 6–27. doi:10.1177/0007650310394323

Palmås, K., Andersson Schwarz, J., & Larsson, S. (2014). The liability of politicalness: Legitimacy and legality in piracy-proximate entrepreneurship. *International Journal of Entrepreneurship and Small Business, 22*(4), 408–425. doi:10.1504/IJESB.2014.064269

Parsons, T. (1964). *Structure and process in modern societies*. Academic Press.

Peng, M. W., & Heath, P. S. (1996). The growth of the firm in planned economies in transition: Institutions, organizations, and strategic choice. *Academy of Management Review, 21*(2), 492–528.

Puffer, S. M., McCarthy, D. J., & Boisot, M. (2010). Entrepreneurship in Russia and China: The impact of formal institutional voids. *Entrepreneurship Theory and Practice, 34*(3), 441-467.

Rockström, J., Lannerstad, M., & Falkenmark, M. (2007). Assessing the water challenge of a new green revolution in developing countries. *Proceedings of the National Academy of Sciences of the United States of America, 104*(15), 6253–6260. doi:10.1073/pnas.0605739104 PMID:17404216

Rottig, D., & Reus, T. H. (2009, August). Institutional distance, organizational legitimacy, and the performance of foreign acquisitions in the United States. In Academy of Management Proceedings (Vol. 2009, No. 1, pp. 1-6). Academy of Management.

Ryszko, A. (2016). Proactive Environmental Strategy, Technological Eco-Innovation and Firm Performance—Case of Poland. *Sustainability, 8*(2), 156. doi:10.3390/su8020156

Schaltegger, S. (2002). A Framework for Ecopreneurship. *Greener Management International, 38*(38), 45–58. doi:10.9774/GLEAF.3062.2002.su.00006

Schepers, J., Voordeckers, W., Steijvers, T., & Laveren, E. (2014). The entrepreneurial orientation–performance relationship in private family firms: The moderating role of socioemotional wealth. *Small Business Economics, 43*(1), 39–55. doi:10.1007/s11187-013-9533-5

Scherer, A. G., Palazzo, G., & Seidl, D. (2013). Managing legitimacy in complex and heterogeneous environments: Sustainable development in a globalized world. *Journal of Management Studies, 50*(2), 259–284. doi:10.1111/joms.12014

Scott, W. R. (1995). *Institutions and organizations* (Vol. 2). Thousand Oaks, CA: Sage.

Shane, S. (2000). Prior knowledge and the discovery of entrepreneurial opportunities. *Organization Science, 11*(4), 448–469. doi:10.1287/orsc.11.4.448.14602

Shane, S., & Venkataraman, S. (2000). The promise of entrepreneurship as a field of research. *Academy of Management Review, 25*(1), 217–226.

Sheng, S., Zhou, K. Z., & Li, J. J. (2011). The effects of business and political ties on firm performance: Evidence from China. *Journal of Marketing, 75*(1), 1–15. doi:10.1509/jmkg.75.1.1

Shepherd, D. A., & Patzelt, H. (2011). The new field of sustainable entrepreneurship: Studying entrepreneurial action linking what is to be sustained with what is to be developed. *Entrepreneurship Theory and Practice, 35*(1), 137–163. doi:10.1111/j.1540-6520.2010.00426.x

Shirokova, G., Bogatyreva, K., Beliaeva, T., & Puffer, S. (2016). Entrepreneurial orientation and firm performance in different environmental settings: Contingency and configurational approaches. *Journal of Small Business and Enterprise Development, 23*(3), 703–727. doi:10.1108/JSBED-09-2015-0132

Silajdžić, I., Kurtagić, S. M., & Vučijak, B. (2015). Green entrepreneurship in transition economies: A case study of Bosnia and Herzegovina. *Journal of Cleaner Production, 88*, 376–384. doi:10.1016/j.jclepro.2014.07.004

Simón-Moya, V., Revuelto-Taboada, L., & Guerrero, R. F. (2014). Institutional and economic drivers of entrepreneurship: An international perspective. *Journal of Business Research, 67*(5), 715–721. doi:10.1016/j.jbusres.2013.11.033

Singh, J. V., Tucker, D. J., & Meinhard, A. G. (1991). Institutional change and ecological dynamics. *The new institutionalism in organizational analysis*, 390-422.

Soto-Acosta, P., Cismaru, D. M., Vătămănescu, E. M., & Ciochină, R. S. (2016). Sustainable Entrepreneurship in SMEs: A Business Performance Perspective. *Sustainability*, *8*(4), 342. doi:10.3390/su8040342

Stam, W., & Elfring, T. (2008). Entrepreneurial orientation and new venture performance: The moderating role of intra-and extraindustry social capital. *Academy of Management Journal*, *51*(1), 97–111. doi:10.5465/AMJ.2008.30744031

Starr, J. A., & MacMillan, I. C. (1990). Resource cooptation via social contracting: Resource acquisition strategies for new ventures. *Strategic Management Journal*, 79–92.

Stevens, C. E., Xie, E., & Peng, M. W. (2015). Toward a legitimacy-based view of political risk: The case of Google and Yahoo in China. *Strategic Management Journal*.

Suchman, M. C. (1995). Managing legitimacy: Strategic and institutional approaches. *Academy of Management Review*, *20*(3), 571–610.

Tabachnick, B. G., Fidell, L. S., & Osterlind, S. J. (2001). *Using multivariate statistics*. Academic Press.

Tan, J. (2001). Innovation and risk-taking in a transitional economy: A comparative study of Chinese managers and entrepreneurs. *Journal of Business Venturing*, *16*(4), 359–376. doi:10.1016/S0883-9026(99)00056-7

Vandaie, R., & Zaheer, A. (2014). Surviving bear hugs: Firm capability, large partner alliances, and growth. *Strategic Management Journal*, *35*(4), 566–577. doi:10.1002/smj.2115

Wagner, M., & Schaltegger, S. (2004). The effect of corporate environmental strategy choice and environmental performance on competitiveness and economic performance: An empirical study of EU manufacturing. *European Management Journal*, *22*(5), 557–572. doi:10.1016/j.emj.2004.09.013

Walder, A. G. (1995). Local governments as industrial firms: An organizational analysis of Chinas transitional economy. *American Journal of Sociology*, *101*(2), 263–301. doi:10.1086/230725

Wales, W. J. (2016). Entrepreneurial orientation: A review and synthesis of promising research directions. *International Small Business Journal*, *34*(1), 3–15. doi:10.1177/0266242615613840

Wang, C. L., & Ahmed, P. K. (2007). Dynamic capabilities: A review and research agenda. *International Journal of Management Reviews*, *9*(1), 31–51. doi:10.1111/j.1468-2370.2007.00201.x

Waylen, G. (2014). Informal institutions, institutional change, and gender equality. *Political Research Quarterly*, *67*(1), 212–223. doi:10.1177/1065912913510360

Westhead, P., Ucbasaran, D., & Wright, M. (2005). Experience and cognition: Do novice, serial and portfolio entrepreneurs differ? *International Small Business Journal*, *23*(1), 72–98. doi:10.1177/0266242605049104

Wiklund, J., & Shepherd, D. (2003). Knowledge-based resources, entrepreneurial orientation, and the performance of small and medium-sized businesses. *Strategic Management Journal*, *24*(13), 1307–1314. doi:10.1002/smj.360

Wu, K. J., Liao, C. J., Tseng, M. L., & Chou, P. J. (2015). Understanding innovation for sustainable business management capabilities and competencies under uncertainty. *Sustainability*, *7*(10), 13726–13760. doi:10.3390/su71013726

Yang, C. S., Lu, C. S., Haider, J. J., & Marlow, P. B. (2013). The effect of green supply chain management on green performance and firm competitiveness in the context of container shipping in Taiwan. *Transportation Research Part E, Logistics and Transportation Review*, *55*, 55–73. doi:10.1016/j.tre.2013.03.005

Zhu, Q., & Sarkis, J. (2004). Relationships between operational practices and performance among early adopters of green supply chain management practices in Chinese manufacturing enterprises. *Journal of Operations Management*, *22*(3), 265–289. doi:10.1016/j.jom.2004.01.005

Zikmund, W. G., Babin, B. J., Carr, J. C., & Griffin, M. (2013). *Business research methods*. Cengage Learning.

Zimmerman, M. A., & Zeitz, G. J. (2002). Beyond survival: Achieving new venture growth by building legitimacy. *Academy of Management Review*, *27*(3), 414–431.

Chapter 12
Improving Renewable Energy Source in Automotive Applications:
Evaluation, Analysis, and Design

Wei Liu
University of Sydney, Australia

Hongbo Shi
Harbin Institute of Technology, China

Hailing Ma
Jinggangshan University, China

Sang-Bing Tsai
University of Electronic Science and Technology of China (Zhongshan Institute), China & Civil Aviation University of China, China

ABSTRACT

With the rapid development of automobile industry, people have become more and more aware of the shortage of the traditional automobile energy and the irreversible pollution to the environment. Sustainable development will be the first place in the automobile industry, but there is no clear evidence to identify the most effective specific solution. This chapter is divided into two parts, the first part aims to analyze the evaluation of automotive sustainable development by synthesis. A new hybrid multiple criteria decision making (MCDM) model that combines DEMATEL technique and DANP method is proposed. By analysis and comparison, the result shows that renewable energy source is the main influencing criteria that affects the automobile sustainable development. Based on this, the second part of this chapter provides an effective solution. An advanced energy storage based on the super capacitor of self-generating wireless charging is designed to be used in automotive applications. It has advantages of short charging time, high power conversion efficiency and low power consumption.

DOI: 10.4018/978-1-5225-3537-9.ch012

INTRODUCTION

The automobile is one of the most important vehicles in the world, and its invention has greatly affected people's lives. Growing demand for automobiles has resulted in a large vehicle population. As of 2015, the total number of vehicles in use worldwide, including cars, trucks and buses, exceeded 1 billion. The number is expected to exceed 2 billion by 2030 (Gao, Hensley and Zielke 2014). The vigorous development of the automobile is providing an ever more convenient mode of transport that saves time, money and improves efficiency.

The automobile industry has become one of the pillar industries of the national economy, and accounts for a large proportion of the output value and sales revenue. As a comprehensive assembly industry, it produces in large quantities. In this way, huge numbers of jobs can be provided in order to help the government solve employment issues. Meanwhile, the technological development of the automotive industry has also been promoting the progress of human science and technology. For example, high-end technology developed for winning the Formula One competition has been successfully transferred to other modes of transportation such as passenger cars and high-speed railway. In addition, the development of the automotive industry has greatly improved the efficiency of human work by saving time. Therefore, the automobile industry provides the benefits from different perspectives and social and technological development is inseparable from the automotive industry.

However, the accelerated development of the global automotive industry has led to a significant increase in greenhouse gas emissions. Environmental pollution is increasing, accompanied by calls for more environmental protection. At the same time, the shortage of international energy supplies has led to the continuing rise of international crude oil prices. Energy saving and environmental protection are the pressing issues dominating the automobile industry around the world. Different countries have gradually paid attention to the field of new energy vehicles in their R & D and industrialization. As the world's largest developing country and also one of the most important countries regarding energy production and consumption, China's rapid economic development has led to increasing demand for energy and dependence on oil imports. Transportation energy consumption is gradually increasing. With the rapid growth of car ownership, this proportion will continue to rise. So in the long run, national energy security issues will be more serious. Energy consumption has brought serious environmental problems. In recent years, air pollution has become a serious problem. An increasing number of people are beginning to realize the impacts of air pollution and the automobile as the primary cause of air pollution in large cities. Therefore, energy security and environmental pollution have become the bottleneck restricting the sustainable development of one country's economy. Under such pressure, the government attaches great importance to renewable energy in the automotive industry and gradually introduced a variety of programs and policies to encourage and promote the development of a new energy automotive industry, and its market penetration.

The promotion of renewable energy vehicles is designed to reduce environmental pollution and energy shortages. Investment in renewable energy sources in automotive applications can avoid damage to the environment due to the use of renewable energy as a driving force to reduce pollutant emissions. However, a large external cost exists in investing in renewable energy. As for automobile manufacturers, the development of renewable energy technologies not only has technical risk and requires large-scale business restructuring, but also needs enormous investment. That is to say, the producers bear a greater cost, but the benefits (environmental protection and energy conservation) can be shared by the whole of society. So this system imbalance means that manufacturers will not take the initiative to invest. In the initial stage, the market environment and mechanism is imperfect and the government needs to regulate resources and implement policies to guide the development of renewable energy use in the automobile industry.

The premise of the development of renewable energy use in the automobile industry is a comprehensive evaluation analysis as a guide to correct decision management, especially in the first stage of development. A comprehensive evaluation analysis aims to find the influencing factors of development and provide a specific method for effective improvement. The existing evaluations of renewable energy use in the automobile mainly focus on three aspects, technical, economic and social, but only reflect on the level of sustainable management and decision-making. As a highly concentrated industrial product, the evaluation based on the technical level is more effective and direct and this is the research gap that we need to fill with this study.

In this study, we argue that a comprehensive evaluation analysis which combines management level and technical level is needed in the development of renewable energy use in the automobile industry, and a new technical design is proposed after the evaluation in order to better solve this problem. In this way, firstly, we review the literature on sustainable management and sustainable development, development of automobile sustainable technology, and theory and method of comprehensive evaluation. Then, criteria are chosen and criteria determination for the evaluation index system is designed. Via the evaluation index system, we design a new hybrid multi criteria model that combines DEMATEL and DEMATEL-based ANP (DANP) for the evaluation, and a case study is produced in order to analyze the development of renewable energy use in the automobile industry. Last, the design of renewable energy technology for improving sustainable development in the automotive industry is proposed. Through a new design of advanced energy storage based on the super capacitor of self-generating wireless charging, the automobile application of renewable energy has the advantages of short charging time, high power conversion efficiency and low power consumption. Therefore, this chapter provides mixed research from management and technology perspectives to provide a more effective comprehensive evaluation method and a technical method of improving renewable energy technology in the automobile industry.

REVIEW OF AUTOMOBILE SUSTAINABLE DEVELOPMENT AND ITS EVALUATION

Review of Automobile Sustainable Development

Transportation energy consumption accounts for a large proportion of the total consumption of petroleum resources. The statistical data from the International Energy Agency (IEA) indicates that 58% of the world's oil consumption was used in the transportation sector in 2014, and is estimated to account for more than 62% of the total oil consumption in 2020. The energy crisis will be more serious in the future, so the traditional traffic energy power system must be reconsidered, and more and more countries have sped up the development of renewable energy technology to achieve the transformation of their transportation energy power systems.

Alternative energy sources that can be used as renewable energy sources include gaseous fuels, coal-based liquid fuels, bio-based liquid fuels, hydrogen, electricity, solar, nuclear, efficient energy storage, etc. In the long term, the main vehicle renewable energy sources will be hydrogen, electricity, solar and efficient energy storage. We will introduce some matured renewable energy technologies in the automobile industry.

Battery Electric Vehicle (BEV)

BEVs use electric power as the driving force, with wheels driven by an electric motor. BEVs do not use the internal combustion engine as a launch device, or use gasoline, diesel or other fuels, only a rechargeable battery. The basic structure is very simple; while an electric generator and battery are the key components, the battery is the most critical.

BEV has the advantages of being pollution-free, low noise, energy efficient and easy to maintain. However, the achieved mileage is not ideal, there is short battery life and high battery cost, so it is difficult to bring BEVs into commercial operation, and they can only be used in some limited places. Currently, BEV technology is in its infancy.

Hybrid Electric Vehicle (HEV)

HEV's system consists of two or more simultaneous operations of a single drive system. The driving power is provided by a single drive system alone or jointly according to the actual vehicle driving state. According to the hybrid-driven connection, general HEVs are divided into three categories: Series Hybrid Electric Vehicle (SHEV), Parallel Hybrid Electric Vehicle (PHEV) and Parallel-series Hybrid Electric Vehicle (PSHEV).

The most prominent feature of HEVs is fuel economy, as they typically use 50% less fuel than traditional vehicles and reduce emissions simultaneously. HEVs also have the problems on high prices, low efficiency and dependence on gasoline or diesel. However, HEVs are the most mature renewable energy technology for automobiles and have been put on the market. Many automobile manufacturers, such as Toyota (Prius, Japan), Honda (Insight, Japan), Buick (LaCROSS, USA), BYD (F3DM, China) have mastered the entire set of hybrid electric technology and consumers can choose them in passenger vehicles. It can be said that HEVs are the best choice in the transition from traditional vehicles to BEVs.

Hydrogen-Powered Vehicle

Hydrogen-powered vehicles get their power from the combustion of hydrogen and air based on a transformation of the traditional engine. Hydrogen-powered vehicles are not only pollution-free and have low emissions but also have lower requirements of hydrogen, high combustion performance, internal combustion engine technology and so on. However, current developmental problems of hydrogen-powered vehicles include efficiency and cost. Vehicles can only use hydrogen energy over the distance of 250 kilometers and it is currently difficult to fill vehicles with hydrogen anywhere. Otherwise, obtaining hydrogen by electrolysis of water is prohibitively expensive. So hydrogen-powered vehicles need to be improved and their commercialization will not come soon.

Fuel Cell Vehicles (FCV)

FCVs are powered by a vehicle fuel cell. The fuel used in the on-board fuel cell device is a high hydrogen-containing reformed gas obtained by reforming high purity hydrogen or hydrogen-containing fuel. Compared with the usual electric vehicle, the difference is that the power used by the FCV comes from the on-board fuel cell device, and the power used by the electric vehicle comes from the battery charged by the grid. So the key to FCV is the fuel cell. FCVs have the advantages of being pollution-free, energy efficient, quiet, powerful and having good control systems. However, current FCV technology is still in the trial stage and it will be a long time before commercialization.

Review of Sustainable Development Theory and Sustainable Evaluation

Sustainable development is defined as development that can not only meet the needs of contemporary people but also not pose a hazard to the ability of future generations to meet their needs (World Commission on Environment and Development 1987). In recent years, as energy crises and environmental problems have become increasingly prominent, the concept of sustainable development is gaining more and more attention. Sustainable development can slow down the human energy crisis, protect the human environment and organically combine environmental issues and development issues to become a comprehensive strategy of social-economic development (Giddings, Hopwood and O'Brien 2002).

The implementation of sustainable development in the automobile industry depends on the design. Automobiles should be designed with the concept of sustainable development in mind in the first stages to ensure renewable and sustainable characteristics. The product design stage is the process to determine the product structure, specifications and production system layout, so it is important in achieving sustainable development. Curtis and Walker (2001) pointed out that sustainable design is a balance of social, ethical, environmental and economic factors involved in the development of a product or service. From an economic perspective, sustainable design of products creates greater market competitiveness and greater market efficiency. From an environmental protection perspective, sustainable design of products can control the impact of products on the environment. From a social perspective, sustainable design of products meets the needs of people in the stages of product manufacturing, circulation, utilization and disposal.

In order to achieve the sustainable development of automobiles, we need to analyze the life cycle of automobiles. Sundin (2004) divided the automobile life cycle into four stages: the raw material extraction stage, the manufacturing stage, the utilize stage and the disposal stage. Ashby (2009) added the transportation phase to the automobile life cycle. Mildenberger and Khare (2000) analyzed the automobile life cycle process from the raw material stage to the disposal stage and found that the key aspect is the design phase which directly determines the sustainability (economy, environment and society) of raw material consumption, sales and transportation, utilization, maintenance and disposal. Therefore, the design stage should take full account of sustainability factors. The sustainable concept of automobiles must be reflected in the design stage and will be of great significance on the impact on economy, environment and society throughout the life cycle.

The United Nations Environment Programme (UNEP) (2012) defined the Life Cycle Sustainability Assessment (LCSA) as the environmental, social and economic impact of the product's entire lifecycle process in the decision-making process for product sustainability. Sustainability assessment is a systematic and cyclic evaluation process that is parallel to planning and strategy development. Through this process, sustainability and sustainability of planning or strategy are defined in terms of environmental, economic and social goals, and ultimately promote the planning, strategy and policy to the direction of sustainable development by a series of reports (Boothroyd and Dewhurst 1994). Therefore, in order to evaluate the sustainability of the product, it is necessary to find a reasonable and scientific method of evaluation, so as to ensure that the designed products are sustainable and conducive to the sustainable development of the industry.

Figure 1. Steps of LCA
Source: ISO 14040

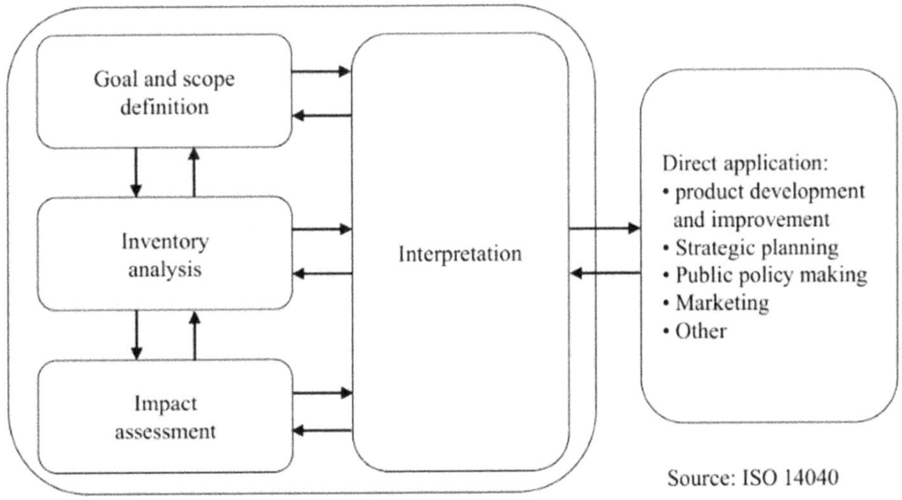

Life Cycle Assessment (LCA) is one of the core elements of sustainable evaluation. Life cycle assessment can be achieved through the use and combination of independent life cycle assessment techniques through environmental, economic and social lifecycle evaluation techniques, and it is possible to integrate these technologies into a comprehensive Life Cycle Sustainability Assessment (LCSA). LCA is divided into three categories: Environment Life Cycle Assessment (E-LCA), Life Cycle Cost (LCC) and Social Life Cycle Assessment (S-LCA). E-LCA is used to evaluate the environmental impact of the product or service from the cradle to the grave, taking into account the potential environmental impact of the product (Hemdi, Saman and Sharif 2009). Kapur et al. (2012) conducted a life cycle assessment of traditional cleaning products and green cleaning products. The result showed that traditional cleaning products are less environmentally friendly than green cleaning products. Warsen, Laumer and Momberg (2011) used the LCA method to evaluate the lifecycle of the new manual and remanufactured manual transmissions, and the result showed that the full life cycle energy consumption of the remanufactured transmission is 33% less than that of the new manual transmission. Puri, Compston and Pantano used the E-LCA method to evaluate the life cycle of three different car doors, illustrating the raw material consumption and environmental emissions associated with the three doors. Hunkeler and Rebitzer (2005) argued that life cycle assessment should not only be environmentally oriented, but also economically and socially oriented. The essence of LCC is to consider the economic cost of the product during the life cycle. Atling (1993) argued that LCC is made up of manufacturer costs, user costs, and social costs. Cost is the sum of the cost of the product's entire life cycle, such as raw materials, assembly, transportation, operation and maintenance and disposal. After comparing the cost efficiencies of different business decisions or policy makers' views, only costs and benefits of the product material life cycle are added in the product or system life cycle (Hu, Fang and Ben 2004). S-LCA mainly examines the potential social impact of each stage of the product's life cycle. Becker and Vanclay (2003) first proposed the concept of social impact assessment. Weidema (2006) also suggested the idea of integrating sociality into life-cycle evaluations, examining the social nature of the product from the perspectives of health, safety, freedom and equality. Labuschagne and Brent (2006) proposed social impact indicators for the evaluation of the processing of industrial projects and technologies. According to the 'Guidelines for Social Life Cycle Assessment of Products' published by UNEP, the evaluation of S-LCA is based on subclasses. Subclasses are important, measurable attributes of sociality. It can usually be divided into five categories according to the different types of stakeholder: workers, local communities, society, consumers and value chain participants. Subclasses can also be divided into 6 according to the type of impact: human rights, working conditions, health and safety, cultural heritage, management and socio-economic impacts. Subclasses can be analyzed by inventory and measured by certain indicators. Therefore, the main task of S-LCA is to collect and calculate the input and output of a product system and its potential social impact.

To sum up, life-cycle sustainability assessments should include an assessment of the potential environmental, economic and social impacts of the product or system life cycle, which will help industry, government, or non-governmental organizations acquire technologies based on sustainable decision-making.

EVALUATION OF AUTOMOBILE SUSTAINABLE DEVELOPMENT

Criteria Determination for Evaluation Index System

Evaluation of automotive sustainable development is a complex engineering system. Evaluation indicators are the basic composition of the evaluation system, so we should follow the following principles in the selection.

Systematic Principle

Indicators should be selected around the core of sustainability and introduce the sustainability of automotive products from economic, environmental, social perspectives. According to the inherent logical relationship between the index hierarchy, the hierarchical structure can be established.

Operational Principle

Indicators should be clear, easy to understand, and facilitate the collection of data. Index classification and content should not be too cumbersome and lengthy. As far as possible indicators should be streamlined to build a reasonable system to achieve the overall optimal function of the target.

Openness Principle

First, the established indicators should allow for appropriate adjustments to the weight of different regions, rather than be immutable. Second, indicators are in dynamic development, and the index system should be amended as people continue to strengthen their consciousness of economic and social development.

In this chapter, we follow the requirements of sustainable development principles and establish the integrated system model with economic, environmental and social coordination. The evaluation index system is shown in Table 1.

Table 1. Evaluation index system

Dimensions	Criteria	Descriptions
Economy (D_1)	• Direct Cost (C_{11}) • Indirect Cost (C_{12}) • Profit (C_{13})	• Refers to the cost that can be directly included in the production process. • Refers to the cost that cannot be directly included in the process of organization and management. • Refers to the operating income of business in a certain period of time.
Environment (D_2)	• Resource Consumption (C_{21}) • Energy Consumption (C_{22}) • Environmental Pollution (C_{23}) • Recycling Disposal (C_{24})	• Refers to the consumption of raw materials used in the production stage. • Refers to the energy consumption of the product during usage. • Refers to the environmental pollution of the product in the process of production and usage. •Refers to whether the product is easily dismantled, whether it can be reused or recycled, ease of separating harmful substances.
Society (D_3)	• Economic Contribution (C_{31}) • Social Welfare (C_{32}) • Environmental Responsibility (C_{33})	• Refers to the promotion of local employment and the development of local economy. • Refers to the social commitments of sustainable development and environmental security. • Refers to the disposal of waste, the level of resource consumption and the development of new energy technologies.

Constructing a New Hybrid MCDM Model for Development Evaluation

This research uses the DEMATEL technique and combines a DANP method to establish a new hybrid MCDM model to address the problems of interdependence and feedback among certain criteria. The DEMATEL technique is used to build an INRM, and the DANP is expected to obtain the influential weights using the basic concept of ANP. The research processes are concisely illustrated in Figure 2.

Model Calculation Process

The DEMATEL Technique for Building INRM

1. Calculate the direct impact matrix of scores

Experts are asked to assess the relationship between the standards of the relationship between the individual in the sense that in the use of an integer range from 4 to 0, the score is expressed by the natural

Figure 2. The New Hybrid Index Model of the Research Process

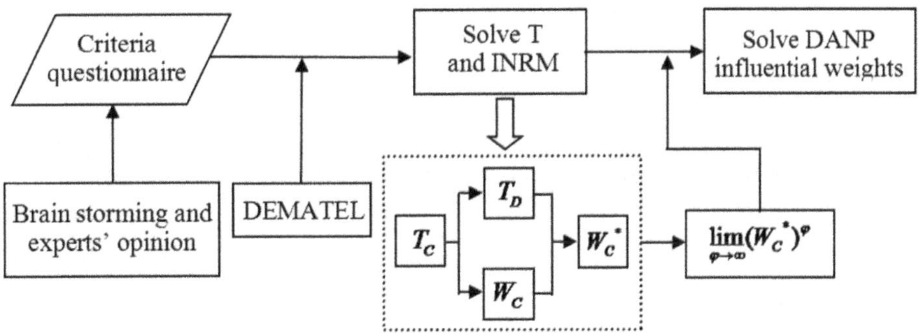

language: "influence (4)", "high impact (3)", "media (2)", "low impact (1)" and "absolutely no effect (0)". The direct effect of being compared by two means means that if the expert thinks that the standard has an impact on the standard, they should show it. Then the direct relationship of the matrix can be expressed as Equation 1.

$$D = \begin{pmatrix} d_{11} \cdots d_{1j} \cdots d_{1n} \\ \vdots \quad \vdots \quad \vdots \\ d_{i1} \cdots d_{ij} \cdots d_{in} \\ \vdots \quad \vdots \quad \vdots \\ d_{n1} \cdots d_{nj} \cdots d_{nn} \end{pmatrix} \tag{1}$$

2. Standardized Direct Impact Matrix R.

Using matrix D, the normalized matrix $R = [r_{ij}]_{nxn}$ can be calculated by Equation 2 and Equation 3 as below.

$$R = v \text{ x } D \tag{2}$$

$$v = \min\left\{\frac{1}{\max_i \sum_{j=1}^{n} d_{ij}}, \frac{1}{\max_j \sum_{i=1}^{n} d_{ij}}\right\}, i, j \in \{1, 2, \ldots n\} \tag{3}$$

3. Obtain the total influence matrix T.

The total influence matrix T for building the influential network relationship map (INRM) can be obtained from Equation 4, in which *I* denotes the identity matrix.

$$\mathrm{T} = \mathrm{R} + \mathrm{R}^2 + \ldots + \mathrm{R}^{\mathrm{X}} = \mathrm{R}(\mathrm{I} - \mathrm{R}^{\mathrm{X}})(\mathrm{I} - \mathrm{R})^{-1} \tag{4}$$

Then,

$$\mathrm{T} = \mathrm{R}(\mathrm{I} - \mathrm{R})^{-1}$$

When $\lim_{X \to \infty} R^X$, $T = [0]_{n \times n}$ where $R = [r_{ij}]_{n \times n}$, $0 \le R_{ij} < 1$, $0 \le \sum_i R_{ij} \le 1$, $0 \le \sum_j R_{ij} \le 1$ and at least one row or one column of the summation, but not each, equals 1. Accordingly, $\lim_{X \to \infty} R^X = [0]_{n \times n}$ is guaranteed naturally.

4. Analyze the results and build the INRM.

After obtaining the total influence matrix T, its sum of rows and columns can be expressed respectively as vector r and vector c which come from Equation 5 and Equation 6.

$$r = \left[\sum_{j=1}^{n} t_{ij}\right]_{n\times 1} = \left[t_i\right]_{n\times 1} = (r_1,\ldots,r_i,\ldots,r_n)',\ j \in \{1,2,\ldots,n\} \tag{5}$$

$$c = \left[\sum_{i=1}^{n} t_{ij}\right]'_{1\times n} = \left[t_j\right]_{n\times 1} = (c_1,\ldots,c_j,\ldots,c_n)',\ j \in \{1,2,\ldots,n\} \tag{6}$$

In the equation the superscript ' denotes the transpose. Because of the same number of elements in vector r and vector c, r+c and r - c can stablish two column vectors as Equation 7 and Equation 8.

$$r + c = (r_1 + c_1,\ldots,\ r_i + c_i,\ldots,r_n + c_n)',\ i \in \{1,2,\ldots,n\} \tag{7}$$

$$r - c = (r_1 - c_1,\ldots,\ r_i - c_i,\ldots,r_n - c_n)',\ i \in \{1,2,\ldots,n\} \tag{8}$$

r_i show standard direct and indirect effects on other criteria, indicating that the sum of the direct and indirect effects of the standard has been received from other criteria. In addition, if the total amount of the impact of the standard in this system is positive, then the standard net effect of other standards, if it is negative, then the standard is subject to other standards. Thus, the causal map can be mapped by the data set that is called INRM. INRM can provide a valuable idea to improve the standard.

Combing the ANP Method for Determining the Criterion's Influence Weights

1. Find the normalized matrix T_c^a by using dimensions and take the normalization of T_c^{a11} for instance.

In this part, we specify the total influence matrix T_C by the criteria and T_D by the dimensions. $T_C = \left[t_D^{ij}\right]_{n\times n}$, $T_D = \left[t^{ij}\right]_{m\times m}$ and T_D can be obtained from T_C. So, the total influence matrix T_C is shown as Equation 9.

After obtaining T_C, the normalized total influence matrix T_c^a for the criteria can be shown easily as Equation 10.

Take the normalization of T_c^{a11} based on $a11$ for instance and it is shown by Equation 11 and Equation 12.

$$d_{ci}^{11} = \sum_{j=1}^{m_1} t_{ij}^{11},\ i = 1,2,\ldots,m_1 \tag{11}$$

Box 1. Equation 9

$$
T_c = \begin{array}{c} \\ D_1 \\ \vdots \\ D_i \\ \vdots \\ D_m \end{array}
\begin{array}{c} \\ c_{11} \; c_{12} \cdots c_{1n_1} \\ \vdots \\ c_{i1} \; c_{i2} \cdots c_{in_i} \\ \vdots \\ c_{m1} \; c_{m2} \cdots c_{mn_m} \end{array}
\overset{\begin{array}{ccccc} D_1 & \cdots & D_j & \cdots & D_m \\ c_{11} \cdots c_{1n_1} & \cdots & c_{j1} \cdots c_{jn_j} & \cdots & c_{m1} \cdots c_{mn_m} \end{array}}{\begin{bmatrix} T_c^{11} & \cdots & T_C^{1j} & \cdots & T_C^{1m} \\ \vdots & & \vdots & & \vdots \\ T_C^{i1} & \cdots & T_C^{ij} & \cdots & T_C^{im} \\ \vdots & & \vdots & & \vdots \\ T_C^{m1} & \cdots & T_C^{mj} & \cdots & T_C^{mm} \end{bmatrix}}
$$

Box 2. Equation 10

$$
T_C^a = \begin{array}{c} \\ D_1 \\ \vdots \\ D_i \\ \vdots \\ D_m \end{array}
\begin{array}{c} \\ c_{11} \; c_{12} \cdots c_{1n_1} \\ \vdots \\ c_{i1} \; c_{i2} \cdots c_{in_i} \\ \vdots \\ c_{m1} \; c_{m2} \cdots c_{mn_m} \end{array}
\overset{\begin{array}{ccccc} D_1 & \cdots & D_j & \cdots & D_m \\ c_{11} \cdots c_{1n_1} & \cdots & c_{j1} \cdots c_{jn_j} & \cdots & c_{m1} \cdots c_{mn_m} \end{array}}{\begin{bmatrix} T_c^{a11} & \cdots & T_C^{a1j} & \cdots & T_C^{a1m} \\ \vdots & & \vdots & & \vdots \\ T_C^{ai1} & \cdots & T_C^{aij} & \cdots & T_C^{aim} \\ \vdots & & \vdots & & \vdots \\ T_C^{am1} & \cdots & T_C^{amj} & \cdots & T_C^{amm} \end{bmatrix}}
$$

2. Obtain the unweighted supermatrix W_c.

The replacement of the supermatrix W_c can be obtained by normalizing the total coefficient of influence matrix T_c^a by using dimensions because of the interdependence between the clusters and dimensions. Based on the basic concept of ANP, supermatrix $W_c = (T_c^a)'$ is shown as Equation 13.

In this equation, the supermatrix W_c has a case that its dimensions and criteria are independent if a blank or zero shows in the matrix.

3. Obtain the weighted normalized supermatrix W_c^*.

In order to obtain the weighted normalized supermatrix W_c^*, the unweighted supermatrix W_c is the basis and the normalized total influence matrix T_D^a is the essence. We can get W_c^*, multiplying T_D^a by

Box 3. Equation 12

$$T_C^{a11} = \begin{bmatrix} \frac{t_{C11}^{11}}{d_{C1}^{11}} & \cdots & \frac{t_{C1j}^{11}}{d_{C1}^{11}} & \cdots & \frac{t_{C1m1}^{11}}{d_{C1}^{11}} \\ \vdots & & \vdots & & \vdots \\ \frac{t_{Ci1}^{11}}{d_{Ci}^{11}} & \cdots & \frac{t_{Cij}^{11}}{d_{Ci}^{11}} & \cdots & \frac{t_{Cim_1}^{11}}{d_{Ci}^{11}} \\ \vdots & & \vdots & & \vdots \\ \frac{t_{Cm_11}^{11}}{d_{Cm_1}^{11}} & \cdots & \frac{t_{Cm_1j}^{11}}{d_{Cm_1}^{11}} & \cdots & \frac{t_{Cm_1m_1}^{11}}{d_{Cm_1}^{11}} \end{bmatrix} = \begin{bmatrix} t_{C11}^{a11} & \cdots & t_{C1j}^{a11} & \cdots & t_{C1m_1}^{a11} \\ \vdots & & \vdots & & \vdots \\ t_{Ci1}^{a11} & \cdots & t_{Cij}^{a11} & \cdots & t_{Cim_1}^{a11} \\ \vdots & & \vdots & & \vdots \\ t_{Cm_11}^{a11} & \cdots & t_{Cm_1j}^{a11} & \cdots & t_{Cm_1m_1}^{a11} \end{bmatrix}$$

Box 4. Equation 13

$$W_C = (T_C^a)' = \begin{array}{c c} & \begin{array}{ccccc} D_1 & \cdots & D_j & \cdots & D_m \\ c_{11} \cdots c_{1n_1} & \cdots & c_{j1} \cdots c_{jn_j} & \cdots & c_{m1} \cdots c_{mn_m} \end{array} \\ \begin{array}{cc} D_1 & \begin{array}{c} c_{11} \\ c_{12} \\ \vdots \\ c_{1n_1} \end{array} \\ \vdots & \vdots \\ D_i & \begin{array}{c} c_{i1} \\ c_{i2} \\ \vdots \\ c_{in_i} \end{array} \\ \vdots & \vdots \\ D_m & \begin{array}{c} c_{m1} \\ c_{m2} \\ \vdots \\ c_{mn_m} \end{array} \end{array} & \begin{bmatrix} W^{11} & \cdots & W^{1j} & \cdots & W^{1m} \\ \vdots & & \vdots & & \vdots \\ W^{i1} & \cdots & W^{ij} & \cdots & W^{im} \\ \vdots & & \vdots & & \vdots \\ W^{m1} & \cdots & W^{mj} & \cdots & W^{mm} \end{bmatrix} \end{array}$$

the unweighted supermatrix W_c. T_D^a can be obtained by the total influence matrix T_D in process which can be shown as Equation 14.

$$T_D = \begin{bmatrix} t_D^{11} \cdots t_D^{1j} \cdots t_D^{1n} \\ \vdots \quad\quad \vdots \quad\quad \vdots \\ t_D^{i1} \cdots t_D^{ij} \cdots t_D^{in} \\ \vdots \quad\quad \vdots \quad\quad \vdots \\ t_D^{n1} \cdots t_D^{nj} \cdots t_D^{nn} \end{bmatrix} \tag{14}$$

After normalizing the total influence matrix T_D, we can obtain a new normalized total influence matrix T_D^a which can be shown as Equation 15.

$$T_D^a = \begin{bmatrix} \frac{t_D^{11}}{d_1} \cdots \frac{t_D^{1j}}{d_1} \cdots \frac{t_D^{1n}}{d_1} \\ \vdots \quad \vdots \quad \vdots \\ \frac{t_D^{i1}}{d_i} \cdots \frac{t_D^{ij}}{d_i} \cdots \frac{t_D^{in}}{d_i} \\ \vdots \quad \vdots \quad \vdots \\ \frac{t_D^{n1}}{d_n} \cdots \frac{t_D^{nj}}{d_n} \cdots \frac{t_D^{nn}}{d_n} \end{bmatrix} \tag{15}$$

$$= \begin{bmatrix} t_D^{a11} \cdots t_D^{a1j} \cdots t_D^{a1n} \\ \vdots \quad \vdots \quad \vdots \\ t_D^{ai1} \cdots t_D^{aij} \cdots t_D^{ain} \\ \vdots \quad \vdots \quad \vdots \\ t_D^{an1} \cdots t_D^{anj} \cdots t_D^{ann} \end{bmatrix}$$

In here, $t_D^{aij} = t_D^{ij} / d_i, d_i = \sum_{j=1}^{n} t_D^{ij}$.

After obtaining the normalized total influence matrix T_D^a, we can calculate and obtain the new weighted supermatrix W_C^* by the unweighted supermatrix W_C. The result is shown by Equation 16.

$$W_C^* = T_D^a \times W$$
$$= \begin{bmatrix} t_D^{a11} \times W^{11} & \cdots & t_D^{a11} \times W^{i1} & \cdots & t_D^{a11} \times W^{n1} \\ \vdots & & \vdots & & \vdots \\ t_D^{ai1} \times W^{1j} & \cdots & t_D^{aij} \times W^{ij} & \cdots & t_D^{ain} \times W^{nj} \\ \vdots & & \vdots & & \vdots \\ t_D^{an1} \times W^{1n} & \cdots & t_D^{anj} \times W^{in} & \cdots & t_D^{ann} \times W^{nn} \end{bmatrix} \tag{16}$$

4. Obtain the DANP.

Limit the weighted supermatrix by raising it to a sufficiently large power φ until it converges and becomes a long-term stable supermatrix to obtain global priority vector, i.e., $\lim_{\varphi \to \infty} \left(W_C^*\right)^{\varphi}$ where φ represents any number of powers when $\varphi \to \infty$ and the influential weights $W = \left(W_1, \ldots, W_1, \ldots, W_n\right)$. Then we can call it the DANP influential weights.

ANALYSIS OF EVALUATION RESULT

Data Collection

The data in this research were collected from an expert forecasting questionnaire. 20 representative experts were selected to complete the questionnaire. After recovering the forecasting questionnaires, the data was processed and integrated and then the average impact score of each criteria was obtained. The initial influence matrix D is shown in Table 2.

Model Calculation Process

Using the initial influence matrix D, the normalized matrix R can be calculated and is shown in Table 3.

The total influence matrix T_C by the criteria and T_D by the dimensions are calculated and shown in Table 4 and Table 5.

Table 2. The initial influence matrix D for the criteria

Criteria	C_{11}	C_{12}	C_{13}	C_{21}	C_{22}	C_{23}	C_{24}	C_{31}	C_{32}	C_{33}
C_{11}	0.000	2.750	2.250	1.500	1.125	1.500	1.250	1.250	1.500	2.750
C_{12}	2.875	0.000	2.000	1.750	1.750	1.750	2.000	1.750	1.875	2.125
C_{13}	2.500	1.375	0.000	1.500	1.375	2.500	1.250	1.250	1.500	2.625
C_{21}	2.000	1.625	1.750	0.000	2.125	1.625	1.750	1.125	1.375	1.625
C_{22}	1.750	1.625	2.000	1.625	0.000	2.750	1.875	1.750	2.250	1.625
C_{23}	1.500	2.250	3.000	1.250	2.875	0.000	1.375	1.875	2.125	1.625
C_{24}	1.250	1.625	1.500	1.500	1.625	1.625	0.000	1.375	1.125	1.375
C_{31}	1.625	1.625	1.250	1.250	3.125	2.250	1.250	0.000	3.500	2.125
C_{32}	2.250	2.375	1.750	1.375	3.125	2.750	1.000	2.875	0.000	1.875
C_{33}	3.000	2.500	2.500	1.500	1.375	1.625	0.875	1.500	1.750	0.000

Table 3. The normalized matrix R for the criteria

Criteria	C_{11}	C_{12}	C_{13}	C_{21}	C_{22}	C_{23}	C_{24}	C_{31}	C_{32}	C_{33}
C_{11}	0.000	0.092	0.076	0.050	0.038	0.050	0.042	0.042	0.050	0.092
C_{12}	0.097	0.000	0.067	0.059	0.059	0.059	0.067	0.059	0.063	0.071
C_{13}	0.084	0.046	0.000	0.050	0.046	0.084	0.042	0.042	0.050	0.088
C_{21}	0.067	0.055	0.059	0.000	0.071	0.055	0.059	0.038	0.046	0.055
C_{22}	0.059	0.055	0.067	0.055	0.000	0.092	0.063	0.059	0.076	0.055
C_{23}	0.050	0.076	0.101	0.042	0.097	0.000	0.046	0.063	0.071	0.055
C_{24}	0.042	0.055	0.050	0.050	0.055	0.055	0.000	0.046	0.038	0.046
C_{31}	0.055	0.055	0.042	0.042	0.105	0.076	0.042	0.000	0.118	0.071
C_{32}	0.076	0.080	0.059	0.046	0.105	0.092	0.034	0.097	0.000	0.063
C_{33}	0.101	0.084	0.084	0.050	0.046	0.055	0.029	0.050	0.059	0.000

Table 4. The total influence matrix T_C by the criteria

Criteria	C_{11}	C_{12}	C_{13}	C_{21}	C_{22}	C_{23}	C_{24}	C_{31}	C_{32}	C_{33}
C_{11}	0.321	0.350	0.370	0.325	0.323	0.325	0.263	0.272	0.303	0.395
C_{12}	0.424	0.281	0.376	0.348	0.358	0.349	0.297	0.301	0.329	0.392
C_{13}	0.401	0.313	0.305	0.327	0.335	0.360	0.265	0.275	0.307	0.395
C_{21}	0.409	0.340	0.379	0.306	0.378	0.353	0.299	0.289	0.321	0.389
C_{22}	0.370	0.316	0.359	0.324	0.287	0.362	0.279	0.287	0.326	0.357
C_{23}	0.370	0.339	0.395	0.319	0.392	0.285	0.269	0.296	0.328	0.365
C_{24}	0.299	0.267	0.290	0.273	0.286	0.279	0.181	0.233	0.245	0.295
C_{31}	0.364	0.316	0.334	0.310	0.383	0.349	0.258	0.233	0.363	0.369
C_{32}	0.401	0.353	0.367	0.330	0.399	0.378	0.265	0.334	0.272	0.381
C_{33}	0.590	0.381	0.420	0.367	0.371	0.370	0.283	0.313	0.348	0.357

Table 5. The total influence matrix T_D by the dimensions

Dimensions	D_1	D_2	D_3
D_1	0.349	0.323	0.330
D_2	0.344	0.304	0.311
D_3	0.377	0.339	0.330

Table 6. The sum of influences given and received on the dimensions

Dimensions	r_i	c_j	$(r_i + c_i)$	$(r_i - c_i)$
D_1	1.746	1.884	3.630	-0.138
D_2	1.665	1.708	3.374	-0.043
D_3	1.797	1.700	3.497	0.097

Then, r, c, $r + c$ and $r - c$ used for building the INRM are calculated and shown in Table 6.

The influence weights for the 10 criteria can be calculated by using DANP, as shown in Table 7, Table 8 and Table 9.

Results and Discussion

Through the stable matrix of DANP $\lim_{\varphi \to \infty} \left(W_C^*\right)^{\varphi}$, the final results are sorted in Table 10.

According to the table, it is easily seen that Direct Cost, Environmental Responsibility and Profit are the main influencing criteria and the results are 0.080, 0.078 and 0.076, respectively. Social Welfare, Economic Contribution and Recycling Disposal are the least important criteria and the results are 0.065, 0.059 and 0.056, respectively. From this method, we can evaluate which criteria are more important, and which criteria are less important. From the results above, automotive applications should be developed through energy technology with the characteristics of environmental responsibility and economy in mind. A method of designing renewable energy technology for improving automobile sustainable development is presented below.

Table 7. The total influence matrix T_C by the criteria

Criteria	C_{11}	C_{12}	C_{13}	C_{21}	C_{22}	C_{23}	C_{24}	C_{31}	C_{32}	C_{33}
C_{11}	0.308	0.392	0.394	0.362	0.354	0.335	0.349	0.359	0.358	0.364
C_{12}	0.337	0.307	0.307	0.301	0.302	0.307	0.312	0.311	0.315	0.302
C_{13}	0.355	0.299	0.299	0.336	0.344	0.358	0.339	0.330	0.327	0.333
C_{21}	0.263	0.254	0.254	0.299	0.259	0.254	0.268	0.239	0.240	0.264
C_{22}	0.61	0.392	0.260	0.283	0.229	0.304	0.281	0.295	0.291	0.267
C_{23}	0.263	0.392	0.280	0.264	0.289	0.227	0.274	0.268	0.276	0.266
C_{24}	0.212	0.392	0.206	0.224	0.223	0.215	0.178	0.199	0.193	0.204
C_{31}	0.281	0.392	0.281	0.289	0.296	0.299	0.301	0.241	0.338	0.307
C_{32}	0.312	0.392	0.314	0.321	0.336	0.331	0.317	0.377	0.276	0.342
C_{33}	0.407	0.392	0.405	0.389	0.368	0.369	0.382	0.382	0.386	0.351

Table 8. The total influence matrix T_C by the criteria

Criteria	C_{11}	C_{12}	C_{13}	C_{21}	C_{22}	C_{23}	C_{24}	C_{31}	C_{32}	C_{33}
C_{11}	0.067	0.086	0.086	0.082	0.080	0.076	0.079	0.082	0.082	0.083
C_{12}	0.074	0.057	0.067	0.068	0.068	0.069	0.071	0.071	0.072	0.069
C_{13}	0.078	0.076	0.066	0.076	0.078	0.081	0.077	0.075	0.075	0.076
C_{21}	0.071	0.069	0.068	0.061	0.069	0.067	0.071	0.065	0.066	0.072
C_{22}	0.070	0.071	0.070	0.075	0.061	0.081	0.074	0.081	0.079	0.073
C_{23}	0.071	0.069	0.075	0.070	0.077	0.060	0.073	0.073	0.075	0.073
C_{24}	0.057	0.059	0.055	0.059	0.059	0.057	0.047	0.054	0.053	0.056
C_{31}	0.058	0.061	0.058	0.059	0.060	0.061	0.061	0.048	0.068	0.061
C_{32}	0.065	0.067	0.065	0.065	0.069	0.068	0.065	0.075	0.055	0.068
C_{33}	0.084	0.080	0.084	0.080	0.075	0.075	0.078	0.077	0.077	0.070

Table 9. The total influence matrix T_C by the criteria

Criteria	C_{11}	C_{12}	C_{13}	C_{21}	C_{22}	C_{23}	C_{24}	C_{31}	C_{32}	C_{33}
C_{11}	0.080	0.067	0.076	0.070	0.073	0.071	0.056	0.059	0.065	0.078
C_{12}	0.080	0.067	0.076	0.070	0.073	0.071	0.056	0.059	0.065	0.078
C_{13}	0.080	0.067	0.076	0.070	0.073	0.071	0.056	0.059	0.065	0.078
C_{21}	0.080	0.067	0.076	0.070	0.073	0.071	0.056	0.059	0.065	0.078
C_{22}	0.080	0.067	0.076	0.070	0.073	0.071	0.056	0.059	0.065	0.078
C_{23}	0.080	0.067	0.076	0.070	0.073	0.071	0.056	0.059	0.065	0.078
C_{24}	0.080	0.067	0.076	0.070	0.073	0.071	0.056	0.059	0.065	0.078
C_{31}	0.080	0.067	0.076	0.070	0.073	0.071	0.056	0.059	0.065	0.078
C_{32}	0.080	0.067	0.076	0.070	0.073	0.071	0.056	0.059	0.065	0.078
C_{33}	0.080	0.067	0.076	0.070	0.073	0.071	0.056	0.059	0.065	0.078

Table 10. Results sort and summary

Criteria	Results	Sort
C_{11}	0.080	1
C_{12}	0.067	7
C_{13}	0.076	3
C_{21}	0.070	6
C_{22}	0.073	4
C_{23}	0.071	5
C_{24}	0.056	10
C_{31}	0.059	9
C_{32}	0.065	8
C_{33}	0.078	2

DESIGN OF RENEWABLE ENERGY TECHNOLOGY FOR IMPROVING AUTOMOBILE SUSTAINABLE DEVELOPMENT

Self-Generating Wireless Charging System

In view of the bad weather, we must take into account the various circumstances, such as the super capacitor in the exhaustion of electricity. We hereby join self-generating devices in this taxi, which known as joint power generation.

Piezoelectric Power Generation Foot Pad

As China is a populous country, the car is an important means of transport. People will always be stepping on the foot pad when they get into the car. So we use piezoelectric ceramic in the foot pad in the front and rear of the car. Electricity can be generated through the pressure and charge the super capacitor with the circuit rectifier and regulator.

Our aim is to use the piezoelectric chip strain to obtain the current, the current from the rectifier circuit into DC, and then after the DC boost circuit output stable DC voltage, and then power management chip to control the lithium battery charge. However, the piezoelectric device produces only a few microamps, and cannot directly drive the DC boost circuit, so we use a circuit rectifier, get a pulse DC voltage, see the voltage of the oil filter circuit output a stable DC voltage, and then super Capacitor charging, the previous part of the role of buffer power. And then drive the DC boost circuit and power management circuit work, by the lithium-ion battery to generate the energy collected. Our ultimate goal is to transform tiny vibrations into possible energy that can be used by us.

In general, there are two modes of operation for piezoelectric material: the length vibration mode and the transverse vibration mode. We use the transverse vibration mode; the direction T of the externally applied stress is parallel to the polarization direction P, and the resulting voltage is a polarized directional piezoelectric element.

Figure 3. Schematic diagram of the piezoelectric device

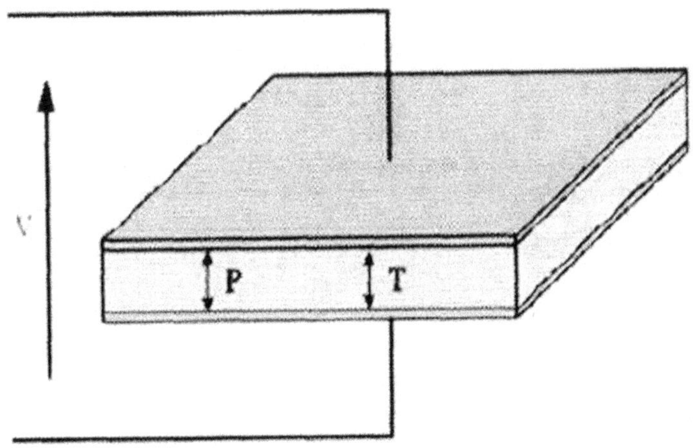

Thermoelectric Conversion

The heat-to-electrical conversion module with the highest power generation capability in the world has been successfully developed this year to convert heat energy directly into electrical energy without having to undergo mechanical energy conversion (turbine or turbine, generator) as in conventional thermal power plants. Its power generation principle is based on the Seebeck phenomenon, that is, in the same loop, by the different bimetallic two connection points between the temperature difference and the voltage, the heat of an electrical semiconductor components of the upper and lower temperature difference between the positive and negative Potential difference, thus generating electricity. This type of power generation device does not exist in the drive part of the static state of the new power generation mode. There will be no fault. Its temperature performance is up to 500 °C, low temperature side is only 20 °C, the maximum output power density up to 1w / cm, more than just one module can generate 15w, an area of only 14.4cm2, for which we will use the Seebeck effect In the car's engine, through the engine's motion heat and the combination of Seebeck effect of the engine heat conversion into electricity will be recycled, and then connected with the super capacitor to achieve the energy storage, easy to use electric vehicles.

In addition to the use of light-volt effect of a direct conversion of light to electricity power generation mode, the other solar power generation method is the use of the Seebeck effect of a precise thermal conversion of electricity, so as to achieve the comprehensive utilization of energy.

Solar Roofs

We have the top of the car covered with solar panels, which can provide for the daily use of electronic equipment within the vehicle, and can also generate electricity for the super capacitor, so that cars can use the super capacitor for electricity, thus allowing them to run further. Solar panels added to the top of the car can be effectively integrated as it is an optimum use of space.

Because of vast geographical regions, China's resources to be too much radiation will be based on regional and weather changes are somewhat different, due to the northern weather conditions change little, we have to the north of solar radiation and combined with different light Angle and the connection

Figure 4. Solar roof

between the solar panels as an example: in the fixed light intensity of the output power of the photovoltaic cell depends on the load resistance R, the output power of solar cells in the best load resistance R reached the maximum power.

$$P = U \times I$$

However, in the actual use of solar cells, the direction of solar panels and the angle of solar light on the power generation efficiency of the battery plate has a profound impact, for which the experimental results are shown in Figure 3. Experiments show that when the incident light is parallel to the normal of the battery plate (i.e., the incident light is irradiated with the solar panel), the voltage generated by the solar panel is the most efficient, and the power is gradually reduced as the irradiation angle increases.

At the same time, the sun has differing intensity throughout the day, the experiment in the sun and solar panels parallel to the normal, that is, vertical sunlight, to explore the efficiency of solar cells throughout the day. The exact trend of the law is shown in Figure 4. It can be seen from the figure, between 12:00 noon and two o'clock in the afternoon, the solar power generation capacity should be the strongest. When the sun tilt angle increases, the efficiency gradually decreases.

Because of the electrical loss makes the combination of solar panels is different, the possible power and its efficiency are also atypical, so in the experiment, for the different series of parallel arrangement of the battery panels were related experiments. There are five combinations of series and parallel. Measurements are perfect for each type of connection. Respectively, the measurement of discrete angles of exposure, measurement time, each connection state is to take the maximum measurement. The following is the five types of connection plate.

For the five connections, the volt-ampere characteristics of the panel, the experiment for each connection are used at different angles of exposure, obtained a unique data. For each connection, in the case of open circuit, measure its output voltage V, and then in the case of short circuit to measure its output current I. According to the volt-ampere characteristics, the power curve under different connection modes is plotted.

Based on the measured data, the solar power generation efficiency of the five connections is calculated and the results are shown in figure.

It can be seen from the experimental results that in the five different combinations, the combination of four parallel combinations of batteries provides the strongest current, the combination of four series of battery voltage to provide the strongest, four-series parallel and four parallel Power is greatest. In order to optimize the efficiency, we chose the four strings in parallel as our bus top connection.

In 2016 the country has about 1 million taxis.
Standard Taxi Size: 4945 * mm1845mm * 1480mm.
Four series of parallel (the latest solar panels): 320 watts per square meter.
Roof area S total = length * width = 4.945 * 1.845 = 9.12 square meters, remove the exhaust fan, assuming the available area is S = 8 square meters, then P too always = S * p = 8 * 320 = 2560 watts (assuming Sunny, appropriate angle of exposure)

Hand-Cranked Generator

Everyone has a smart phone and they have a short battery life. Therefore, we have added in each side of the tables and chairs a hand-cranked generator. There is a mobile phone charging line interface, so that people in the car can charge their phones. The idea takes into account the "people-oriented" dimension.

In addition, smart phones generally run on 5V charges. We designed the hand-cranked generator to stabilize the original voltage to about 5V to provide a stable charge environment.

Resonant Electric Vehicle Car Wireless Charging System

Electromagnetic resonant coupling radio power transmission technology is the use of magnetic resonance coupling wireless power transmission. Power transmission is a cutting-edge field. It is a new technology and it can be efficient and flexible in achieving wireless transmission of electricity. Electromagnetic resonant radio power transmission technology as a modern energy transmission, charging convenient and quick, to stop the station can quickly charge the request.

The Principle and Application of Super Capacitor

Super capacitors, also known as large capacity capacitors, gold capacitors, energy storage capacitors, Faraday capacitors, electrochemical capacitors or electric double layer capacitors, rely on polarized electrolytes to store electrical energy. The super capacitor is based on the theory of electrochemical double

Figure 5. Hand-cranked generator model

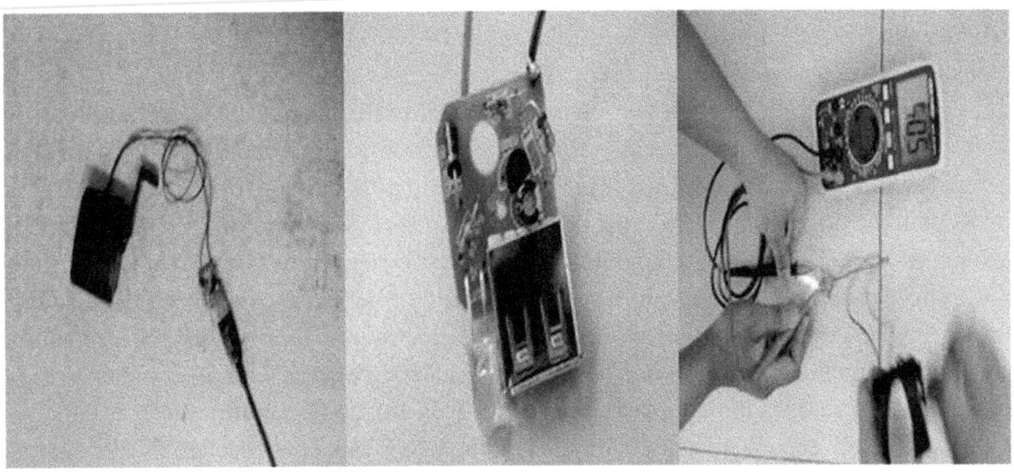

Figure 6. Electric vehicle wireless charging device

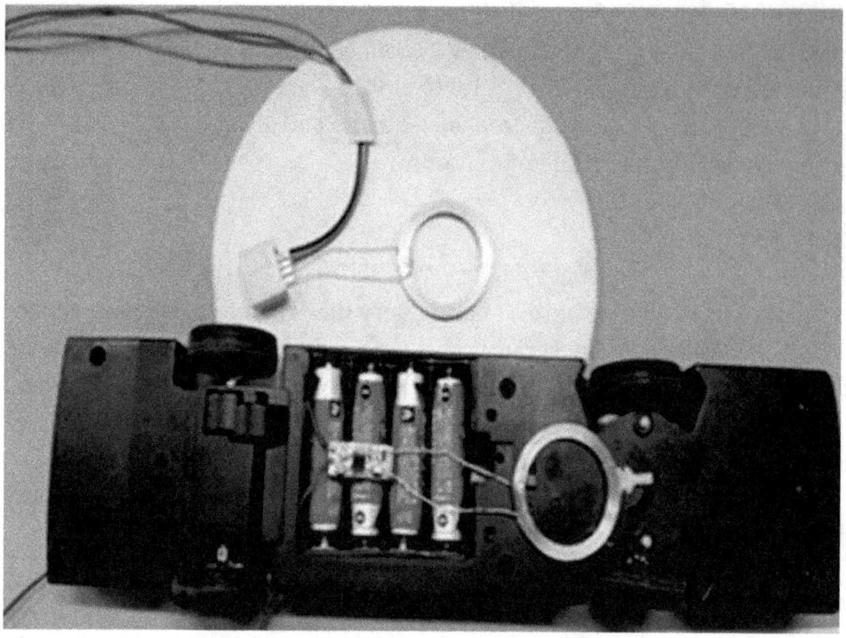

layer, which belongs to the electric double layer capacitor, which is the largest capacity in the electric double layer capacitor which has been put into mass production in the world. In principle, the two types of conductive plates can be used to hold the insulating layer between the plates. The voltage is stored on the plates. When the voltage is applied between the plates, the charge is stored on the plates. The stored energy can be expressed for the We, where the diaphragm is the insulating medium, ΔV is the volume of the insulating medium, and E is the electric field strength of the insulating medium. Energy consumption per unit volume, i.e. energy storage density, can be expressed as

$$\omega e = \frac{1}{2} \varepsilon E^2$$

It can be seen that increasing the dielectric constant and the electric field strength increases the energy storage density of the capacitor. The super capacitor structure is shown in Figure 7.

A Super capacitor usually contain two electrodes, electrolyte, current collector, and four separate components which make use of activated carbon porous electrode and electrolyte composition of the double layer structure to obtain large capacity. When the voltage is applied to the two plates of the super capacitor, as with the ordinary capacitor, the positive electrode of the plate stores a positive charge, and the negative plate stores a negative charge. Under the electric field generated by the charge on the two plates of the super capacitor, the opposite charge is formed at the interface between the electrolyte and the electrode to balance the internal electric field in the electrolyte. This positive charge and negative charge are arranged on the contact surface between two different phases with a very small gap between positive and negative charges in the opposite position. This charge distribution layer is called an electric double layer, so the capacitance is very large. As the super capacitor discharges, the charge on the

Figure 7. Super capacitor structure

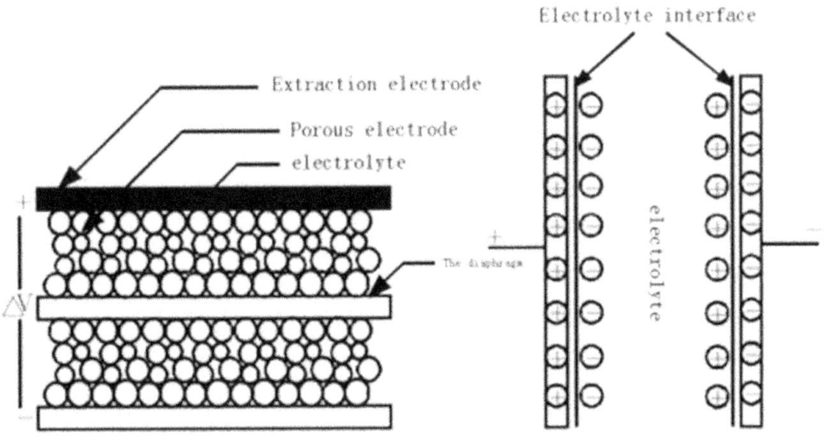

positive and negative plates is discharged by the external circuit, and the charge on the interface of the electrolyte is reduced accordingly. The super capacitor charging and discharging process is always a physical process. There is no chemical reaction. Therefore, unlike batteries using chemical reactions, the performance is stable.

According to the basic framework, we have developed a small renewable energy grid-connected power generation system. The power generation system in the system may be a solar or wind power generation system in which the respective power generation devices are connected together by respective converters to form a DC bus. At the same time, the DC bus through the power electronic devices and inertia flywheel and super capacitor connected to the energy storage device through the corresponding supplement or absorption to adjust the output current tends to a stable level, that is, within a certain range of fluctuations. Then the DC bus through the grid inverter device to get AC, through the switch can be attached to the public grid to achieve power generation, can also be connected to the AC load. This system principle can be implemented to microgrids.

The flywheel energy storage device is mainly used in the regulation of renewable energy power generation systems, in real-time detection of changes in the momentum of the power grid, real-time control flywheel energy storage device compensation momentum, in order to achieve renewable energy power generation system and flywheel energy storage system energy state of a reasonable description, Real-time computing and energy exchange control, thus achieving the premise of grid. At present, recent research on the generalized momentum compensation control of renewable energy power generation systems based on flywheel energy storage technology solves this problem.

The Generalized Momentum Model of Flywheel Energy Storage Systems

$$P_s = \frac{\partial T_s}{\partial \omega} = J\omega$$

Control the flywheel energy storage system in a certain period of time with the renewable energy power generation grid energy exchange, energy storage system from one operating state changes to another operating state, with the interaction of the power grid because of the energy or lost energy, Its operating state will change. This process is characterized as: the flywheel energy storage system collides with the grid, and the total momentum remains constant during the collision. With real-time measurement of flywheel energy storage system state parameter ω size, you can quickly calculate the flywheel energy storage system, instantaneous momentum and a certain amount of time change in momentum.

The Generalized Momentum Model of Renewable Energy Power Generation System

$$P_{rg} = \frac{\partial T_{rg}}{\partial U_{rg}} = C_{rg} U_{rg}$$

The use of Hall voltage sensors and program-controlled microprocessors can mean real-time measurement and rapid calculation of renewable energy power generation system center point voltage Urg, which can quickly calculate the instantaneous momentum of renewable energy power generation systems and a certain period of time changes in momentum the amount.

Mathematical Model of Flywheel Energy Storage

Assuming that the three-phase potential of the motor is Ea, Eb, Ec, the three-phase winding current is ia, ib, ic, the terminal voltage is Ua, Ub, Uc, and the three-phase stator winding adopts star connection, the three-phase winding resistance is Ra = Rb = Rc = R, the three-phase winding self-sensation is La = Lb = Lc = L, the three-phase winding resistance is the same as the parameters, the three-phase winding interaction Lab = Lba = Lac = Lc = Lbc = Lcb = M, D is the differential operator, and Un is the common connection point voltage, then the voltage equation can be expressed as follows when running in generator mode:

$$\begin{bmatrix} U_a \\ U_b \\ U_c \end{bmatrix} = \begin{bmatrix} E_a \\ E_b \\ E_c \end{bmatrix} - \begin{bmatrix} R & 0 & 0 \\ 0 & R & 0 \\ 0 & 0 & R \end{bmatrix} \begin{bmatrix} i_a \\ i_b \\ i_c \end{bmatrix} - \begin{bmatrix} L & M & M \\ M & L & M \\ L & M & L \end{bmatrix} D \begin{bmatrix} i_a \\ i_b \\ i_c \end{bmatrix} - U_n \begin{bmatrix} 1 \\ 1 \\ 1 \end{bmatrix}$$

When running in motor mode, the voltage equation can be expressed as:

$$\begin{bmatrix} U_a \\ U_b \\ U_c \end{bmatrix} = \begin{bmatrix} E_a \\ E_b \\ E_c \end{bmatrix} + \begin{bmatrix} R & 0 & 0 \\ 0 & R & 0 \\ 0 & 0 & R \end{bmatrix} \begin{bmatrix} i_a \\ i_b \\ i_c \end{bmatrix} + \begin{bmatrix} L & M & M \\ M & L & M \\ L & M & L \end{bmatrix} D \begin{bmatrix} i_a \\ i_b \\ i_c \end{bmatrix} + U_n \begin{bmatrix} 1 \\ 1 \\ 1 \end{bmatrix}$$

According to this equation, the equivalent circuit of the flywheel energy storage device can be established.

In the case of a phase, when the motor rotates at an angular velocity ω, the single-phase electromagnetic torque Ta can be expressed as

$$T_a = E_a i_a / \omega$$

The total electromagnetic torque Te is the sum of the three-phase torque

$$\mathrm{Te} = (\mathrm{E_a i_a} + \mathrm{E_b i_b} + \mathrm{E_c i_c}) / \omega$$

When the flywheel energy storage device is operated in the discharge mode, the total torque of the flywheel overcomes the load torque TL, the damping torque Bω and the generator total electromagnetic torque Te, the relationship can be expressed as follows:

$$-\mathrm{J}\frac{\mathrm{d}\omega}{\mathrm{dt}} = T_L + B\omega + T_e$$

When the motor is running in the motor mode, its mechanical energy equation can be expressed as

$$\mathrm{T_e} = T_L + B_\omega + J\frac{d\omega}{dt}$$

where J is the flywheel moment of inertia. For cylindrical flywheels, the moment of inertia is:

$$\mathrm{J} = \mathrm{mr^2} = \rho\pi \mathrm{r^2 l r^2}$$

where ρ is the flywheel material density, r is the flywheel radius of rotation, and l is the flywheel thickness.

Generalized Momentum Compensation Control Method

The generalized momentum conservation theory provides a reliable basis for the energy exchange control between the flywheel energy storage device and the renewable energy power generation system. When the external environmental factors change or load changes, the renewable energy power generation system can be considered outside of the energy collision; the original balance of the system has changed, the momentum has changed, the voltage Urg has also undergone a corresponding change, thus affecting the supply voltage quality. Similarly, when the grid due to a short circuit, open circuit and other reasons caused by sudden changes in the grid structure, equivalent inertia parameter Crg changes, the system to maintain its generalized momentum conservation, Urg will inevitably change accordingly, may exceed the allowable supply voltage range. Each time the central point voltage does not meet the requirements, first a flywheel energy storage device and the power grid between the small momentum collision, measuring the small momentum collision caused by the flywheel speed change Δω and power grid. The central point of the voltage changes ΔUrg, after a series of calculations, and finally the required torque is obtained.

$$T = J\beta = \frac{mr^2}{2}\frac{\Delta\omega}{\Delta t} = \frac{\rho\pi r^2 l r^2}{2}\frac{\Delta n \cdot 2\pi}{\Delta t}$$

According to the torque obtained, the change value of the voltage at the output end of the flywheel motor can be obtained by reversing equations to compensate the bus voltage of ΔUrg, and then the controllable variable of the control system can be obtained.

Flywheel Energy Storage Device Control Flow Chart

The method of energy colliding with the renewable energy power generation system changes the momentum of the power grid by collision, so that the voltage of the grid center point can the method of energy colliding with the renewable energy power generation system changes the momentum of the power grid by collision, so that the voltage of the grid center point can be restored to the level of power quality requirements. The control flow is shown in Fig. If the grid voltage is reduced, the control system quickly calculates the amount of power loss in the grid, and calculates the same magnitude of the flywheel output to reduce the required speed and time and adjust the flywheel energy storage device. If the grid voltage increases, the control system quickly calculates the amount of power fluctuations in the grid, while calculating the flywheel to obtain the same size of the momentum needs to increase the speed and time and the flywheel energy storage device to accelerate energy storage adjustment.

Experiments show that, regardless of whether the solar photovoltaic power generation system is subject to external interference or internal network changes, flywheel energy storage devices can be real-time to the photovoltaic power generation system to provide appropriate compensation momentum; the photovoltaic power generation system voltage quickly and accurately adjust to ensure that the central point voltage is stable.

An inverter is a photovoltaic device, fan, fuel cell, energy storage components through the power electronic switch and micro-network system connected to the crucial equipment. The converter may include both a rectifier and an inverter, or only an inverter. The converter has the characteristics of fast speed, low inertia and weak overcurrent capacity. At the same time, it is suitable for micro-network inverter in addition to the need for conventional inverter function and can run in parallel, but being also dependent on the special needs of micro-network system with some control functions. The structure and operation of the inverter are the major technologies of the microgrid.

Widely used in pure electric vehicle energy supply systems, super capacitor batteries, compared toother batteries, charging time is short, charge and discharge cycle long life, high quality than high power, high reliability, in the vehicle design and development process, its Single cell arrangement of the connection not only to meet the performance requirements of the vehicle, but also in the energy recovery process to ensure that the battery safe, as much as possible to store more energy. In the energy recovery process, the pure electric vehicle super capacitor energy storage system's principal role is to quickly and effectively absorb more braking energy. Be returned to the level of power quality requirements. If the grid voltage is reduced, the control system quickly calculates the amount of power loss in the grid, and calculates the same magnitude of the flywheel output to reduce the required speed and time and adjust the flywheel energy storage device. If the grid voltage increases, the control system quickly calculates the amount of power fluctuations in the grid, while calculating the flywheel to obtain the same size of the momentum needed to increase the speed and time and the flywheel energy storage device to accelerate energy storage adjustment.

In order to meet the requirements, the super capacitor energy storage system must have sufficient capacity to store the maximum braking energy. This is also necessary in order to meet the safety requirements of the super capacitor energy storage process, but also to consider the regenerative braking process peak power. Electric taxi model using rear wheel drive, model taxi motor for permanent magnet DC brush motor, model electric bus power supply for the 4 1.2V lithium battery, super capacitor consists of two 2.5V600F super capacitor series, each Super capacitor at both ends and a voltage regulator circuit to ensure that the super capacitor can be stable and efficient absorption of energy, super capacitor and power supply between the lithium battery through the chopper circuit and the use of rated current charging method of lithium battery charging device phase connection.

Innovation Advantage

The use of a super capacitor and flywheel energy storage of these two advanced energy storage methods as a new energy power generation system in the energy storage means, not only real-time detection, compared with other means of storage, the more accurate to add momentum or absorb momentum to achieve a smooth Transmission. And in high efficiency, safety, conditions of use, life and so on, there is irreplaceable superiority. In addition, these two energy storage methods are clean and pollution-free, and the technology is mature, which is very good for development prospects.

Solar thermal power generation on the vehicle roof and two total generations in power co-generation unit of up to 12KW, fully provide a real-time source of energy for the electric taxis. To ensure that the electric taxi operation of the safe and reliable, to ensure the real-time influx of energy, green, energy saving.

CONCLUSION

In this chapter, we contribute to the field of renewable energy sources in the automobile industry and sustainable development in the following ways. First, we propose a new method of comprehensive evaluation analysis which combines the management level and technical level by using a new hybrid multi criteria method that combines DEMATEL and DEMATEL-based ANP. This method is a good way to evaluate and understand the important criteria which can improve sustainable development most. We obtained the data from experts from the automobile industry and evaluated it through the new model. The results show that automotive applications should be developed keeping environmental responsibility and economy in mind. Second, we propose a design of renewable energy technology for improving automobile sustainable development in order to solve such problems as derived from the evaluation above. Through a new design of advanced energy storage based on the super capacitor of self-generating wireless charging, the automobile application of renewable energy source has the advantages of short charging time, high power conversion efficiency and low power consumption.

There are several limitations to our study that future work might seek to address. First, the number of selected experts which are used for data collection needs to be expanded. More data samples can improve the credibility of the research. Second, the selected experts should be more representative. The experts' quality has a strong positive correlation with the results of the research. Finally, in the process of selecting dimensions and criteria, future researches need to find a scientific method to establish a more practical evaluation index system which can improve the accuracy of the results.

REFERENCES

Alting, L. (Ed.). (1993). *Life-cycle design of products: a new opportunity for manufacturing enterprises, concurrent engineering: Automation, tools and techniques*. New York, NY: Wley.

Alting, L., & Jorgensen, J. (1993). Life cycle concept as a basic for sustainable industrial production. *CIRP Annals-Manufacturing Technology.*, *42*(1), 163–167. doi:10.1016/S0007-8506(07)62417-2

Becker, H. A., & Vanclay, F. (2003). *The international handbook of social impact assessment*. Cheltenham, UK: Edward Elgar. doi:10.4337/9781843768616

Boothroyd, G., & Dewhurst, P. (1994). *Adecade of DFMA Research, Conference Proceedings, The 1994 International Forum on Design for Manufacture and Assembly.* Newport.

Curtis, H., & Walker, J. (2001). *Design council European survey of manufacturing companies' attitudes towards design for sustainability*. London: Design Council.

Gao, P., Hensley, R., & Zielke, A. (2014). A road map to the future for the auto industry. *The McKinsey Quarterly*, *10*, 1–11.

Giddings, B., Hopwood, B., & OBrien, G. (2002). Environment, economy and society: Fitting them together into sustainable development. *Sustainable Development*, *10*(4), 187–196. doi:10.1002/sd.199

Hemdi, A. R., Saman, M. Z., & Sharif, S. (2009). Proposed decision making framework for sustainable product design evaluation. *Asia Pacific Industrial Engineering & Management Systems Conference,* 1601-1609.

Hu, Z., Fang, F., Ben, D., Pu, G., & Wang, C. (2004). Net Energy, CO2 Emission and Life-cycle Cost Assessment of Cassava-based Ethanol as an Alternative Automotive Fuel in China. *Applied Energy*, *78*(3), 247–256. doi:10.1016/j.apenergy.2003.09.003

Hunkeler, D., & Rebitzer, G. (2005). The Future of Life Cycle Assessment. *International Journal of Life Cycle Assessment.*, *10*(5), 305–308. doi:10.1065/lca2005.09.001

Kapur, A., Baldwin, C., Swanson, M., Wilberforce, N., McClenachan, G., & Rentschler, M. (2012). Comparative life cycle assessment of conventional and Green Seal-compliant industrial and institutional cleaning products. *International Journal of Life Cycle Assessment.*, *17*(4), 377–387. doi:10.1007/s11367-011-0373-8

Labuschagne, C., & Brent, A. C. (2006). Social Indicators for Sustainable Project and Technology Life Cycle Management in the Process Industry. *International Journal of Life Cycle Assessment*, *11*(1), 3–15. doi:10.1065/lca2006.01.233

Mildenberger, U., & Khare, A. (2000). Planning for an environment-friendly car. *Technovation*, *20*(4), 205–214. doi:10.1016/S0166-4972(99)00111-X

Puri, P., Compston, P., & Pantano, V. (2009). Life cycle assessment of Australian automotive door skins. *International Journal of Life Cycle Assessment*, *14*(5), 420–428. doi:10.1007/s11367-009-0103-7

Sundin, E. (2004). *Product and process design for successful remanufacturing*. Linköping, Sweden: Linköping University.

United Nations Environment Programme. (2012). *Towards a Life Cycle Sustainability Assessment*. New York, NY: United Nations.

Warsen, J., Laumer, M., & Momberg, W. (2011). Comparative Life Cycle Assessment of Remanufacturing and New Manufacturing of a Manual Transmission, Globalized Solutions for Sustainability in Manufacturing. *Proceedings of the 18th CIRP International Conference on Life Cycle Engineering*. Braunschweig: Technische Universitat Braunschweig.

Weidema, B. P. (2006). The Integration of Economic and Social Aspects in Life Cycle Impact Assessment. *International Journal of Life Cycle Assessment, 11*(1), 89–96. doi:10.1065/lca2006.04.016

World Commission on Environment and Development (WCED). (1987). *Our common future*. Oxford, UK: Oxford University Press.

Chapter 13
Risk Implications of Heavy Metal Contamination in Agricultural Soil and Crop Productivity

Archana
M. J. P. Rohilkhand University, India

Ajai Kumar Jaitly
M. J. P. Rohilkhand University, India

ABSTRACT

Heavy metals especially lead, nickel, cadmium, copper, cobalt, chromium and mercury are more toxic and chief contaminants of the environment. Agricultural soils in many parts of the world are slightly to moderately polluted with heavy metals due to increase in geologic and anthropogenic activities (use of phosphate fertilizers, sewage sludge application, dust from smelters, industrial waste). Plants growing on these contaminated soils showed toxicity symptoms that results in reduce growth and activity which declined the productivity and posing threats to agro-ecosystems. They put plants under stress and affect their physiology. In this chapter, we have summarized the effects of heavy metals on plants including both symptoms and productivity.

INTRODUCTION

Heavy metals are mostly metalloids and cannot be degraded, thus accumulated in soils and water system. Heavy metal pollution exerts an unpropitious effect on agricultural soil and hence become a serious concern. Continuous use of fertilizers, application of sewage sludge, dust from smelters, industrial effluents and wrong irrigation process may lead to the accumulation of heavy metals in agricultural land (Bell et al., 2001; Schwartz et al., 2001; Passariello et al., 2002). Though heavy metal plays a significant role in physiological process of plants in trace amounts but higher concentration of heavy metal than permissible level slow down their growth and biomass yield which further decline the crop productivity.

Plants uptake heavy metals from contaminated soil and accumulate them throughout the food chain. It is a potential threat to animal and human health (Jordao et al., 2006; Sprynskyy et al., 2007; Oves, 2016).

DOI: 10.4018/978-1-5225-3537-9.ch013

The heavy metals can cause some adverse effect in the body, such as depletion of some vital nutritional minerals from their original place, cancer development, kidney and liver diseases, damage of lung, neurological abnormalities, depletion of immunological defense, malnutrition etc. (Khan et al., 2008).

Heavy metals also affect soil fertility by modifying micro-organism through their toxic effects. The microbial population and their activity in soil influences the fertility of soil (Ashraf and Ali, 2007; Rathnayake et al., 2013). Agricultural runoff containing heavy metals enter in aquatic environment and caused toxicity to plants and animals inhabiting the habitat. Heavy metal contamination of sewage sludge, municipal solid waste and pig manure affects the process of composting by inhibiting bacterial growth. Heavy metals affect the life cycle of Earth-worm in vermicomposting. These impacts of heavy metals adversely affect the bio-community and thus the productivity (Figure 1)

In these respects, heavy metal contamination of agricultural soils has always been considered a critical challenge. Heavy metals are normally present at low levels in agricultural soil but due to their cumulative behavior and toxicity, they have a potential harmful effect, not only on crop plants but also on human health. Therefore, this chapter has been focused on the impact of the heavy metals to determine the loss of agricultural productivity and their possible management.

BACKGROUND

Heavy Metals

Heavy metals exhibit metallic properties and belong to the group of loosely-defined subset of elements. Most of the heavy metals are transition elements having incomplete filled d-orbitals. These d-orbitals provide heavy metal cations that have ability to form complex compounds which may or may not be redox active (Nies, 1999). It mainly comprises the transition metals, some metalloids, lanthanides, and

Figure 1. Pathways for heavy metals in soil, plants, animals and humans

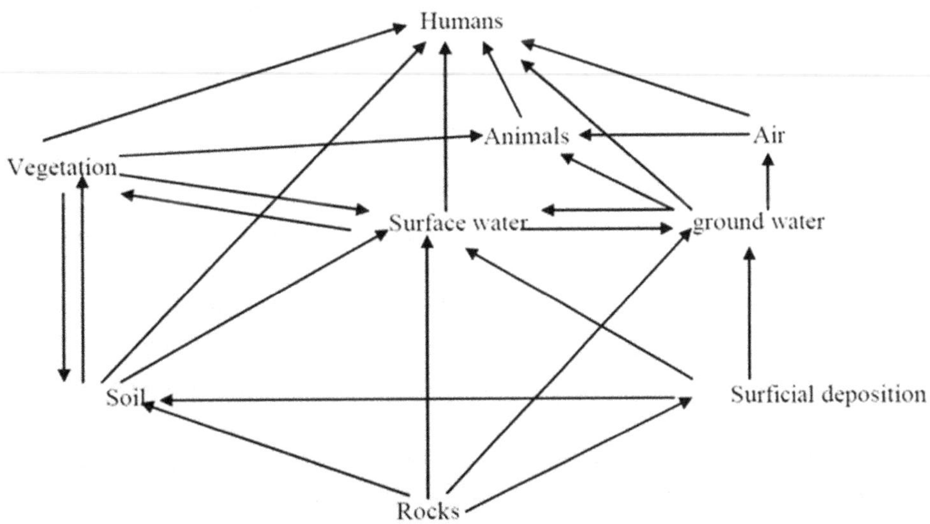

actinides. Many different definitions have been proposed, some based on density, some on atomic number or atomic weight, and some on chemical properties or toxicity. Accordingly, some of the definitions are cited as under:

- Heavy metals are elements having atomic weights between 63.546 and 200.590 (Kennish, 1992).
- According to Weast (1984) heavy metals are elements those have density above 5 g/cm^3, accordingly the transition elements from Vanadium 'V' (Sc and Ti not included) to the half-metal Arsenic 'As', from Zirconium 'Zr' (Y not included) to Antimony 'Sb', from Lanthanum 'La' to Polonium 'Po', the lanthanides and the actinides are known to as heavy metals. Out of 90 naturally occurring elements, 21 are non-metals, 16 are light metals and the remaining 53 (As included) are heavy metals.
- Lenntech (2004) said heavy metals are any metallic element that has a relatively high density (perhaps > 5 g/cm^3, Weast, 1984) and is toxic even at low concentration.

There are various sources of heavy metal contamination of soil, such as metal mining and smelting, agricultural and horticultural products (application of fertilizer and pesticide), sewage sludge, combustion of fossil fuel, metallurgical industries, electroplating, chemical and other industrial sources and disposal of waste. The agricultural land of many countries have been contaminated by As, Cd, Zn and Pb due to extensive use of fertilizers (Singh and Agrawal, 2007; Alkhader, 2015) and industrial practices (Papazoglou et al., 2005; Yang et al., 2016) and according to Lugon-Moulin et al. (2004) Cd in phosphorus- fertilizers is of great concern. Copper is often used as a fungicide in agriculture that increase the concentration of copper in the cultivated soils. Moreover, the heavy metals are found naturally at various high background levels at different places throughout the world, due to various concentrations in the bedrock.

Roles of Heavy Metals in Plants

Heavy metals have been divided into two classes on the basis of their role in living systems: essential and non-essential. Essential heavy metals are required by living organisms for their growth, development and physiological functions, while non-essential heavy metals are those, which are not required by living organisms for any physiological functions. The elevated concentrations of both essential and nonessential heavy metals in the soil can lead to metal stress thus inhibiting growth in most of the plants (Hall, 2002; Shahid et al., 2015). They cannot be degraded and can cause direct or indirect toxicity, when accumulated above certain threshold levels. In general, three different molecular mechanisms (Figure 2) of metal toxicity can be distinguished: (a) blocking of essential functional groups in biomolecules, (b) dislocation of essential metal ions from biomolecules and (c) production of reactive oxygen species by autoxidation and the Fenton reaction, which is typical for transition metals (Schutzendubel and Polle, 2002).

As, Cd, Hg, Pb or Se are non-essential metals, since they do not accomplish any physiological function in plants, while, Co, Cu, Fe, Mn, Mo, Ni and Zn are essential metals for plant growth. However, essential elements cause poisoning when their concentration is greater than permissible values. Copper (Cu) plays a key role in CO_2 assimilation and ATP synthesis and also is an essential component of various proteins such as plastocyanin (a component of photosynthetic system) and cytochrome oxidase (enzyme used in respiratory electron transport system) (Demirevska-kepova et al., 2004; Arunakumara,

Figure 2. Molecular mechanism of heaby metal toxicity in plants (Peralta-Videaa et al., 2009). Purple spheres indicate redox actove metals, and red and blue spheres are redox inactive metals. The green spheres is a metal center that is displaced by a heavy metal (red).

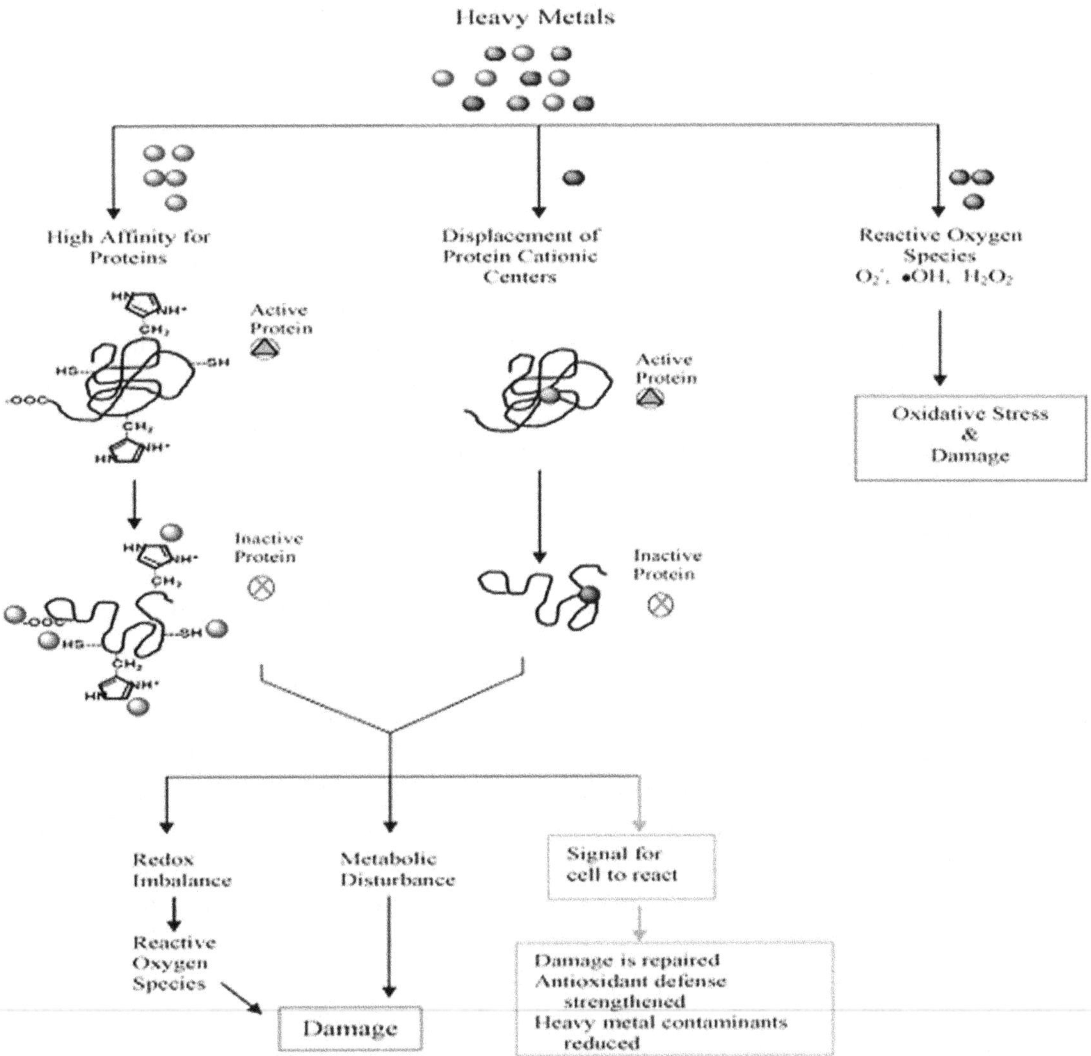

2013). Zinc is a constituent of metallo-enzyme or a cofactor for several enzymes such as anhydrases, dehydrogenases, oxidases and peroxidases and plays an important role in regulating the nitrogen metabolism, cell multiplication, photosynthesis and auxin synthesis in plants ((Hewitt, 1983; Shier, 1994; Makoi, 2016). It also plays an important role in the synthesis of nucleic acid and proteins, and helps in the utilization of phosphorous and nitrogen during seed development. Nickel is a component of the enzyme urease, microbial dehydrogenases, hydrogenases and methyl reductase and plays a significant role in urea and ureide metabolism, iron absorption, nitrogen fixation and seed development (Rao et al., 2006, Khoshgoftarmanesh et al., 2011). Manganese act as cofactor in many enzymes and also is a part of the oxygen evolving complex in chloroplast. Manganese (Mn) perform an important part in oxidation-reduction processes, electron transport in photosynthesis, as an activator of more than 35 different enzymes involved in oxidation reactions. Mn is also a structural component in enzymes evolved in carboxylation, carbohydrates metabolism and citric acid cycle (Makoi, 2016). Molybdenum is a key component of nitrogen metabolism and is required by enzymes catalyzing important reactions within the cell (Salisbury and Ross, 1992; Mendel, 2011).

Accessibility of Heavy Metals to Plants: Uptake Mechanism

The root system of plants has very large surface area that absorbs and accumulates water and essential nutrients with other non-essential contaminants (Raskin and Ensley, 2000; Thakur et al., 2016). The metals are transported through apoplast within the cortex according to their concentration gradient and also accumulate in the cell walls (Schutzendubel and Polle, 2002; Jan and Parray, 2016). Access of heavy metals to bare roots is confined to the first few millimeters of the root tip. Uptake and transport across cell membrane of root is an important process which initiates metal absorption into plant tissues. There are two routes of heavy metal uptake in plants: (a) passive uptake: which is driven by the concentration gradient across the membrane, and (b) active uptake: which is an energy dependent uptake facilitated by transporter proteins within the membrane (Nies, 1999; Williams et al., 2000; Jan and Parray, 2016).

Effects of Various Heavy Metals on Plants

Heavy metal toxicity of plants leads to chlorosis, reduction of yield, nutrient uptake, disorders in metabolism and reduced ability to fix molecular nitrogen in plants (Guala et al., 2010). At high concentration, direct toxic effects of heavy metal involve inhibition of cytoplasmic enzymes and destruction of cell structures through oxidative stress. An example of indirect toxic effect is the replacement of essential nutrients at cation exchange sites of plants. Furthermore, heavy metals have adverse effect on the growth and activities of soil microorganisms that may also ultimately affect the growth of plants. For instance, a reduction in the number of useful soil microorganisms due to high metal concentration may lead to decrease in organic matter decomposition leading to a decline in soil nutrients. Enzyme activities that are useful for plant metabolism may also be inhibit due to heavy metal interference with activities of soil microorganisms. These toxic effects (both direct and indirect) sometimes may lead to death of plant (Chibuike and Obiora, 2014). One of the most destructive effects induced by heavy metals is lipid peroxidation, which can cause deterioration of cell membrane. Malondialdehyde (MDA) is a decomposition products of polyunsaturated fatty acids of membrane and it is considered as a positive indicator of oxidative stress (Demiral and Türkan, 2005). Some of the potential impacts of heavy metal are as follows and have been summarized in Table 1.

Effect of Heavy Metals on Antioxidative System

Production of reactive oxygen species (ROS) and free radicals are basic responses of plants to high levels of heavy metal stress (Dietz et al., 1999; Baycu et al., 2017). Various metals either generate ROS (reactive oxygen species) directly through Haber-Weiss reactions or they can overproduce ROS. The occurrence of oxidative stress in plants could be the indirect result of heavy metal toxicity (Wojtaszek, 1997; Mithofer et al., 2004; Baycu et al., 2017). Oxidative stress can seriously interrupt normal metabolism through oxidative damage of cellular components. Plants have developed a complex antioxidant system to repair the damage against oxidative stress and inducing resistance to metals by protecting labile macromolecules (Zhang and Kirkham, 1994; Kapoor et al., 2016).

The antioxidative system encompasses several enzymes and compounds of low molecular weight. Antioxidants like cysteine, proline, ascorbic acid and non-protein thiol (sulfhydryl) play a central role in detoxification of toxic metal ions (Noctor and Foyer, 1998; Singh and Sinha, 2005; Sharma et al., 2012). For e.g., ascorbic acid is an ubiquitous soluble antioxidant in photosynthetic organisms and the most important reducing substrate for H_2O_2 detoxification (Singh et al., 2005; Vidović et al., 2016). It also serves as a free radical scavenger against O_3 (El-Khatib, 2003). Tocopherols, the other antioxidative compound protect lipids and other membrane components by physically quenching and reacting chemically with singlet oxygen (Hare et al., 1999; Sharma et al., 2012). Accumulation of free proline in response to heavy metal exposure seems widespread among plants. Proline increases the stress tolerance of plants through the mechanisms like osmoregulation, protection of enzymes against denaturation, and stabilization of protein synthesis ((Costa and Morel, 1994; Kuznetsov and Shevyakova, 1997; Hossain et al., 2014).

Lead and mercury have been reported to boost the ascorbic acid and α-tocopherol level in two cultivars of *Oryza sativa* (Mishra and Choudhuri, 1999). Mercury exposure was found to increase the ascorbic acid levels in *Bacopa monnieri* (Sarita et al., 1996). Zengin and Munzuroglu (2005) explored the effects of lead, copper, cadmium and mercury on the level of proline, retinol, α-tocopherol and ascorbic acid in 17-day-old seedlings of *Phaseolus vulgaris* L (bean). After ten-day exposure to heavy metals, primary leaves showed a considerable increase in proline, retinol, α -tocopherol and ascorbic acid level. The strongest effect on proline, retinol, α -tocopherol and ascorbic acid level was found in plants exposed to mercury (Hg^{++}), followed by the sequence $Cd^{++} > Cu^{++} > Pb^{++}$. Increased proline level has been observed at lower concentrations of Cd and Pb but at higher concentrations it was found to be decreased in *Brassica juncea* (Jhon et al., 2009).

Malar et al. (2014) studied the effect of Pb on the activity of superoxide dismutase (SOD) at the concentrations of 100, 200, 400, 600, 800, 1000 mg/L in *Eichhornia crassipes* seedlings. SOD activity was found to increase significantly at a concentration dependent manner, up to 800 mg/L, although, the SOD activity has decreased slightly at 1000 mg/L Pb concentration. Treatment of different concentrations of Cd (0, 0.2, 0.4, and 0.6 mM) for 30 days has been given to plants of *Brassica juncea* (var. RLC-1) to analyze its effect on anti-oxidative enzymes (superoxide dismutase, polyphenol oxidase, glutathione-S transferase, and glutathione peroxidase). Activities of all these enzymes were enhanced with the increased dose of Cd as compared to control plants (Kapoor et al., 2014). Sidhu et al. (2016) conducted an experiment to assay the effect of lead (Pb) on anti-oxidative response in *Coronopus didymus*. Exposure of Pb stimulated the activities of antioxidants (SOD, CAT, APX, GPX and GR) after 6 weeks. However, at the highest Pb concentration, the activities of SOD and CAT declined (Sidhu et al., 2016).

Effect of Heavy Metals on Chlorophyll Content

Chlorophyll is an important pigment necessary for plant growth. Chlorophyll content is often measured in plants in order to assess the impact of environmental stress, as changes in pigment content are linked to visual symptoms of plant illness and photosynthetic productivity (Parekh et al., 1990). Increase in the concentration of heavy metals decrease the chlorophyll content and causes deleterious effect in the metabolism especially in inhibition of the enzymes responsible for chlorophyll biosynthesis.

Table 1. Affects of some heavy metals on plants at their higher concentrations

Heavy Metal	Effect on Plants	References
Cd	Reduction in photosynthesis, growth and also in water and nutrient uptake.	(Wójcik and Tukiendorf, 2004; Kapoor et al., 2014; Asati et al., 2016).
	Symptoms of injury reflected as chlorosis and browning of root tips.	
Zn	Inhibit many plant metabolic functions; result in retarded growth (limit the growth of both root and shoot) and cause senescence.	(Ebbs and Kochian, 1997; Asati et al., 2016).
	It causes chlorosis in the leaves at their high level and give rise the manganese (Mn) and copper (Cu) deficiencies in plant shoots.	(Ebbs and Kochian, 1997; Asati et al., 2016).
	Appearance of purplish-red color in leaves, which is ascribed to phosphorus (P) deficiency.	(Lee et al., 1996; Asati et al., 2016).
Cu	Perform a cytotoxic role, induces stress and causes injury to plants that leads to plant growth retardation and leaf chlorosis.	(Lewis et al., 2001; Asati et al., 2016).
	Excess of Cu generates oxidative stress and causes disturbance of metabolic pathways and damage to macromolecules.	(Hegedus et al., 2001; Asati et al., 2016).
Hg	Induce visible injuries and physiological disorders in plants For example, it can bind to water channel proteins, consequently inducing leaf stomata to close and hinder the flow of water.	(Zhou et al., 2007; Asati et al., 2016).
	High level of Hg^{2+} interfere the mitochondrial activity and induces oxidative stress by promoting the generation of ROS (reactive oxygen species).	(Zhang and Tyerman, 1999; Asati et al., 2016).
Co	Adversely affect the shoot growth and biomass.	(Li et al., 2009).
	Extremity of Co restricted the concentration of Fe, chlorophyll, protein and catalase activity in leaves of cauliflower and also affect the translocation of P, S, Mn, Zn and Cu from roots to top.	(Chatterjee and Chatterjee, 2000; Asati et al., 2016).
Ni	Causes various physiological disturbance and toxicity symptoms such as chlorosis and necrosis in different plant species.	(Pandey and Sharma, 2002; Rahman et al., 2005; Asati et al., 2016).
	Excess of Ni^{2+} in soil showed impairment of nutrient balance and resulted in alternation of cell membrane functions (alter lipid composition and H-ATPase activity of membrane) as reported in shoots of *Oryza sativa*.	(Ros et al., 1992; Asati et al., 2016).
	It can decrease water uptake at their higher level.	(Pandey and Sharma, 2002; Gajewska et al., 2006; Asati et al., 2016).
Pb	A very low level, it may inhibit some vital plant processes i.e. photosynthesis, mitosis and water absorption. The other symptoms includes: dark green leaves, wilting of older leaves, stunted foliage and brown short leaves and roots.	(Bhattacharyya et al., 2008; Asati et al., 2016).

Oncel et al. (2000) gave Cd and Pb treatments in two varieties of wheat (*Triticum aestivum* Gerek 79 and Bolal 2973) and observed that the total chlorophyll content decreased by 50% in "Gerek 79" and by 70% in "Bolal 2973". Cadmium was reported to affect the chlorophyll development and to inhibit the production of protochlorophyll reductase and aminolevulinic acid (Stobart et al., 1985). The effects of lead, copper, cadmium and mercury on the chlorophyll content of *Phaseolus vulgaris* has been studied by Zengin and Munzuroglu (2005) and it was observed that total chlorophyll content in leaves significantly reduced on application of Pb (1.5 mM), Cu (0.3 mM), Cd (0.08 mM) and Hg (0.06, mM). Jhon et al. (2009) also showed the toxic effect of Pb and Cd on chlorophyll 'a' and chlorophyll 'b' level in *Brassica juncea.* Pb proved to be less toxic as compared to Cd.

Christofaki (2011) studied the effect of the different concentrations of Ni (0.1, 0.5 and 1 mM) on total chlorophyll content of lower and upper leaves of *Nicotiana tabacum* and *Nicotiana glauca* (Figure 3). The harmful effect of Ni became more prominent with increasing concentrations. Chlorophyll a and b were gradually reduced up to about 28% and 37%, respectively when exposed to 1mM Ni, compare to control plants.

Liu et al. (2014) observed the effects of Cd on Chl a, Chl b, and Chl (a + b) in cotton seedlings after 7 days. They found that the chlorophyll content of the cotton seedlings was significantly inhibited at low concentration of Cd (25 μM), and the inhibitory effects were enhanced as the Cd concentration was increased. Dey et al. (2014) studied the effect of high concentration of copper stress on chlorophyll content in two different cultivars of *Camellia sinensis* (tea plant) *viz.* TV-23, TV-17. A significant decrease in chlorophyll content was detected at concentrations above 300 μM. They also found that photosynthetic activity of TV-17 was more sensitive (retardation level is high) in comparison to TV-23, for Cu stress.

The different concentrations of Cd (0, 0.2, 0.4, and 0.6mM) has been given to the plants of *Brassica juncea* (var. RLC-1) for 30 days to analyze the effect on level of photosynthetic pigments. There was a significant decrease in total chlorophyll content after 30-day treatment (Kapoor et al., 2014). Plants of *Brassica juncea* (var. Pusa Jaikisan) grew on soil supplemented with different concentrations of nickel and chromium (50 and 100 mg kg^{-1} soil), lead (100 and 200 mg kg^{-1} soil), cadmium (5 and 10 mg kg^{-1} soil) and mercury (25 and 50 mg kg^{-1} soil) to observe their effect on chlorophyll content. A significant decrease in chlorophyll a and b content at higher concentration of these metals was observed (Sheetal et al., 2016).

Figure 3. Symptoms of Ni Stress: a) on leaves of Nicotiana tabacum nd b) on leaves of Nicotiana glauca; Symptoms of Cd stress: a) on leaves of Nicotiana tabacum and b) on leaves of Nicotiana glauca (Christofaki, 2011)

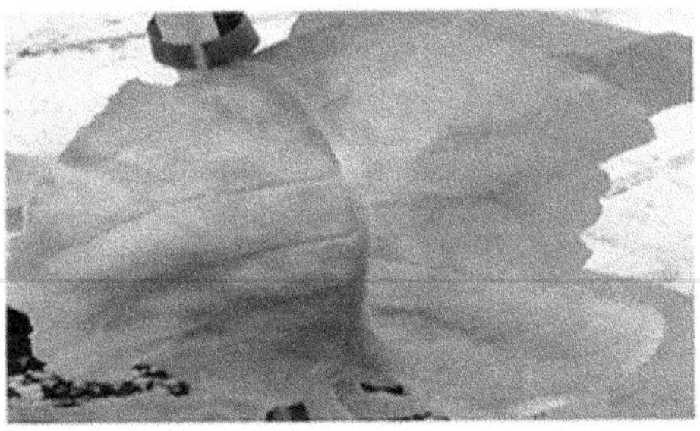

(a) (b)

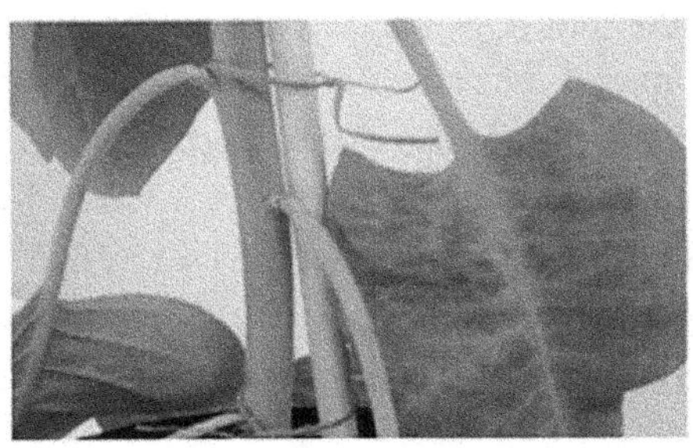

(a) (b)

Effect of Heavy Metals on Seed Germination

Heavy metals also affect seed germination in plants. At high concentration of metal, seed germination was gradually delayed (Ashraf and Ali, 2007). Jadia and Fulekar (2008) studied a concentration dependent inhibition of the seed germination in *Helianthus annuus* and observed that 5 ppm of Cd, Ni and Pb had very low toxic effects on seed germination, while 5 ppm of copper increased seed germination. Cd, Cu, Ni and Pb at 10, 20, 40 and 50 ppm level were found most deleterious for seed germination. Delayed germination was observed at higher concentration i.e., 40 and 50 ppm. The decreasing order of toxicity for metals on seed germination observed was Cd > Cu > Ni > Pb >Zn (Jadia and Fulekar, 2008). Mahmood et al. (2007) found that increasing level of Cu significantly inhibited the seeds germination in rice, wheat and barley plants. The inhibitory effect of Cu on seed germination was more distinct on rice than wheat and barley. At 10 μM of Cu, seed germination of wheat and rice was reduced more than 35 and 60%, respectively over control. They also reported that Pb and Zn did not considerably affect seed germination.

Abraham et al. (2013) studied the effect of cadmium, lead, and copper on seed germination of *Arachis hypogaea*. They found that higher concentration of Cd (at 75 and 100 mg/L), Pb (at 75 and 100 mg/L), Cu (at 100 mg/L) reduced seed germination of groundnut as compared to control. Cadmium produced more significant effect on seed germination than lead and copper. Effect of several concentrations (1, 10, 20, and 30 mM) of Cd, Cr, Zn, Cu, Pb and Hg on seed germination of *Jatropha curcas* was studied. After 10 days, they found a significant reduction in seed germination as metal concentrations increased (Acharya and Sharma, 2014). Bae et al. (2016) conducted an experiment to measure the effects of Zn, Pb, Ni, Cu, and Cd on seed germination of legumes (*Coronilla varia*, *Lotus corniculatus*, and *Trifolium arvense*). Seed germination of *T. arvense* inhibited by all the metals but *A. artemisiifolia* showed least effect. Pb and Ni were found to stimulate initiation of seed germination in *A. artemisiifolia*. Zn, Pb, and Ni did not affect the germination of *L. corniculatus,* while Cu and Cd inhibited the germination. Germination of *C. varia* was decreased by Ni, Cu, and Cd and delayed by Zn and Pb.

Effect of Heavy Metals on Growth and Yield

The first symptoms related to heavy metal toxicity include stunted growth, reduced root growth and deformation of roots. Typically roots are more affected than leaves because they have direct contact with the metal. The toxicity of Cu on *Brassica juncea* plants was studied by Wang et al. (2004). They observed a significant decrease in root biomass at 4 μM concentration of Cu. Zaier et al. (2010) reported decreased plant biomass of the same species at 200 μM and higher concentration of Pb. Nguyen (2015) studied the effect of Cu, Pb and Zn on growth and biomass yield of *Brassica juncea*. 50 ppm, 100 ppm 200 ppm concentrations of Cu and Pb were added in soil, while those for Zn were added as 100 ppm, 300 ppm, 500 ppm. The height of the plant decreased significantly when concentrations of Cu, Pb and Zn were increased. The amount of added heavy metal not only reduced plant height but also affected the yield. Jhon et al. (2009) also exhibited the toxic effects of Cd and Pb on growth of *Brassica juncea*. They found decreased root length from 17% (300 μM Cd) to 54% (900 μM Cd) at the pre-flowering stage. Similarly, stem height declined from 4% (300 μM Cd) to 51% (900 μM Cd) at the post-flowering stage. Only the higher concentration of Pb (1500 μM) was toxic and it was maximum at the post-flowering stage (1500 μM Pb declined root length from 28.21 cm to 13.97 cm after 60 days).

Lead (Pb) has been reported to significantly affect the *Brassica napus* plants growing in hydroponic solution at their 50 and 100 μM concentrations (Shakoor et al., 2014). The seedling growth of *Gossypium hirsutum* was intensely decreased by Cd treatments and with the increased level of Cd, plant height and leaf area were also found to decrease (Liu et al., 2014). Jadia and Fulekar (2008) conducted an experiment to determine the effect of heavy metals (Cd, Cu, Ni, Pb and Zn) on the growth of Sunflower (*Helianthus annuus*). At lower concentrations (5 and 10 ppm) of these metals, root and shoot growth were found to increase except for Cd which resulted in reduced root growth, however the growth of roots and shoots were decreased at higher concentrations of metals (40 and 50 ppm).

Mahmood et al. (2007) studied the effect of Cu, Zn, Pb, Mg and Na (1, 5 and 10 μM concentrations) on seedling growth, root length and shoot height at barley (*Hordeum vulgare* L.), rice (*Oryza sativa* L.) and wheat (*Triticum aestivum* L.). Cu showed the most adverse effects on growth of seedlings followed by Zn and Pb, while Mg and Na did not affect seedling growth of these plants. Cu at 10 μM reduced seed germination of wheat and rice by 35 and 60%, respectively over the control. Shoot height of wheat seedlings decreased significantly, however, barley shoot height increased at the same concentration (at 10 μM of Cu). The shoot height of barley was increased, while that of wheat decreased by 15% when level of Pb was raised. The effect of Zn on shoot height of barley seedlings were same to that of the Pb. Root length of all crop seedlings was significantly affected by increasing Cu, Pb and Zn. The effect of Cu on root length was more prominent than Pb and Zn. At 10 μM of Cu, the root length of both barley and wheat was decreased by 50% and rice by 40% over control.

Christofaki (2011) studied the effect of Zn on weight of leaves, shoot and root of *Nicotiana tabacum* and *Nicotiana glauca* followed a dose-dependent manner (0.5, 2.5 and 5 mM). Reduction of fresh weight of *N. glauca* was more prominent than that of the *N. tabacum*, at the higher concentrations of Zn. At 2.5 and 5 mM of Zn, weight of leaves and shoot of *N. glauca* was decreased by 28% and 44%, respectively, while in case of *N. tabacum* the addition of 2.5 and 5 mM Zn led to significant reduction by 26% and 39%, respectively, over control.

Dey et al. (2014) studied the effect of high concentration of copper on growth of *Camellia sinensis*. The weight of shoot and root decreased progressively with increased Cu concentration. Plants showed diverse symptoms at the different concentration of Cu. In comparison to control, there was a significant decrease in the growth of plants treated with Cu. Leaves were least affected than roots and shoots at lower concentration (200 μM of Cu). However increased concentration of Cu (500 μM) severely inhibited the root growth with accumulation of Cu in leaves (Figure 4). Black brown spots were observed on the

Figure 4. Cu toxicity on two different cultivars of Camellia sinesis (TV-23 and TV-17) at concentrations of 200 μm and 500 μm: (a) small yellow patches on leaves of TV-23 (at 200 μm of Cu) (b) Effects of leaves on TV-23 (at 500 μm of Cu) (c) Effects of leaves of TV-17 (at 500 μm of CU) (Dey et al., 2014)

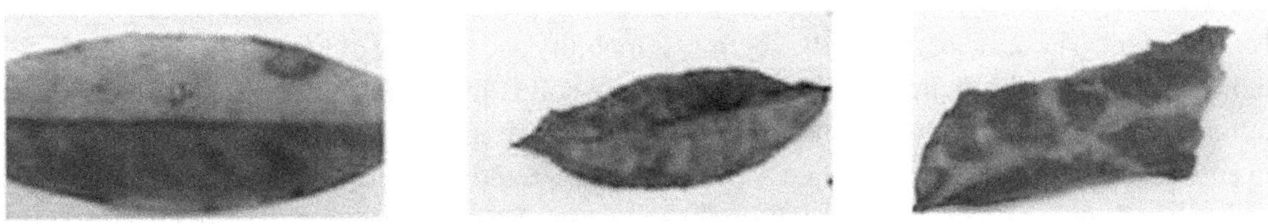

(a) (b) (c)

leaves at higher concentration (500 μM) (Figure 4b and 4c). The root showed reduced length. Formation of new leaves and stem branches also got reduced. As a whole the growth gets stunted and this resulted in decreased biomass.

Malar et al. (2014) observed the effect of Pb on growth and biomass of seedling of *Eichhornia crassipes* (water hyacinth) at the concentrations of 100, 200, 400, 600, 800, 1000 mg/L. Growth retardation was a common response to heavy metal toxicity. Inhibitory effect was more on Shoot than root at different concentrations of Pb after 10 days' treatment. Plants did not show any visible toxicity symptoms up to 800 mg/L of Pb, however, 1000 mg/L of Pb displayed toxicity symptoms like chlorosis and drying edges in seedlings. At this concentration biomass of seedlings reduced by 33.28% as compared to control.

Effect of different concentration of Cd, Cr, Zn, Cu, Pb and Hg (1, 10, 20, and 30 mM) on growth of *Jatropha curcas* has been studied (Acharya and Sharma, 2014). 1 mM-dose of Cd and Cu and 1 and 10 mM-dose of Zn favor the growth of roots of the plants over control, while for all the treatments of Cr, Pb and Hg, root length was significantly decreased. Exposures of 1 mM of Cd, Cr, Pb and Hg reduced the shoot length as compared to control, while Cu and Zn increased the shoot length by 21.1 and 29.5% respectively. Production of biomass of *Brassica juncea on* different concentrations of nickel and chromium (50 and 100 mg kg^{-1} soil), lead (100 and 200 mg kg^{-1} soil), cadmium (5 and 10 mg kg^{-1} soil) and mercury (25 and 50 mg kg^{-1} soil) has been studied by Sheetal et al. (2016). They found significant reduction in biomass of the plant with the increased level of metal ions over control.

Effect of Heavy Metals on Water Content

Heavy metal stress resulted in decreased water content in plants which in turn increased toxic effects of heavy metals (Ali et al., 2015). Zaier et al. (2010) reported that the water content of *Brassica juncea* plants drastically decreased under Pb toxicity, although this species has been considered as tolerant species. Some authors reported that soluble sugar, proline and other amino acids can have a protective role, that maintained osmotic potential and also helped in detoxification of reactive oxygen species (ROS) (Sharma and Dietz, 2006; Keunen et al., 2013). Seedlings of *Eichhornia crassipes* (water hyacinth) were treated with different concentrations of Pb (100, 200, 400, 600, 800, 1000 mg/L) to observe the effect on water content. Relative water content (RWC) in water hyacinth increased slightly, up to 400 mg/L of Pb thereafter it decreased at higher concentrations (Malar et al., 2014)

Management of Contaminated Soil

Plant uptake metals from the soil. Soil is a major conveying media for heavy metals because both soil and heavy metals have rich and diverse binding properties. Soil may adsorb, exchange, oxidize, reduce, catalyze or precipitate metal ions. This depends on several factors such as pH, temperature, water content, nature of metal, distribution of particle size, cation exchange capacity, nutrient balance, organic matter and clay content. These factors affect the solubility, mobility and toxicity of heavy metals (Morel, 1997; Rutkowska et al., 2015). Removal of heavy metal from soil is important for the improvement of crop health and productivity. Phyto- and microbial remediation are major way to remove the heavy metals from soil or metal may be modified in a form that metals could not harm the plant productivity. Following methods may help in preventing the uptake of metals by plants, leaving them in the soil, besides phyto-remediation and microbial remediation:

1. High pH (6.5 or higher) increased immobility of heavy metals in soil. Cationic metals are more soluble at lower pH. Increased pH of soil makes them less available to plants, consequently less likely to be incorporated in their tissues and ingested by humans. Rutkowska et al. (2015) reported that the total Zn concentration in the soil solution decreased with increasing pH. This may be due to the greater adsorptive capacity of the soil solid surfaces resulting from increased pH-dependent negative charge, the formation of hydrolyzed forms of Zn, chemisorption on calcite and co-precipitation in Fe oxides (Alloway, 2009). Increased pH has an opposite effect on anionic metals. Liming has been reflected to be an economically acceptable method which reduced the transport of heavy metals as it induces an increase in soil pH (McLaughlin, 2002).
2. Organic matter in soil (humus) bear negatively charged sites on carboxyl and phenol groups, to bind metals, thus form metal-organic complex leads to reduce metal uptake (Stevenson, 1982; Violante et al., 2010). High amounts of insoluble organic matter in soil is negatively correlated with uptake of metals by plant, as often observed on peat soils with Cu (Miller et al., 1976). Soil components differ greatly in their sorption capacities, their cation and anion exchange capacities, and the binding energies of their sorption sites (Violante et al., 2010).
3. Redox potential was found to affect solubility of heavy metals in soil. (Patrick et al, 1990; Masscheleyn et al.,1991). Yaron et al. (1996) showed that under same pH values, metal solubility increases as redox potential decreased. Chuan et al. (1996) also reported that solubility of Pb, Cd and Zn in soil increased at reduced redox potential. The reductive transformation of some heavy metals may also take place chemically, for example Cu(II) reduction to Cu(I) by Fe2+ or H2S, reduction of Cu(II), Ag(II), and Hg(II) to elemental forms by Fe(II)- bearing green-rust, thus reducing solubility of metals (Violante et al. 2010). Microorganisms may directly reduce toxicity heavy metals (e.g., Cr, Hg, U) via detoxification pathways. Microbial reduction of certain metals to a lower redox state may result in reduced mobility and toxicity. Aerobic and anaerobic reduction of Cr (VI) to Cr (III) is widespread in microorganisms (Violante et al., 2010).
4. Drainage improve soil aeration, thus allow metals to oxidize and makes them less soluble. Therefore, when aerated, these metals are less available to plants; however, chromium is more available in oxidized forms. Zhao et al., (2016) Investigated effect of irrigation and drainage modes on the residual characteristics of heavy metals (As, Cd, Cr, Cu, Mn, and Pb) in soil. The results showed that different water-saving modes in farmland had a notable effect on the residual concentration of heavy metals: Water-saving irrigation > flooding irrigation > dry field. The bioavailability fractions of Cd showed the most significant difference between water-saving irrigation (8.55%) and flooding irrigation (6.16%), and for bioavailability fractions of non-rhizosphere (10.53–10.78%) it was higher compared with rhizosphere fractions (3.26–7.10%). Therefore, bioavailability fractions of heavy metals in soil were closely associated with the input intensity of pollutants and the leaching intensity of different water-saving modes. The potential bioavailability fractions (Fe/Mn oxide fraction and organic matter fraction) of heavy metals in paddy fields (28.89%) were lesser than in dry fields (30.37%), the potential bioavailability fractions of heavy metals in flooding irrigation (30.14%) were greater than in water-saving irrigation (27.64%), and for rhizosphere soils (25.73–37.64%) it was greater compared with non-rhizosphere soils (27.64–31.18%); the most significant response on different water-saving modes showed Cu. Thus, the increase of crystalline forms of Fe/Mn oxide induced by a high redox potential and dissolved oxygen in dry fields and the rhizosphere soils played an important role for the high percentage of potential bioavailability fractions of heavy metals (Zhao et al., 2016).

5. Plant micronutrient requirement also control the ability to take up or exclude toxic elements. Some plants are well adapted for survival in stressful environmental conditions. These plants can hold heavy metals at a level more than 1% and up to 25% on a dry matter basis. These plants are called 'hyperaccumulators'. Grasses take up fewer trace elements than fast-growing plants, e.g. lettuce, spinach and carrots. Accumulation of Cd by different plant species have been found in the decreasing order as: leafy vegetables > root vegetables > grain crops (Morel, 1997).
6. There are some remediation technologies for metal in soils (Sidhu, 2016). However, they are expensive when applied to large area of soil. Treatment can be done *in-situ* (on-site), or *ex-situ* (removed and treated off-site). Some available treatments include:
   a. High temperature treatments (produce a vitrified, granular, non-leachable material).
   b. Washing process (leaches out contaminants).
   c. Excavation (Physical removal of the contaminants)
   d. Stabilization: Heavy metals treated in a way that reduced their availability to plants, for e.g. application of phosphate on large scale reduced the availability of cationic metals but posed the opposite effect on anionic compound like arsenic (As).
7. Stimulation of microbial population in the contaminated soil can reduce heavy metal concentrations in soil. The increased microbial biomass enhances accumulation of certain heavy metals thus reducing their availability for plants.

REFERENCES

Abraham, K., Sridevi, R., Suresh, B., & Damodharam, T. (2013). Effect of heavy metals (Cd, Pb, Cu) on seed germination of *Arachis hypogeae*. L. *Asian Journal of Plant Science and Research*, *3*, 10–12.

Acharya, S., & Sharma, D. K. (2014). Study on the effects of heavy metals on seed germination and plant growth on *Jatropha curcas. International Journal of Agricultural Science Research*, *3*, 31–34.

Ali, B., Gill, R. A., Yang, S., Gill, M. B., Farooq, M. A., Liu, D., & Zhou, W. et al. (2015). Regulation of cadmium-induced proteomic and metabolic changes by 5-aminolevulinic acid in leaves of *Brassica napus* L. *PLoS ONE*, *10*(4), 1–23. doi:10.1371/journal.pone.0123328 PMID:25909456

Alkhader, A. M. F. (2015). The impact of phosphorus fertilizers on heavy metals content of soils and vegetables grown on selected farms in Jordan. *Agrotechnology*, *5*(01), 1–15. doi:10.4172/2168-9881.1000137

Alloway, B. J. (2009). Soil factors associated with zinc deficiency in crops and humans. *Environmental Geochemistry and Health*, *31*(5), 537–548. doi:10.1007/s10653-009-9255-4 PMID:19291414

Arunakumara, K. K. I. U., Walpola, B. C., & Yoon, M.-H. (2013). Alleviation of Phyto-toxicity of Copper on Agricultural Plants. *Journal of the Korean Society for Applied Biological Chemistry*, *56*(5), 505–517. doi:10.1007/s13765-013-3126-5

Asati, A., Pichhode, M., & Kumar, N. (2016). Effect of heavy metals on plants: An overview. *International Journal of Application or Innovation in Engineering & Management*, *5*, 56–66.

Ashraf, R., & Ali, T. A. (2007). Effect of heavy metals on soil microbial community and mung beans seed germination. *Pakistan Journal of Botany*, *39*, 629–636.

Bae, J., Benoit, D. L., & Watson, A. K. (2016). Effect of heavy metals on seed germination and seedling growth of common ragweed and roadside ground cover legumes. *Environmental Pollution, 213*, 112–118. doi:10.1016/j.envpol.2015.11.041 PMID:26882164

Baycu, G., Rognes, S. E., Hakan, O., Gören-Saglam, N., Csatári, I., & Szabo, S. (2017). Abiotic stress effects on the antioxidative response profile of *Albizia julibrissin* Durazz. (Fabaceae). *Brazilian Journal of Botany, 40*(1), 21–32. doi:10.1007/s40415-016-0318-3

Bell, F. G., Bullock, S. E. T., Halbich, T. F. J., & Lindsay, P. (2001). Environmental impacts associated with an abandoned mine in the Witbank Coalfield, South Africa. *International Journal of Coal Geology, 45*(2-3), 195–216. doi:10.1016/S0166-5162(00)00033-1

Bhattacharyya, P., Chakrabarti, K., Chakraborty, A., Tripathy, S., & Powell, M. A. (2008). Fractionation and bioavailability of Pb in municipal solid waste compost and Pb uptake by rice straw and grain under submerged condition in amended soil. *Geosciences Journal, 12*(1), 41–45. doi:10.1007/s12303-008-0006-9

Chatterjee, J., & Chatterjee, C. (2000). Phytotoxicity of cobalt, chromium and copper in cauliflower. *Environmental Pollution, 109*(1), 69–74. doi:10.1016/S0269-7491(99)00238-9 PMID:15092914

Chibuike, G. U., & Obiora, S. C. (2014). Heavy Metal Polluted Soils: Effect on Plants and Bioremediation Methods. *Applied and Environmental Soil Science, 2014*, 1–12. doi:10.1155/2014/752708

Christofaki, M. I. (2011). *Effect of heavy metal stress in plant metabolism of solanaceous plant species with emphasis on nitrogen assimilation* (PhD thesis). Cranfield University.

Chuan, M. C., Shu, G. Y., & Liun, J. C. (1996). Solubility of heavy metals in a contaminated soil: Effects of redox potential and pH. *Water, Air, and Soil Pollution, 90*(3-4), 543–556. doi:10.1007/BF00282668

Costa, G., & Morel, J. L. (1994). Water relations, gas exchange and amino acid content in Cd-treated lettuce. *Plant Physiology and Biochemistry, 32*, 561–570.

Demiral, T., & Türkan, I. (2005). Comparative lipid peroxidation, antioxidant defense systems and proline content in roots of two rice cultivars differing in salt tolerance. *Environmental and Experimental Botany, 53*(3), 247–257. doi:10.1016/j.envexpbot.2004.03.017

Demirevska-kepova, K., Simova-Stoilova, L., Stoyanova, Z., Holzer, R., & Feller, U. (2004). Biochemical changes in barely plants after excessive supply of copper and manganese. *Environmental and Experimental Botany, 52*(3), 253–266. doi:10.1016/j.envexpbot.2004.02.004

Dey, S., Mazumder, P. B., & Paul, S. B. (2014). Effect of copper on growth and chlorophyll content in tea plants (*Camellia sinensis)* (l.) O. Kuntze. *International Journal of Research in Applied Natural and Social Sciences, 2*, 223–230.

Dietz, K. J., Baier, M., & Krämer, U. (1999). *Free radicals and reactive oxygen species as mediators of heavy metal toxicity in plants: from molecules to ecosystems*. Berlin: Springer- Verlag. doi:10.1007/978-3-662-07745-0_4

Ebbs, S. D., & Kochian, L. V. (1997). Toxicity of zinc and copper to Brassica species: Implications for phytoremediation. *Journal of Environmental Quality, 26*(3), 776–781. doi:10.2134/jeq1997.00472425002600030026x

El-Khatib, A. A. (2003). The response of some common Egyptian plants ozone and their use a biomonitors. *Environmental Pollution, 124*(3), 419–428. doi:10.1016/S0269-7491(03)00045-9 PMID:12758022

Gajewska, E., Sklodowska, M., Slaba, M., & Mazur, J. (2006). Effect of nickel on antioxidative enzyme activities, proline and chlorophyll contents in wheat shoots. *Biologia Plantarum, 50*(4), 653–659. doi:10.1007/s10535-006-0102-5

Guala, S. D., Vega, F. A., & Covelo, E. F. (2010). The dynamics of heavy metals in plant–soil interactions. *Ecological Modelling, 221*(8), 1148–1152. doi:10.1016/j.ecolmodel.2010.01.003

Hall, J. L. (2002). Cellular mechanisms for heavy metal detoxification and tolerance. *Journal of Experimental Botany, 53*(366), 1–11. doi:10.1093/jxb/53.366.1 PMID:11741035

Hare, P. D., Cress, W. A., & Van Staden, J. (1999). Proline synthesis and degradation: A model system for elucidating stress-related signal transduction. *Journal of Experimental Botany, 50*, 413–434.

Hegedus, A., Erdei, S., & Horvath, G. (2001). Comparative studies of H_2O_2 detoxifying enzymes in green and greening barley seedlings under cadmium stress. *Plant Science, 160*(6), 1085–1093. doi:10.1016/S0168-9452(01)00330-2 PMID:11337065

Hewitt, E. J. (1983). Metals and micronutrients: uptake and utilization by plants. London: Academic Press.

Hossain, M. A., Hoque, M. A., Burritt, D. J., & Fujita, M. (2014). *Proline Protects Plants Against Abiotic Oxidative Stress: Biochemical and Molecular Mechanisms. In Oxidative damage to plants* (pp. 477–52). Academic Press.

Jadia, C. D., & Fulekar, M. H. (2008). Phytoremediation: The application of vermicompost to remove zinc, cadmium, copper, nickel and lead by sunflower plant. *Environmental Engineering and Management Journal, 7*, 547–558.

Jan, S., & Parray, J. A. (2016). *Heavy metal uptake in plants approaches to heavy metal tolerance in plants* (pp. 1–18). Singapore: Springer. doi:10.1007/978-981-10-1693-6

John, R., Ahmad, P., Gadgil, K., & Sharma, S. (2009). Heavy metal toxicity: Effect on plant growth, biochemical parameters and metal accumulation by *Brassica juncea* L. *International Journal of Plant Production, 3*, 65–76.

Jordao, C. P., Nascentes, C. C., Cecon, P. R., Fontes, R. L. F., & Pereira, J. L. (2006). Heavy metal availability in soil amended with composted urban solid wastes. *Environmental Monitoring and Assessment, 112*(1-3), 309–326. doi:10.1007/s10661-006-1072-y PMID:16404548

Kapoor, D., Kaur, S., & Bhardwaj, R. (2014). Physiological and Biochemical Changes in *Brassica juncea* plants under Cd-Induced Stress. *BioMed Research International, 2014*, 1–13. doi:10.1155/2014/726070 PMID:25133178

Kapoor, D., Rattan, A., Kaur, S., & Bhardwa, R. (2016). Influence of cadmium on antioxidative defence system, photosynthesis, level of osmolytes and ions uptake in *Brassica juncea*. *International Journal of Pharmacy and Pharmaceutical Sciences, 8*, 1–13.

Kennish, L. (1992). Toxicity of heavy metals: Effect of Cr and Se on human health. *Journal of Indian Public Health Education*, *2*, 36–64.

Keunen, E., Peshev, D., Vangronsveld, J., Wim, V. D. E., & Cuypers, A. (2013). Plant sugars are crucial players in the oxidative challenge during abiotic stress: Extending the traditional concept. *Plant, Cell & Environment*, *36*(7), 1242–1255. doi:10.1111/pce.12061 PMID:23305614

Khan, S., Cao, Q., Zheng, Y. M., Huang, Y. Z., & Zhu, Y. G. (2008). Health risks of heavy metals in contaminated soils and food crops irrigated with wastewater in Beijing, China. *Environmental Pollution*, *152*(3), 686–692. doi:10.1016/j.envpol.2007.06.056 PMID:17720286

Khoshgoftarmanesh, A. H., Hosseini, F., & Afyuni, M. (2011). Nickel supplementation effect on the growth, urease activity and urea and nitrate concentrations in lettuce supplied with different nitrogen sources. *Scientia Horticulturae*, *130*(2), 381–385. doi:10.1016/j.scienta.2011.07.009

Kuznetsov, V. V., & Shevyakova, N. I. (1997). Stress responses of tobacco cells to high temperature and salinity: Proline accumulation and phosphorylation of polypeptides. *Physiologia Plantarum*, *100*(2), 320–326. doi:10.1111/j.1399-3054.1997.tb04789.x

Lee, C. W., Choi, J. M., & Pak, C. H. (1996). Micronutrient toxicity in seed geranium (*Pelargonium*×*hortorum* Baley). *Journal of the American Society for Horticultural Science*, *121*, 77–82.

Lenntech. (2004). *Water Treatment and Air Purification*. Retrieved from www.excelwater.com/thp/filters/Water-Purification.htm

Lewis, S., Donkin, M. E., & Depledge, M. H. (2001). Hsp70 expression in *Enteromorpha intestinalis* (Chlorophyta) exposed to environmental stressors. *Aquatic Toxicology (Amsterdam, Netherlands)*, *51*(3), 277–291. doi:10.1016/S0166-445X(00)00119-3 PMID:11090890

Li, H. F., Gray, C., Mico, C., Zhao, F. J., & McGrath, S. P. (2009). Phytotoxicity and bioavailability of cobalt to plants in a range of soils. *Chemosphere*, *75*(7), 979–986. doi:10.1016/j.chemosphere.2008.12.068 PMID:19217640

Liu, L., Sun, H., Chen, J., Zhang, Y., Li, D., & Li, C. (2014). Effects of cadmium (Cd) on seedling growth traits and photosynthesis parameters in cotton (*Gossypium hirsutum* L.). *Plant Omics Journal*, *7*, 284–290. doi:10.1111/tpj.12387

Lugon-Moulin, N., Zhang, M., Gadani, F., Rossi, L., Koller, D., Krauss, M., & Wagner, G. J. (2004). Critical review of the science and options for reducing cadmium in tobacco (*Nicotiana tabacum* L.) and other plants. *Advances in Agronomy*, *83*, 111–180. doi:10.1016/S0065-2113(04)83003-7

Mahmood, T., Islam, K. R., & Muhammad, S. (2007). Toxic effects of heavy metals on early growth and tolerance of cereal crops. *Pakistan Journal of Botany*, *39*, 451–462.

Makoi, J. H. J. R. (2016). Status of Fe, Cu, Zn and Mn and their relationship with some physical chemical properties in the soils of Sonjo and Lupiro alluvial/flood plains in Kilombero and Ulanga Districts, Morogoro, Tanzania. *Academia Journal of Biotechnology*, *4*, 76–84.

Malar, S., Sahi, S. V., Favas, P. J. C., & Venkatachalam, P. (2014). Lead heavy metal toxicity induced changes on growth and antioxidative enzymes level in water hyacinths. *Botanical Studies (Taipei, Taiwan), 55*, 1–11. PMID:28597420

Masscheleyn, P., Pardue, J., Delaune, R., & Patrick, J. (1991). Effect of redox potential and pH on As speciation and solubility in a contaminated soil. *Environmental Science & Technology, 25*(8), 1414–1419. doi:10.1021/es00020a008

McLaughlin, M. (2002). Heavy metals. In R. Lal (Ed.), *Encyclopedia of soil science* (pp. 650–653). New York: Marcel Dekker.

Mendel, R. R. (2011). Cell biology of molybdenum in plants. *Plant Cell Reports, 30*(10), 1787–1797. doi:10.1007/s00299-011-1100-4 PMID:21660547

Miller, J. E., Hassett, J. J., & Koeppe, D. E. (1976). Uptake of cadmium by soybean as influenced by soil cation exchange capacity, pH and available phosphorus. *Journal of Environmental Quality, 5*(2), 157–160. doi:10.2134/jeq1976.00472425000500020009x

Mishra, A., & Choudhuri, M. A. (1999). Monitoring or phyto-toxicity of lead and mercury from germination and early seedling growth induces two rice cultivars. *Water, Air, and Soil Pollution, 114*(3/4), 339–346. doi:10.1023/A:1005135629433

Mithofer, A., Schulze, B., & Boland, W. (2004). Biotic and heavy metal stress response in plants: Evidence for common signals. *FEBS Letters, 566*(1-3), 1–5. doi:10.1016/j.febslet.2004.04.011 PMID:15147858

Morel, J. L. (1997). Bioavailability of trace elements to terrestrial plants. In J. Tarradellas, G. Bitton, & D. Rossel (Eds.), *Soil ecotoxicology* (pp. 141–176). CRS Press.

Nguyen, X.C. (2015). Effect of Heavy Metals on plant growth and ability to use fertilizing substances to reduce heavy metal accumulation by *Brassica juncea* L. Czern. *Global Journal of Science Frontier Research: D Agriculture and Veterinary, 15*, 35-40.

Nies, D. H. (1999). Microbial heavy-metal resistance. *Applied Microbiology and Biotechnology, 51*(6), 730–750. doi:10.1007/s002530051457 PMID:10422221

Noctor, G., & Foyer, C. H. (1998). Ascorbate and glutathione: Keeping active oxygen under control. *Annual Review in Plant Physiology and Molecular Biology, 49*(1), 249–279. doi:10.1146/annurev.arplant.49.1.249 PMID:15012235

Oncel, I., Keles, Y., & Ustun, A. S. (2000). Interactive effects of temperature and heavy metal stress on the growth and some biochemical compounds in wheat seedlings. *Environmental Pollution, 107*(3), 315–320. doi:10.1016/S0269-7491(99)00177-3 PMID:15092977

Oves, M., Khan, S., Qari, H., Felemban, N., & Almeelbi, T. (2016). Heavy metals: Biological importance and detoxification strategies. *Journal of Bioremediation and Biodegradation, 7*, 1–15.

Pandey, N., & Sharma, C. P. (2002). Effect of heavy metals Co2+, Ni2+, and Cd2+ on growth and metabolism of cabbage. *Plant Science, 163*(4), 753–758. doi:10.1016/S0168-9452(02)00210-8

Papazoglou, E. G., Karantounias, G. A., Vemmos, S. N., & Bouranis, D. L. (2005). Photosynthesis and growth responses of giant reed (*Arundo donax* L.) to the heavy metals Cd and Ni. *Environment International, 31*(2), 243–249. doi:10.1016/j.envint.2004.09.022 PMID:15661290

Parekh, D., Puranik, R. M., & Srivastava, H. S. (1990). Inhibition of chlorophyll biosynthesis by cadmium in greening maize leaf segments. *Biochemie und Physiologie der Pflanzen, 186*(4), 239–242. doi:10.1016/S0015-3796(11)80078-8

Passariello, B., Giuliano, V., Quaresima, S., Barbaro, M., Caroli, S., Forte, G., & Iavicoli, I. et al. (2002). Evaluation of the environmental contamination at an abandoned mining site. *Microchemical Journal, 73*(1-2), 245–250. doi:10.1016/S0026-265X(02)00069-3

Patrick, H., Ronald, D., & Patrick, J. (1990). Transformations of Se, As affected by sediment oxidation–reduction potential and pH. *Environmental Science & Technology, 24*(1), 91–96. doi:10.1021/es00071a010

Rahman, H., Sabreen, S., Alam, S., & Kawai, S. (2005). Effects of nickel on growth and composition of metal micronutrients in barley plants grown in nutrient solution. *Journal of Plant Nutrition, 28*(3), 393–404. doi:10.1081/PLN-200049149

Rao, K. V. M., Raghavendra, A. S., & Reddy, K. J. (Eds.). (2006). *Physiology and Molecular Biology of Stress Tolerance in Plants*. Springer.

Raskin, I., & Ensley, B. D. (2000). *Phytoremediation of toxic metals: Using plants to clean up the environment*. New York: John Wiley & Sons.

Rathnayake, I. V., Megharaj, M., Krishnamurti, G. S., Bolan, N. S., & Naidu, R. (2013). Heavy metal toxicity to bacteria- are the existing growth media accurate enough to determine heavy metal toxicity. *Chemosphere, 90*(3), 1195–1200. doi:10.1016/j.chemosphere.2012.09.036 PMID:23040649

Ros, R., Cooke, D. T., Martínez-Cortina, C., & Picazo, I. (1992). Nickel and cadmium-related changes in growth, plasma membrane lipid composition, ATPase hydrolytic activity and proton-pumping of rice (*Oryza sativa* L. cv. Bahia) shoots. *Journal of Experimental Botany, 43*(11), 1475–1481. doi:10.1093/jxb/43.11.1475

Rutkowska, B., Szulc, W., Bomze, K., Gozdowski, D., & Spychaj-Fabisiak, E. (2015). Soil factors affecting solubility and mobility of zinc in contaminated soils. *International Journal of Environmental Science and Technology, 12*(5), 1687–1694. doi:10.1007/s13762-014-0546-7

Salisbury, F. B., & Ross, C. W. (1992). *Plant Physiology* (4th ed.). Wadsworth Publisher.

Sarita, S., Manisha, G., & Prakash, C. (1996). Bioaccumulation and biochemical effects of mercury in the plant *Bacopa monnieri* L. *Environmental Toxicology and Water Quality, 11*(2), 105–112. doi:10.1002/(SICI)1098-2256(1996)11:2<105::AID-TOX5>3.0.CO;2-D

Schützendübel, A., & Polle, A. (2002). Plant responses to abiotic stresses: Heavy metal-induced oxidative stress and protection by mycorrhization. *Journal of Experimental Botany, 53*, 1351–1365. PMID:11997381

Schwartz, C., Gerard, E., Perronnet, K., & Morel, J. L. (2001). Measurement of *in situ* phytoextraction of zinc by spontaneous metallophytes growing on a former smelter site. *The Science of the Total Environment, 279*(1-3), 215–221. doi:10.1016/S0048-9697(01)00784-7 PMID:11712599

Shahid, M., Khalid, S., Abbas, G., & Dumat, C. (2015). Heavy metal stress and crop productivity. In K. R. Hakeem (Ed.), *Crop Production and Global Environmental Issues* (pp. 1–25). Springer International Publishing. doi:10.1007/978-3-319-23162-4_1

Shakoor, M. B., Ali, S., Hameed, A., Farid, M., Hussain, S., Yasmeen, T., & Abbasi, G. H. et al. (2014). Citric acid improves lead (Pb) phytoextraction in *Brassica napus* L. By mitigating Pb-induced morphological and biochemical damages. *Ecotoxicology and Environmental Safety*, *109*, 38–47. doi:10.1016/j.ecoenv.2014.07.033 PMID:25164201

Sharma, P., Jha, A. B., Dubey, R. S., & Pessarakli, M. (2012). Reactive Oxygen Species, Oxidative Damage, and Antioxidative Defense Mechanism in Plants under Stressful Conditions. *Le Journal de Botanique*, *2012*, 1–26.

Sharma, S. S., & Dietz, K. J. (2006). The significance of amino acids and amino acid-derived molecules in plant responses and adaptation to heavy metal stress. *Journal of Experimental Botany*, *57*(4), 711–726. doi:10.1093/jxb/erj073 PMID:16473893

Sheetal, K. R., Singh, S. D., Anand, A., & Prasad, S. (2016). Heavy metal accumulation and effects on growth, biomass and physiological processes in mustard. *Indian Journal of Plant Physiology*, *21*(2), 219–223. doi:10.1007/s40502-016-0221-8

Shier, W. T. (1994). Metals as toxins in plants. *Journal of Toxicology. Toxin Reviews*, *13*(2), 205–216. doi:10.3109/15569549409089960

Sidhu, G. P. S. (2016). Heavy metal toxicity in soils: Sources, remediation technologies and challenges. *Advances in Plants & Agriculture Research*, *5*, 1–3.

Sidhu, G. P. S., Singh, H. P., Batish, D. R., & Kohli, R. K. (2016). Effect of lead on oxidative status, antioxidative response and metal accumulation in *Coronopus didymus*. *Plant Physiology and Biochemistry*, *105*, 290–296. doi:10.1016/j.plaphy.2016.05.019 PMID:27214085

Singh, A., Agrawal, S. B., & Rathore, D. (2005). Amelioration of Indian urban air pollution phytotoxicity in *Beta vulgaris* L. by modifying NPK nutrients. *Environmental Pollution*, *134*(3), 385–395. doi:10.1016/j.envpol.2004.09.017 PMID:15620584

Singh, R. P., & Agrawal, M. (2007). Effects of sewage sludge amendment on heavy metal accumulation and consequent responses of *Beta vulgaris* plants. *Chemosphere*, *67*(11), 2229–2240. doi:10.1016/j.chemosphere.2006.12.019 PMID:17289111

Singh, S., & Sinha, S. (2005). Accumulation of metals and its effects in *Brassica juncea* (L.) Czern. (cv. Rohini) grown on various amendments of tannery waste. *Ecotoxicology and Environmental Safety*, *62*(1), 118–127. doi:10.1016/j.ecoenv.2004.12.026 PMID:15978297

Sprynskyy, M., Kosobucki, P., Kowalkowski, T., & Buszewsk, B. (2007). Influence of clinoptilolite rock on chemical speciation of selected heavy metals in sewage sludge. *Journal of Hazardous Materials*, *149*(2), 310–316. doi:10.1016/j.jhazmat.2007.04.001 PMID:17513045

Stevenson, F. J. (1982). *Humus chemistry, genesis, composition, reactions*. New York: John Wiley & Sons.

Stobart, A. K., Griffiths, W. T., Ameen-Bukhari, I., & Sherwood, R. P. (1985). The effect of Cd^{+2} on the biosynthesis of chlorophyll in leaves of barley. *Physiologia Plantarum*, *63*(3), 293–298. doi:10.1111/j.1399-3054.1985.tb04268.x

Thakur, S., Singh, L., Wahid, Z. A., Siddiqui, M. F., Atnaw, S. M., & Din, M. F. M. (2016). Plant-driven removal of heavy metals from soil: Uptake, translocation, tolerance mechanism, challenges, and future perspectives. *Environmental Monitoring and Assessment*, *188*(4), 206. doi:10.1007/s10661-016-5211-9 PMID:26940329

Vidović, M., Morina, F., Milić-Komić, S., Vuleta, A., Zechmann, B., Prokić, Lj., & Jovanović, S. V. (2016). Characterisation of antioxidants in photosynthetic and non-photosynthetic leaf tissues of variegated *Pelargonium zonale* plants. *Plant Biology*, *18*(4), 669–680. doi:10.1111/plb.12429 PMID:26712503

Violante, A., Cozzolino, V., Perelomov, L., Caporale, A. G., & Pigna, M. (2010). Mobility and bioavailability of heavy metals and metalloids in soil environments. *Journal of Soil Science Plant Nutrition*, *10*(3), 268–292. doi:10.4067/S0718-95162010000100005

Wang, S. H., Yang, Z. M., Yang, H., Lu, B., Li, S. Q., & Lu, Y. P. (2004). Copper-induced stress and antioxidative responses in roots of *Brassica juncea* L. *Botanical Bulletin of Academia Sinica*, *45*, 203–212.

Weast, R. C. (1984). *CRC handbook of chemistry and physics* (64th ed.). Boca Raton, FL: CRC.

Williams, L. E., Pittman, J. K., & Hall, J. L. (2000). Emerging mechanisms for heavy metal transport in plants. *Biochimica et Biophysica Acta*, *1465*(1-2), 104–126. doi:10.1016/S0005-2736(00)00133-4 PMID:10748249

Wójcik, M., & Tukiendorf, A. (2004). Phytochelatin synthesis and cadmium localization in wild type of *Arabidopsis thaliana*. *Plant Growth Regulation*, *44*, 71–80. doi:10.1007/s10725-004-1592-9

Wojtaszek, P. (1997). Oxidative burst: An early plant response to pathogen infection. *The Biochemical Journal*, *322*(3), 681–692. doi:10.1042/bj3220681 PMID:9148737

Yang, Y., Mei, Y., Zhang, C., Zhang, R., Liao, X., & Liu, Y. (2016). Heavy metal contamination in surface soils of the industrial district of Wuhan, China. *Human and Ecological Risk Assessment: An International Journal*, *22*(1), 126–140. doi:10.1080/10807039.2015.1056291

Yaron, B., Calvet, R., & Prost, R. (1996). *Soil pollution processes and dynamics*. Berlin: Springer-Verlag. doi:10.1007/978-3-642-61147-6

Zaier, H., Mudarra, A., Kutscher, D., de la Campa, M. R. F., Abdelly, C., & Sanz-Medel, A. (2010). Induced lead binding phytochelatins in *Brassica juncea* and *Sesuvium portulacastrum* investigated by orthogonal chromatography inductively coupled plasma-mass spectrometry and matrix assisted laser desorption ionization-time of flight-mass spectrometry. *Analytica Chimica Acta*, *671*(1-2), 48–54. doi:10.1016/j.aca.2010.04.054 PMID:20541642

Zengin, F. K., & Munzuroglu, O. (2005). Effects of some heavy metals on content of chlorophyll, proline and some antioxidant chemicals in bean (*Phaseolus vulgaris* l.) seedlings. *Acta Biologica Cracoviensia. Series; Botanica*, *47*, 157–164.

Zhang, J., & Kirkham, M. B. (1994). Drought stress induced changes in activities of superoxide dismutase, catalase and peroxide in wheat species. *Plant & Cell Physiology*, *35*(5), 785–791. doi:10.1093/oxfordjournals.pcp.a078658

Zhang, W. H., & Tyerman, S. D. (1999). Inhibition of water channels by $HgCl_2$ in intact wheat root cells. *Plant Physiology*, *120*(3), 849–857. doi:10.1104/pp.120.3.849 PMID:10398721

Zhao, Z., Zhang, Y., Lei, X., Xia, L., Jiang, X., Hussain, J., & Fawundu, E. (2016). Effects of irrigation and drainage modes on the residual characteristics of heavy metals in soil. *Soil Air Water*, *44*(3), 291–298. doi:10.1002/clen.201400368

Zhou, Z. S., Huang, S. Q., Guo, K., Mehta, S. K., Zhang, P. C., & Yang, Z. M. (2007). Metabolic adaptations to mercury-induced oxidative stress in in roots of *Medicago sativa* L. *Journal of Inorganic Biochemistry*, *101*(1), 1–9. doi:10.1016/j.jinorgbio.2006.05.011 PMID:17084899

KEY TERMS AND DEFINITIONS

Agriculture: It is a science of cultivating and growing of useful plants at large scale.

Contamination: Contamination is used in term of presence of undesirable substances in environment that affect in a negative manner.

Heavy Metals: Heavy metals are elements that exhibit metallic properties such as ductility, malleability, conductivity, cation stability, and ligand specificity. They are characterized by relatively high density and relative high atomic weight and atomic number greater than 20.

Permissible Level: It is a regulatory limit for the amount or concentration of a substance in the environment, which is not harmful.

Productivity: Productivity is production of substance or thing per unit time. It is expressed as output per unit of input.

Risk Implications: It is defined as the probability of exposure to a hazard substance that will lead to a harmful effect.

Soil: The surface of earth which supports the life. It contains mixture of organic and inorganic substances, water and air (gases) along with the microorganism.

Symptoms: It is a visible effect of any contaminant on host. It is sign of disease/abnormality.

Toxicity: It refers to harmful effects posed by a substance to living organism (animal or plants) and has a dose dependent relationship.

Chapter 14
Four Strategic Postures for Sustainability in the Project-Based Organization

Gilbert Silvius
LOI University of Applied Sciences, The Netherlands

Ron Schipper
Van Aetsveld, The Netherlands

ABSTRACT

Sustainability is one of the most important challenges of our time. How can we develop prosperity, without compromising the life of future generations? Companies are integrating sustainability in their strategies, processes and actions. In the implementation of strategies, projects play an essential role. Delivering business strategy through projects is even more eminent for project based organizations (PBOs) that execute projects as their core business. This chapter will discuss the question What strategic postures a PBO can take on the consideration of sustainability in its projects, while aligning with both the strategy of the client organization and its own strategy. Based upon an analytical analysis, the authors develop four possible strategic postures on sustainability in the projects of a PBO that provide guidance for the consideration of sustainability in these projects.

INTRODUCTION

In the last 10 to 15 years, the concept of sustainability has grown in recognition and importance. How can humans develop prosperity without compromising the future? The pressure on companies to broaden reporting and accountability from economic performance for shareholders, to sustainability performance for all stakeholders has increased substantially (Visser, 2002). Industry leaders realize that 'greenwashing' of current business practices is not a solution. Integration of the principles of sustainability requires rethinking and redevelopment of business strategies, products/services, processes and resources (Silvius et al., 2012). This integration of sustainability into core business functions is considered one of the most important leadership challenge facing business today (BSR/GlobeScan, 2012).

DOI: 10.4018/978-1-5225-3537-9.ch014

In the implementation of strategies, projects play an instrumental but essential role (Shenhar and Patanakul, 2014). It is estimated that today, more than 20% of global economic activity takes place as projects (Turner et al., 2013). The integration of sustainability into business strategies therefore also requires the integration of sustainability in projects and project management. A growing number of publications address this integration, as reported by Silvius and Schipper (2014) and Aarseth et al. (2016), and Silvius and Tharp (2013) conclude that "the relationship between sustainability and project management is … picking up momentum" (Silvius and Tharp, 2013: xix).

Delivering business strategy through projects is even more eminent for those organizations that execute projects as their core business. These organizations, characterized as "project based organizations" (PBOs) (Huemann, 2015) are executing projects as a way of delivering products and services to external clients. The PBO typology differs from the "project oriented organization" (POO), in which projects are primarily used an instrument to change the own organization.

For a PBO, integrating sustainability implies making the projects it performs more sustainable. Several studies on the integration of sustainability into project management focus on the operationalizing the concepts of sustainability into sets of indicators that can be applied to the project's execution process and output, in order to assess the sustainability of the project (For example Bell and Morse, 2003; Fernández-Sánchez and Rodríguez-López, 2010; Keeble et al., 2003; Labuschagne and Brent, 2006; Martens and Carvalho, 2016). These indicators normally follow a more generic framework of sustainability indicators, for example Global Reporting Initiative's sustainability reporting guidelines, in representing the "Triple Bottom Line" perspectives of social, environmental and economic performance (Elkington, 1997). However, specialists question whether a generic list of sustainability indicators is possible, or even desirable, given the wide variety of conditions and the differences in values in different contexts (Hardi and Zdan, 1997). In line with Mettler and Rohner (2009), Silvius and Schipper (2015) therefore conclude that the assessment of sustainability in projects and/or project management should be configurable to the characteristics and context of the project at hand.

This context relates, amongst other factors, to the industry of the organization and the strategy of the organization. For a PBO, however, there are two contexts to consider: the industry and strategy of the 'client' that commissions the project, and the industry and strategy of the PBO itself, as 'contractor' in the project. The PBO therefore is faced with the challenge to balance its commitment to sustainability between the strategy of the client organization and its own sustainability ambitions. How can this balancing act be approached? Should the PBO follow the sustainability strategy of the client or should it stick to its own strategy and select the assignments and clients that fit that strategy?

This chapter will discuss the question 'What strategic postures a PBO can take on the consideration of sustainability in its projects?', while aligning with both the strategy of the client organization and its own strategy. Based upon our analysis, we will develop four possible strategic postures on sustainability in the projects of a PBO that provide guidance for the consideration of sustainability in these projects.

The remainder of this chapter is organized as follows. In the following section, we will discuss the concepts of the project based organization, the different strategic postures towards sustainability and the relationship between sustainability, project and project management. Based on the concepts found in the literature, the following section will develop a framework of four possible strategic postures for sustainability in the projects of the PBO. These four postures will be presented further and their implications for the PBO will be discussed.

BACKGROUND

This paragraph present a brief discussion of the main concepts addressed in this chapter. First the model of the project based organization will be discussed, followed by the different strategic postures towards sustainability. The last section of this paragraph will discuss sustainability in the context of project and project management.

The Project Based Organization

The introduction to this chapter already introduces the organizational typology of the PBO. This paragraph will elaborate on that.

Companies and organizations in all types of industries are undertaking projects as a growing part of their business activities even when their core business processes are repetitive (Keegan and Turner 2002). In order to identify the emerging role of projects in organizations, characterizations such as the project led organization (Hobday, 2000), project based organization (Turner and Keegan, 1999) and project oriented organization (Gareis, 1990), started to get used. And although the exact meaning of these constructs has been subject to academic debate, for example by Sydow et al. (2004), the discussion seems to have settled on the terms project based organization (PBO) and project oriented organization (POO). And although these labels are sometimes still used synonymously, the literature is now indicating a difference between the two typologies. As indicated in the introduction, a POO performs projects primarily for improving the performance of its own organization. In this typology, projects are temporary organizations, that realize changes of business processes, strategies, resources, products or services in the non-temporary 'permanent' part of the organization (Silvius et al., 2012). Projects are an organizational construct that gather resources across organizational structures in order to make a non-routine 'unique' result. The core of the organization, however, is formed by the permanent organization that performs business processes in a repetitive way.

In a PBO, the core of the organization is organized in projects. These projects do not deliver change to the internal organization, but to a client organization that specifies the required deliverable. The effectuation of the change is normally not included in the scope of the project and left to the client organization. The PBO typology has its roots in the construction industry, but can also be recognized in other industries, such as event management, consulting, marketing communication and information technology.

Table 1 provides an overview of the characteristics of both PBO and POO typologies.

The POO and PBO typologies help us to understand the different relationships between strategy and projects. Where a POO performs projects in order to realize its own goals and strategies, the PBO performs projects that relate to its client's goals and strategies, as well as its own goals. A frequently found situation is that a POO acts as a client of a PBO, for example in construction, consulting, information technology or event management projects. The POO in this situation outsources the realization of the project to a more specialized PBO. This situation is illustrated by Turner (2014), using the terminology project 'owner' for the POO that commissions the project and 'contractor' for the PBO (Figure 1).

The POO (owner) and PBO (contractor) each value of the project from their own perspective (Turner, 2014). For the PBO this confirms the earlier identified challenge of balancing the integration of sustainability in the project between the strategy of the client and its own ambitions.

Table 1. Comparison of the 'project based' and the 'project oriented' organization typologies

	Project Based Organization (PBO)	Project Oriented Organization (POO)
Reason for projects	·Projects are performed per force because of the customized nature of the project	·Projects are a strategic choice ·Projects are one option for the organizational design
Relation	·Projects relate to production processes	·Projects relate to business processes
Type of projects	·Mainly external projects	·External (if adequate) and internal projects
Management logic	·Predominantly functional line organizations ·Projects are forced to fit functional logic, as disturbances are not allowed, large projects are turned into temporary functions	·Permanent and temporary organizations have different management logics ·Organization can deal with these contradictions
Understanding of project management	·Operational capability ·Project management as tools and control	·Operational and strategic capability ·Project management as leadership
Paradigm	·Prevailing mechanistic planning paradigm ·Project is considered as complex task or system	·Systemic-constructivist ·Project is considered as a temporary organization

(Huemann, 2015: 63)

Figure 1. Different perspectives on the value of the project (Turner, 2014)

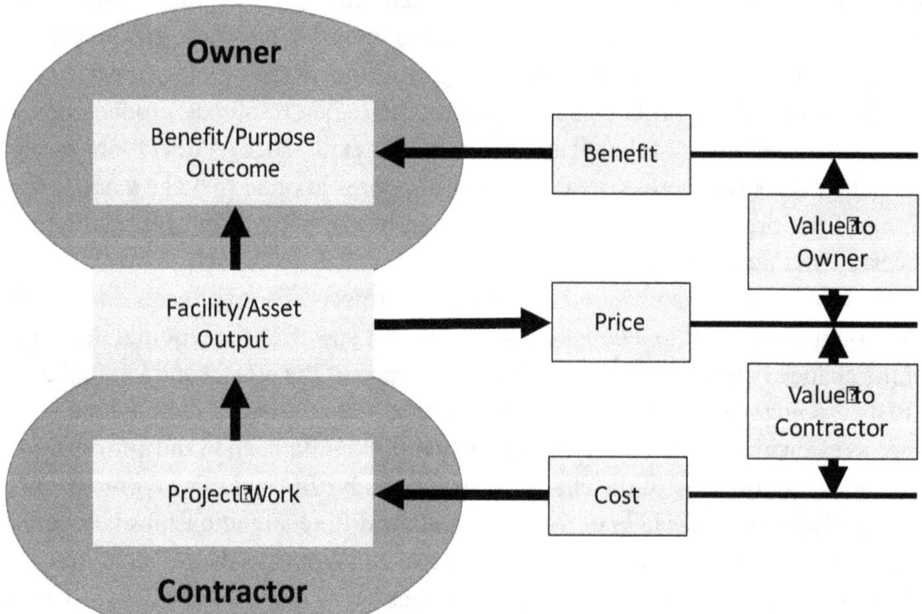

Sustainability in the Context of Strategy

The BSR/GlobeScan (2012) study referenced in the introduction of this chapter is one of many studies that show that corporations are not blind for the societal issues that we are faced with. And although this may have been the case already in past decades, it is not too long ago that issues, such as poverty reduction, protection of the environment and fair working conditions, were considered primarily the responsibility of governments and other regulatory bodies (Tulder et al., 2014). If a company operated within the boundaries of laws and regulations, it could expect societal acceptance. In fact, Friedman argued that a company's only social responsibility is to generate profits for its shareholders (Friedman, 1970). In today's society such a strategy would quite likely face criticism from societal organizations that scrutinize the behavior of larger organizations, thereby influencing public opinion. Examples of successful mobilization of public opinion are plentiful. For example the Brent Spar case of 1995, in which Greenpeace organized a high-profile media campaign around the planned sinking of the oil storage and tanker loading buoy Brent Spar, that successfully influenced Shell. Giving in to public pressure, Shell changed their plan to sink the Brent Spar and instead dismantled it on-shore.

Shell's behavior at the time of the Brent Spar case may be characterized as reactive to public opinion. Today, many organizations proactively accept some level of responsibility for sustainability as part of their 'license-to-operate' and integrate sustainability goals in their corporate strategy. Willard (2005), depicts these different strategic postures as a growth model of 'sustainability stages' a company may develop on (Figure 2).

The stages move from reactive to proactive and describe to what extent a company is committed to sustainability principles. The first stage is when companies fail even to comply with prevailing regulations. They are opportunistic and not engaged with the concept of sustainability . When a company complies with all environmental and social regulations it moves up to stage 2, 'compliance'. In stage 3 'beyond compliance' a company starts to not only react on regulations, but it starts introducing sustainability activities. Yet, these activities are not concerted but are carried out in different departments. Companies who understand the importance of sustainability and the value-added they can gain from sustainable activities e.g. energy-efficient production or eco-friendly products and integrate sustainability into their corporate strategy are in stage 4 'integrated strategy'. The highest stage 5 'purpose and passion' is attained when companies are not just driven by profits but also by a sense of responsibility to improve society and environment and contribute to a better world.

The stages of sustainability model illustrates how companies can move from a reactive approach to a proactive one. If we disregard the pre-compliance stage 1, stages 2 and 3 can be considered as reactive, whereas stages 4 and 5 are considered proactive. This distinction between approaches distinguishes

Figure 2. The stages of sustainability integration. (Adapted from Willard, 2005)

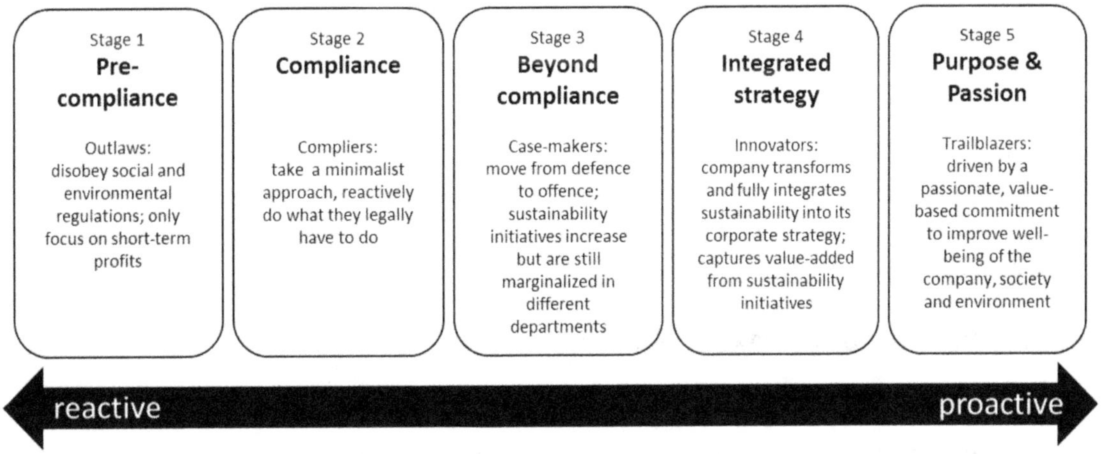

companies that try to 'do no harm', by trying to minimize the negative effects of their business, from companies that try to proactively 'make a positive contribution', by integrating sustainability considerations in their core business strategies.

A similar dichotomy can be found with Baumgartner and Ebner (2010). In their study of the sustainability maturity of corporate strategies, they differentiate between introverted strategies and extroverted strategies. Introverted strategies consider sustainability as an additional aspects of doing business and concentrate on a basic level of sustainability consideration such as conformity and compliance with sustainability-related rules and guidelines. These strategies could be characterized as 'risk avoidance' or 'defensive'.

Extroverted strategies consider sustainability as an integrated part of the business and perhaps even as the reason for doing business. Within the extroverted strategies, Baumgartner and Ebner (2010) distinguish a more conventional and a more transformative approach. In the conventional extroverted strategy, the organization uses its sustainability as a differentiator in the marketplace. Given the 'competitive edge' that these organizations experience from proactively considering sustainability in their business and activities, it makes sense to engage more in sustainability that it is obliged to do by laws and regulations. In these organizations, the marketing of their sustainability efforts plays an important role. Organizations that adopt a transformative extroverted strategy, go beyond the 'competitive edge' perspective on sustainability. For these organizations, their activities are a driver for sustainability in society. Sustainability is integrated in all aspects of business and strategy, and gains therefore much higher credibility.

Tulder et al. (2014) identify the basic attitude of an organization towards its societal role as one of the key variables in the classification of the strategic postures on sustainability. The reactive company sees sustainability as a liability to its business strategy and operations, where a proactive company takes responsibility for societal issues.

Next to the basic attitude, Tulder et al. (2014) recognize another meaning of the reactive-proactive dichotomy: the company's initial drive to consider sustainability. A reactive company acts upon pressure from, mostly external, stakeholders, whereas a proactive company makes the first moves based upon its own leadership or believes.

Based upon these two key variables, Tulder et al. (2004) construct a framework that can be used to characterize the sustainability strategies of organizations and the transition from one strategy to another. Figure 3 shows this framework.

Concluding this part of our literature review, we conclude that consensus is emerging on a basic distinction between inactive/reactive strategies on sustainability, in which the organization views sustainability as a liability, and active/proactive strategies, in which the organization views sustainability as a responsibility.

Provide a normative opinion or judgment on which strategic posture should be considered 'right' or most fitting goes beyond the scope of this chapter. As we are aiming to provide direction for the strategic posture for the PBO, we need to understand how the strategy of the client context of the PBO interacts with the strategy of the PBO itself. For this goal, the dichotomy of inactive/reactive strategies versus active/proactive strategies suffices.

Sustainability in the Context of Projects

The seminal Brundtland report of the UN World Commission on Development and Environment, linked sustainability to change, by stating "In essence, sustainable development is a process of change" (World

Figure 3. Development framework of sustainability strategies. (Tulder et al., 2014)

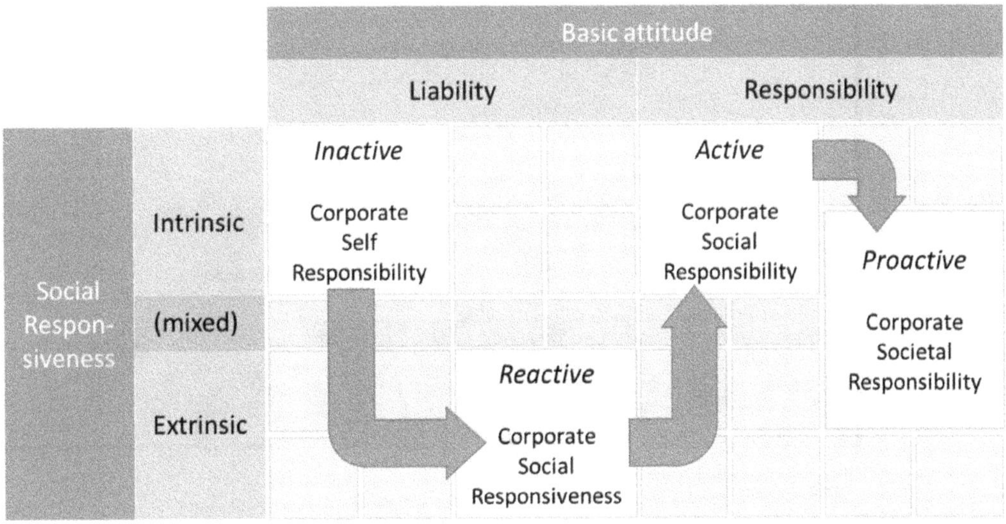

Commission on Development and Environment, 1987). Marcelino-Sádaba et al. (2015) observe that "projects are the ideal instrument for change" and "the necessary change that we require towards sustainability will be boosted by applying the project management discipline to sustainability." (Marcelino-Sádaba et al., 2015). This view elaborates on projects as temporary organizations (Lundin and Söderholm, 1995; Turner and Müller, 2003) that realize change in organizations or across organizational boundaries (Silvius et al., 2012). In according with this, Turner defines project management as "the means by which the work of the resources assigned to the temporary organization is planned, managed and controlled to deliver the beneficial change" (Turner, 2014: 29).

With the growing attention for sustainability, its concepts are also being related to project management (For example by Labuschagne and Brent, 2006; Edum-Fotwe and Price, 2009; Maltzman and Shirley, 2011; Silvius et al., 2012; Martens and Carvalho, 2016). Pasian and Silvius (2016) identify sustainability as one of the evolving schools of thought in project management. They characterize the 'sustainability school of project management' as considering projects in a societal perspective, adopting a management *for* stakeholders approach and considering all aspects of the projects through the triple bottom-line perspectives of economic, environmental and social interests. criteria for the business case and project success. The following section discussed these three characteristics of sustainable project management further.

Projects in a Societal Perspective

As is suggested by Marcelino-Sádaba et al. (2015)., the sustainability school adopts a societal perspective on projects and considers projects as instruments to realize societal change. In their literature review on publications on sustainability in project management, Silvius and Schipper (2014:72) therefore identify the 'recognition of the context of the project' as the starting point of considering sustainability in project management. "Integrating the dimensions of sustainability in project management inevitably implies a broader consideration of the context of the project." (Silvius and Schipper, 2014:72). This enlarged context is illustrated by Silvius et al. (2017) as shown in Figure 4.

Figure 4. The enlarged scope of sustainable project management (Silvius et al., 2017)

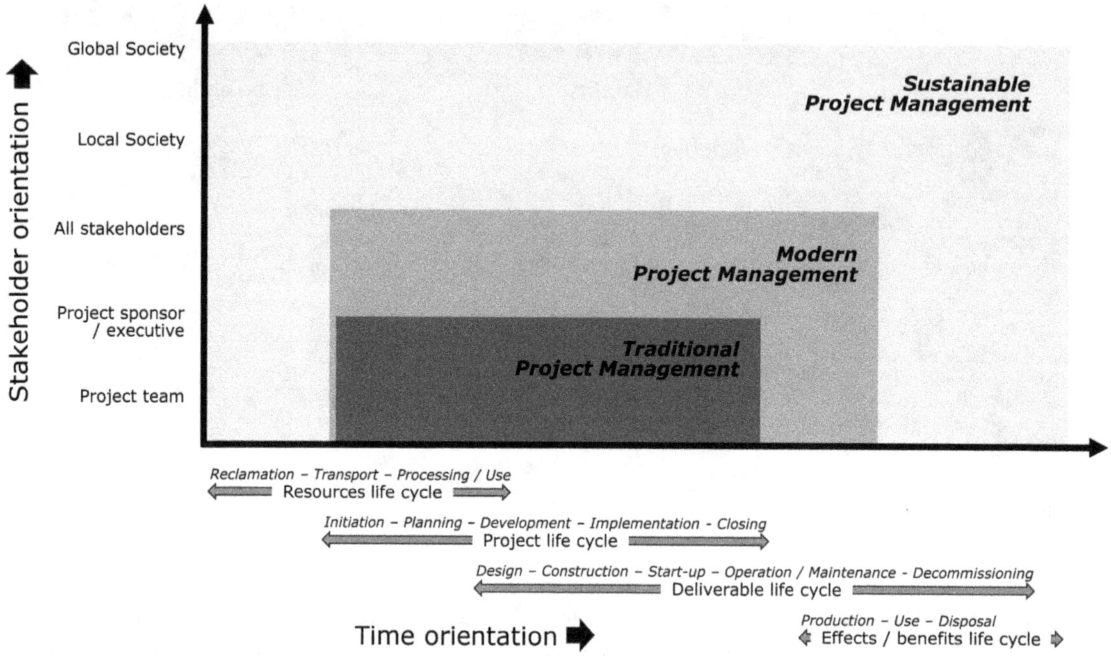

Management for Stakeholders

Several authors (AlWaer et al., 2008; Eskerod and Huemann, 2013; Labelle and Leyrie, 2013) recognize the need for a more proactive engagement of stakeholders as a consequence of integrating sustainability into project management. Eskerod and Huemann (2013) conclude that the current standards of project management guide practitioners towards the recognition of a rather limited group of stakeholders and to "selling the project to the most important stakeholders rather than involving them and their interests into the creation of project objectives" (Eskerod and Huemann, 2013: 43. In this approach, the 'management-of-stakeholders' approach, stakeholders are seen primarily as providers of resources that should be prevented from hindering the project. "In contrast, the 'management-for-stakeholders' approach (Freeman et al., 2007, 2010) takes the point of departure that all stakeholders have the right and legitimacy to receive management attention (Julian et al., 2008). Stakeholders are not means to specific aims in the organization but valuable in their own rights." (Eskerod and Huemann, 2013: p.40). Labelle and Leyrie (2013) conclude that this implies that stakeholder communication becomes stakeholder participation. The information flow between project and stakeholders is no longer unidirectional but transformed into a dialogue that allowed participants to take part in developing the project (Libaert, 1998).

Triple Bottom-Line Criteria for all Aspects of the Project

Integrating sustainability in project management will influence the specifications and requirements of the project's deliverable or output, and the criteria for project success (Eid, 2009; Maltzman and Shirley, 2011). For example the inclusion of environmental or social aspects in the project's objective and intended output and outcome (Silvius et al., 2012). Integrating sustainability into project management suggests that the content, intended output/outcome and success criteria are based on a holistic view of the project (Gareis et al., 2013) and developed together with a broad group of stakeholders (Eskerod and Huemann, 2013).

Integrating sustainability also implies that the definition and perception of project success take into account the 'triple bottom line' of economic, social and environmental benefits as laid out in the business case, both in the short term as in the long term (Silvius and Schipper, 2014). However, studies on sustainability in project management differ in their consideration of the different triple bottom line perspectives. Studies that focus on sustainable or 'green', construction projects and project management mostly discuss the combination of the economic and the environmental dimensions, whereas papers that focus on sustainable development projects tend to discuss mainly the social dimensions (Silvius and Schipper, 2014).

Several studies on the integration of sustainability into project management focus on operationalizing the Triple Bottom Line concept by developing sets of indicators on the different perspectives (For example Bell and Morse, 2003; Fernández-Sánchez and Rodríguez-López, 2010; Keeble et al., 2003; Labuschagne and Brent, 2006; Martens and Carvalho, 2016). This approach bears the risk of lacking the holistic approach of the integration of the economic, environmental and social perspectives. Martens and Carvalho (2016) reported an empirical study on the consideration of Triple Bottom Line variables by project managers from diverse industries in Brazil. One of their conclusion was that the consideration of different sustainability aspects of the Triple Bottom Line is related to the strategy and context of the project. A universal set of sustainability indicators for projects may therefore be illusive.

Addressing this strategic context, Silvius and Schipper (2015) develop a maturity model, SPM3, for the assessment of sustainability in project management with which the degree to which a project considers the different triple bottom line criteria can be assessed on an ordinal scale ranging from 'compliant' to 'purpose' (Table 2). In these maturity levels, the earlier found distinction between an inactive/reactive approach towards sustainability and an active/proactive approach can be recognized. Specific for the SPM3 model, however, is that the model considers how sustainability is approached on the level of a project, and not on the level of the organization or strategy of the organization.

For the measurement of the consideration of sustainability, the SPM3 maturity model defines 22 indicators, organized in the three triple bottom line perspectives (Silvius and Schipper, 2015). During the maturity assessment, each indicator is assessed on the scale of the maturity levels of Table 2. The resulting project sustainability maturity report provides a visual overview on the level of consideration of each indicator (Figure 5).

Table 2. The maturity levels of the SPM3 model

Approach	Maturity Level	Description
'do no harm'	Level 1: Compliant	(This aspect of) Sustainability is considered minimalistic and implicit, and (only) with the intention to comply with laws and regulations.
'positive contribution'	Level 2: Reactive	(This aspect of) Sustainability is considered explicitly, with the intention to reduce negative impacts of the project.
	Level 3: Proactive	(This aspect of) Sustainability is explicitly considered as one of the areas that the project contributes to.
	Level 4: Purpose	Making a contribution to (this aspect of) sustainability is one of the drivers behind the project and sustainability considerations are included in the justification of the project.

(Silvius and Schipper, 2015)

Figure 5. SPM3 framework for the assessment of sustainability consideration (Silvius and Schipper, 2015)

			Integration of sustainability *'do no harm'* → *'positive contribution'*			
			Level 1 Compliant	Level 2 Reactive	Level 3 Proactive	Level 4 Purpose
Sustainability indicators	Economic sustainability	Return on Investment				
		Business agility				
		Competitive potential				
		(Business) Continuity				
		Motivation and incentives				
		Risk reduction				
	Environmental sustainability	Transport				
		Energy				
		Water				
		Eco system				
		Waste and Packaging				
		Materials and resources				
		Emissions				
		Spatial planning				
		Nuisance				
	Social sustainability	Labor practices and decent work				
		Human rights				
		Ethical behaviour				
		Soc. cust and prod responsibility				
		Participation				
		Human capital development				
		Corporate governance				

For each project a logical ambition should be that the project 'does no harm' and therefore addresses all sustainability criteria on at least the level 1 'compliant' level. For the more (pro)active levels of consideration, levels 3 and 4, addressing *all* criteria is not a realistic ambition for a project. Therefore the project needs to select on which aspects of sustainability it aims to make a more active contribution. Guidance for the selection of relevant criteria logically comes from the (sustainability) strategy of the organization. However, for the PBO, a complicating factor is that there are multiple strategic project contexts to consider. Both the 'client' and the PBO itself, the 'supplier', provide relevant strategic contexts to consider in the project.

FOUR STRATEGIC POSTURES FOR PROJECTS OF THE PBO

The basic strategic postures, inactive/reactive versus active/proactive, we derived in the preceding paragraph provides a foundation for the analysis of the possible strategic postures of the projects in a PBO. Earlier in this chapter we established that for the PBO two sustainability strategies provided a relevant context for the projects: the strategy of the client and the strategy of the PBO itself.

Figure 6. Four strategic postures of the PBO

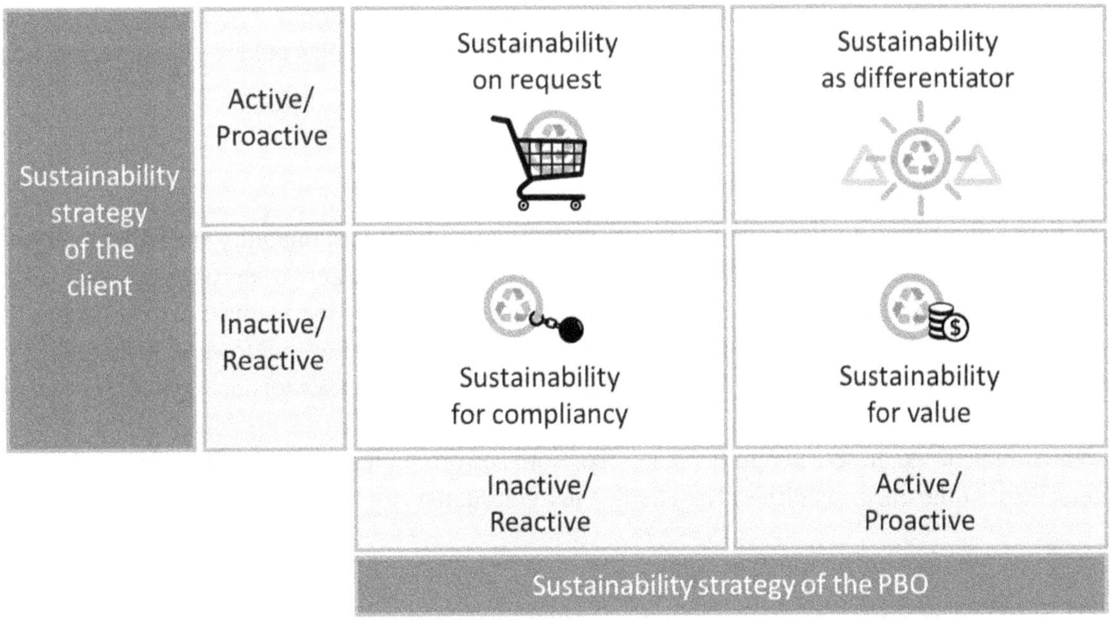

By using the two basic strategic postures for both organizations, we constructed the framework presented in Figure 6.

This framework now allows us to characterize four strategic contexts of the projects of the PBO:

1. An inactive/reactive sustainability strategy of the PBO with an inactive/reactive strategy of the client.
2. We labelled the strategic posture of this situation *Sustainability for compliancy.*
3. An inactive/reactive sustainability strategy of the PBO with an active/proactive strategy of the client.
4. We labelled this situation *Sustainability on request.*
5. An active/proactive sustainability strategy of the PBO with an active/proactive strategy of the client.
6. We labelled this situation *Sustainability as differentiator.*
7. An active/proactive sustainability strategy of the PBO with an inactive/reactive strategy of the client.
8. We labelled this situation *Sustainability for value.*

The following sections discuss the strategic postures of these four situations.

Sustainability for Compliancy

The sustainability for compliancy posture is the most minimalistic of the four strategic postures. In this strategy, both the PBO and the client company have an inactive/reactive approach towards sustainability and therefore sustainability is addressed in projects only with the intention to comply with laws and regulations. In the absence of a more (pro)active sustainability strategy, the different triple bottom line sustainability criteria are addressed on a 'compliance' level.

A sustainability assessment of a project, according to the SPM3 model, would in this strategic posture typically deliver the following result (Figure 7).

Sustainability on Request

In this strategic posture, the PBO has an inactive/reactive approach to sustainability and will only address those sustainability aspects of the project that the client, who has a more active approach to sustainability, requests in the project assignment. Figure 8 illustrates an example of the maturity assessment in this strategic posture. All indicators are considered on the level of compliancy, but towards some indicators a more proactive approach is taken. In Figure 8 this is shown for the indicators energy and transport as an example.

Which indicators are selected to be addressed more actively in the project is determined by the requirements that the client specifies. The client is in the lead, the PBO reacts and follows.

Figure 7. Typical assessment of sustainability consideration in case of the sustainability for compliancy posture

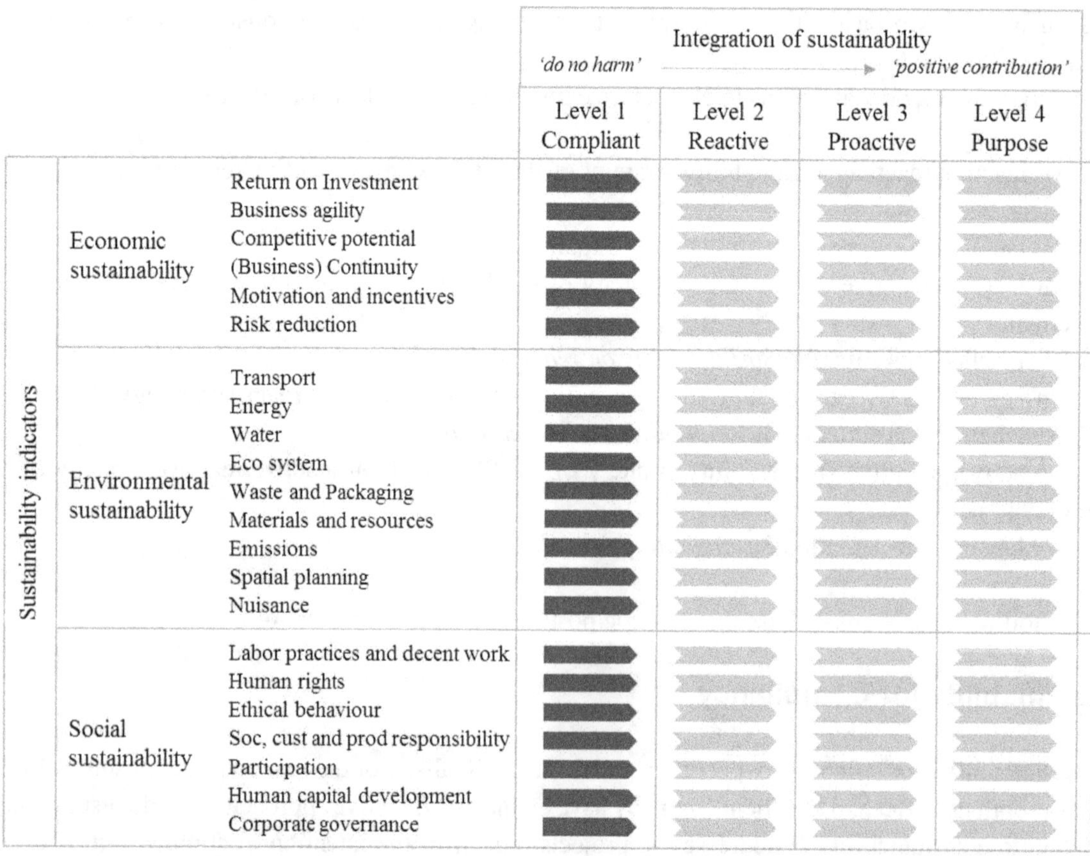

Figure 8. Typical assessment of sustainability consideration in case of the sustainability on request posture. (Note that the level 3 and 4 scored indicators are examples)

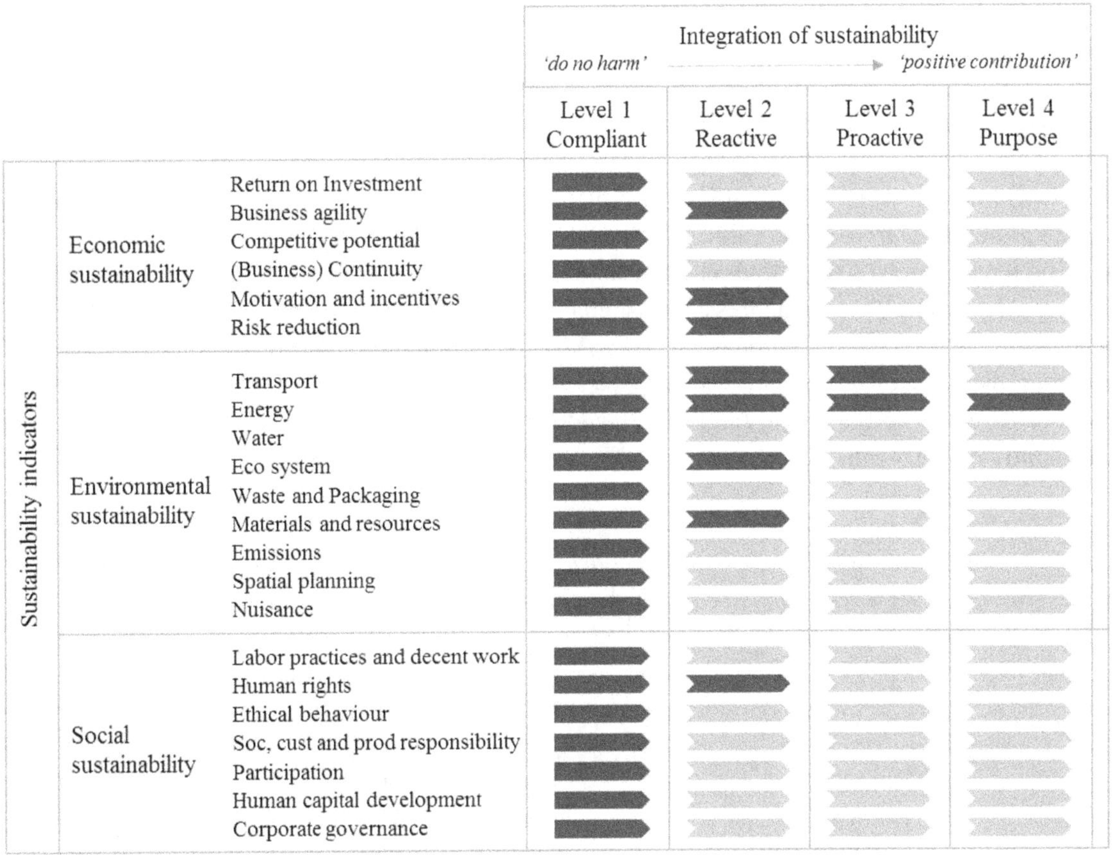

Sustainability as Differentiator

In the sustainability as differentiator and sustainability for success strategic postures, the PBO takes a more (pro)active approach towards sustainability. The PBO has formulated strategic sustainability ambitions and logically aims to realize these through its core business: projects. The sustainability ambitions are therefore communicated and marketed to prospective clients. One reason for this is to allow clients to preselect themselves for the PBO in terms of strategic fit. Another reason is that by communicating its sustainability ambitions, the PBO aims to create a competitive advantage over competitive PBOs with clients that value sustainability.

This strategic posture bears the risk that the sustainability ambitions of the PBO and the client are not aligned. For the project this implies that it will probably have to address multiple criteria, some related to the sustainability strategy of the client and some related to the sustainability strategy of the PBO itself, parallel. In itself this may not be a problem, as the criteria logically not exclude other criteria, however, it may increase the complexity of the project.

Figure 9 shows a logical SPM3 maturity analysis of a project in this strategic posture. Which indicators are selected to be addressed more actively in the project is in this posture a mixture of indicators derived from the sustainability strategy of the client and indicators that are provided by the PBO itself. These indicators may overlap, but they may also be completely different indicators.

Sustainability for Value

Paying attention to sustainability aspects in projects may be perceived as costing time or money and therefore as not supportive of the value or success of a project. However, several studies, such as Tiron-Tudor and Ioana-Maria (2013), Martens and Carvalho (2017) and Silvius and Schipper (2016), point out a positive relationship between addressing sustainability in the management of a project and the success of that project. For example by eliminating waste (Ma, 2011), reducing risks (Silvius and Schipper, 2016) and/or improved stakeholder relationships (Eskerod and Huemann, 2013). Therefore also without an explicit requirement of the client, the PBO may also address sustainability, simply because it expects sustainability delivers value.

Figure 9. Typical assessment of sustainability consideration in case of the sustainability as differentiator posture. (Note that the level 3 and 4 scored indicators are examples)

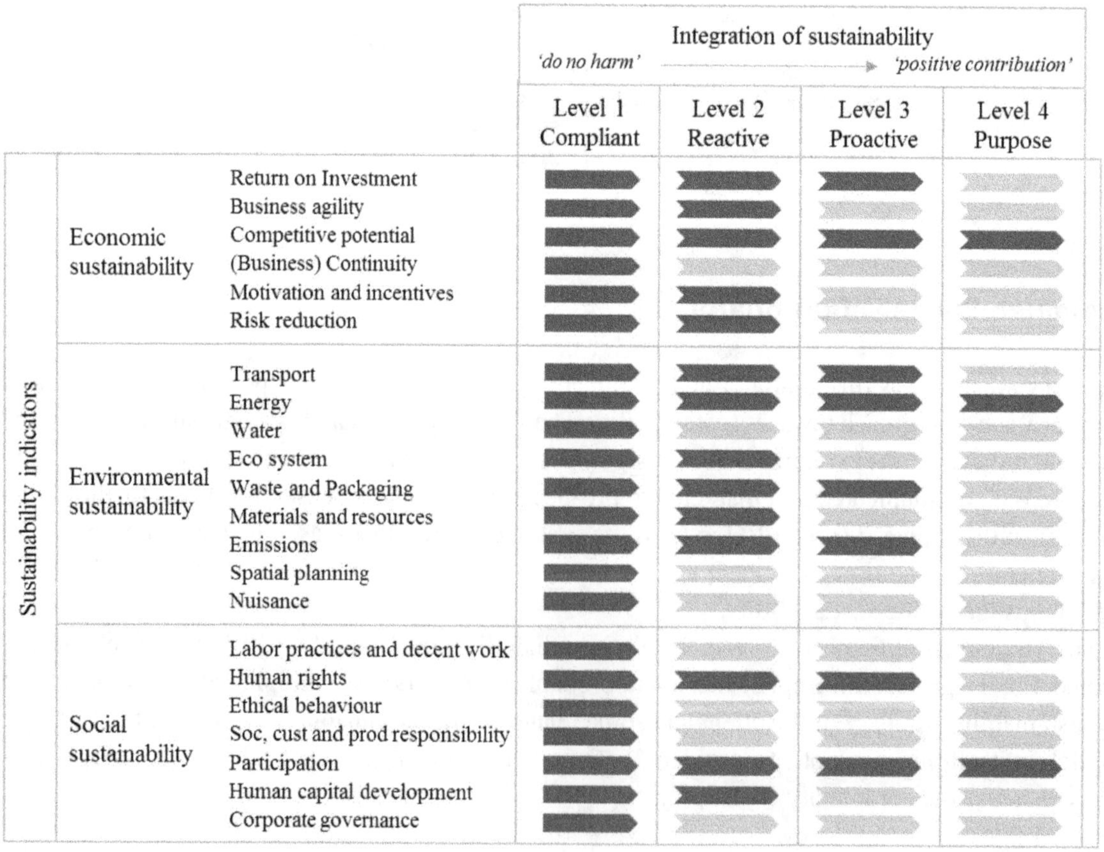

In this strategic posture, it can be expected that the projects score overall better on the active consideration of sustainability aspects, with the highest level scores on a limited set of indicators that relate to the (sustainability) strategy of the PBO. Figure 10 shows how an SPM3 maturity analysis of a project in this strategic posture may look like.

Overview

Table 3 summarizes the four strategic postures for sustainability in the projects of the PBO.

IMPLICATIONS

Our analysis of the four potential strategic postures of a PBO, showed that for three of the four postures, sustainability will be addressed on a level that goes beyond the level of being compliant with laws and regulations. This logically has implications for the way projects are managed and governed within the PBO. This paragraph highlights the implications for the PBO that aims to move towards a more sustainable management of their projects.

Figure 10. Typical assessment of sustainability consideration in case of the sustainability for value posture. (Note that the level 3 and 4 scored indicators are examples)

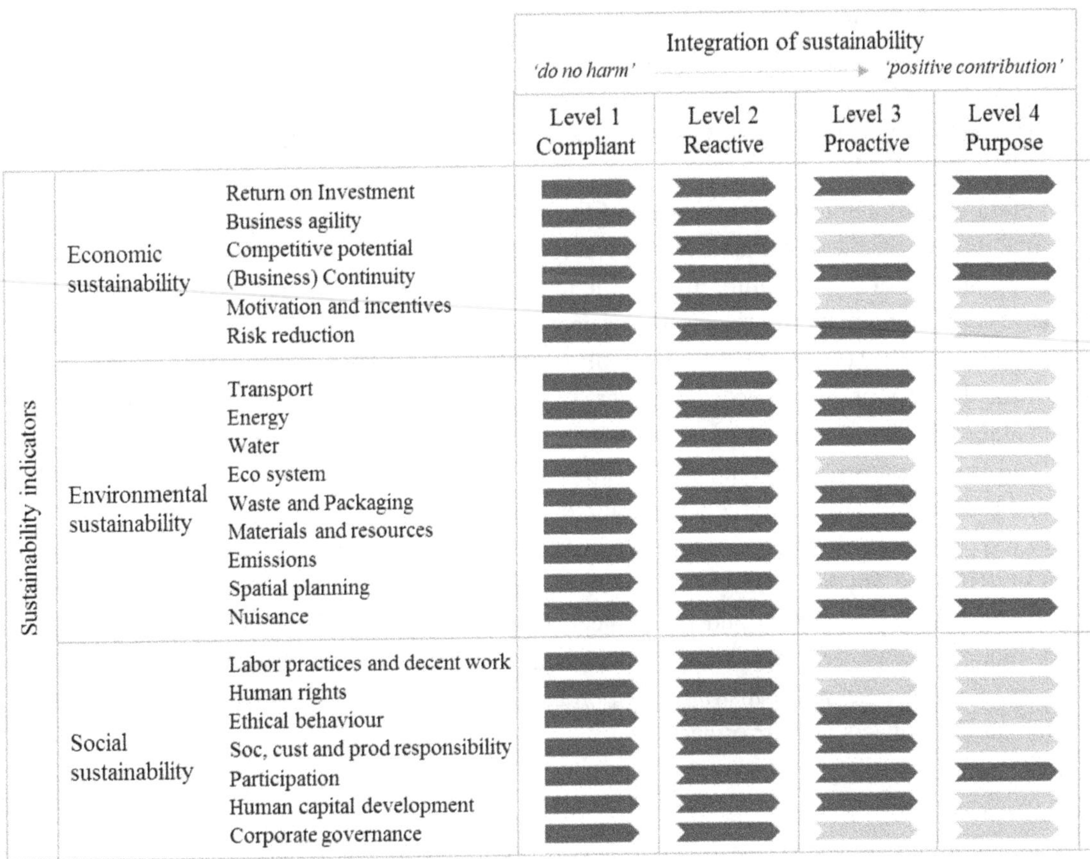

Table 3. Overview of the four strategic postures

Strategic Posture	Sustainability for Compliancy	Sustainability on Request	Sustainability as Differentiator	Sustainability for Value
Main conviction of the PBO	Sustainability brings additional costs and risks and should therefore be kept to a minimum.	Sustainability brings additional costs and the customer should therefore be willing to pay for it.	Sustainability brings opportunities for creating competitive advantage and should therefore be exploited.	Sustainability brings opportunities for creating more valuable projects and should therefore be thrived for.
Main driver	Risk	Sales	Marketing	Value
Main business function involved in sustainability	Risk and compliancy officer	Sales department and project planners	Marketing department and product development	Quality department and project planners

Adopt a Sustainable Project Management Process

Based on an analysis of the most used standards for project management, Silvius (2015a) concludes that on the logical areas of impact of sustainability in project management, the standards of project management processes (PMBOK Guide, PRINCE2 and ISO 21500) fail to refer convincingly to sustainability considerations. Enhancing the project management methodology that is used in the PBO with processes, templates and roles that explicitly address the sustainability aspects of a project therefore should be recommended. At this moment, only the ISO 21500 based PRiSM methodology (Carboni et al., 2013) explicitly addresses sustainability and introduces a 'Sustainability Management Plan' (GPM Global, 2014; Silvius, 2015b) as new element in the project management documentation.

Develop Sustainability Measurement and Monitor Systems

An important condition for the management of the sustainability aspects of a project is a practical tool or instrument to assess the sustainability of the project. And although numerous academic publications include assessments of sustainability impact, the instruments used for these assessments fail to address the specifics of sustainability in projects and project management. Sustainability in projects can be considered as sustainability of the project deliverable (the product that the project realizes) and the sustainability of the delivery of the project (the process of realizing the project) (Silvius and Schipper, 2015). Gareis et al. (2013) observe that the first perspective prevails in studies on sustainability in the business context and that, because of the temporary nature of projects, the second perspective, the sustainable management of projects, is still largely unexplored.

Specific project sustainability assessment instruments range from simple checklists, for example Knoepfel (2010) and Taylor (2008), assessment methodologies, for example the Sustainable Footprint Methodology (Oehlmann, 2011) and the Project Sustainability Impact Assessment (Tam, 2017), to maturity models, the earlier discussed SPM3 (Silvius and Schipper, 2015). A PBO would be wise to develop or select an assessment instrument in order to be able to assess, measure, monitor and report the sustainability impacts of their projects.

Develop 'Sustainable Project Management' Competencies

Addressing sustainability in projects requires adequate competencies. The recently published IPMA Individual Competence Baseline version 4 (International Project Management Association, 2015) includes an explicit reference to sustainability in the competence "Identify, and ensure that the project complies with relevant sustainability principles and objectives". The description of this key competence states that the project manager should be able to "assess the impact of the project on the environment and society" and that he/she "researches, recommends and applies measures to limit or compensate negative consequences". Based upon our analysis of the strategic postures for projects in the PBO this competence makes sense, however, what 'sustainability competencies' does a project manager need to develop in order to fulfill the responsibility for sustainability that the ICB4 puts on him or her? And are these competencies today included in the development path of project managers in the PBO?

Drawing from the literature on 'Education for Sustainable Development', Silvius (2016) derived five key competences for sustainability: systems thinking competences, anticipatory competences, normative competences, strategic competences and interpersonal competences. His specification of these key competences provide a foundation for answering the before mentioned questions and the development of specific training programs. However, his analysis also emphasized the importance of 'attitude' as the foundation on which skills and knowledge are developed. Integrating sustainability in project management therefore a 'mind shift' of the project manager. From executing 'ordered' projects, to accepting a professional responsibility for contributing to a better society (Silvius, 2016).

Develop Organizational Maturity

The PBO also needs to develop an organizational competence on sustainability. In addition to the actions mentioned above on project management process, sustainability measurement instruments and individual competence development, the development of the organizational competence also requires sharing knowledge and experiences, developing good examples and best practices on sustainability in projects, creating supporting structures, engaging stakeholders in the process and committing to ambitions and continuous improvement on the topic.

CONCLUSION

The PBO is an organizational typology that executes projects primarily for external clients. This complicates the alignment of projects and strategy, as two strategies come into play: the one of the PBO and the one of the client. Given this dualistic situation, this chapter set out to discuss what strategic postures a PBO can take on the consideration of sustainability in its projects, while aligning with both the strategy of the client organization and its own strategy. Based on our analysis of basic strategic postures, we concluded that consensus is emerging on a basic distinction between inactive/reactive strategies on sustainability, in which the organization views sustainability as a liability, and active/proactive strategies, in which the organization views sustainability as a responsibility. In the case of the PBO, this classification applies to both the strategy of the client as the strategy of the PBO itself, thereby creating four possible strategic contexts for the projects the PBO executes: (1) an inactive/reactive sustainability strategy of the PBO with an inactive/reactive strategy of the client, (2) an inactive/reactive sustainability strategy of the PBO with an active/proactive strategy of the client, (3) an active/proactive sustainability strategy of the PBO with an active/proactive strategy of the client and (4) an active/proactive sustainability strategy of the PBO with an inactive/reactive strategy of the client.

For each of the four situations we developed a fitting strategic posture and an expectation how the consideration of sustainability in projects for each of the strategic postures logically looks like.

The conclusion of our analysis of the four strategic postures is that in three of the four postures, sustainability needs to be considered beyond the level of compliancy. Only in the situation in which both the PBO and the client have an inactive/reactive approach to sustainability, can the consideration of sustainability be limited to being compliant. However, this situation is rapidly becoming outdated exception. As was indicated in the introduction of this chapter, businesses are integrating sustainability into their business strategies, products/services, processes and resources. And where a PBO, by the nature of its business, may be reactive to the requirements of its clients, it should be clear from our analysis that it does not suffice to take a reactive approach to sustainable project management. The PBO needs to develop competency in sustainable project management, in order to be competitive and to contribute to the strategy of its clients. This competency refers both to the individual competence of the project managers, as to the organizational competence in terms of project management processes, project management documentation, project reporting, supporting organizational structures and organizational learning and knowledge development.

FUTURE RESEARCH DIRECTIONS

Future research may elaborate on the strategic postures we derived in this chapter. What fraction of PBOs adopt a certain posture and how are these implemented? Does the size of the PBO play a role in this?

Also the implications of the PBO forms an interesting direction for future research. How do PBOs develop their organizational competence in sustainable project management? And the competences of the individual project managers?

REFERENCES

Aarseth, W., Ahola, T., Aaltonen, K., Økland, A., & Andersen, B. (2016). (in press). Project sustainability strategies: A systematic literature review. *International Journal of Project Management*.

AlWaer, H., Sibley, M., & Lewis, J. (2008). Different Stakeholder Perceptions of Sustainability Assessment. *Architectural Science Review*, *51*(1), 48–59. doi:10.3763/asre.2008.5107

Baumgartner, R. J., & Ebner, D. (2010). Corporate Sustainability Strategies: Sustainability Profiles and Maturity Levels. *Sustainable Development*, *18*(2), 76–89. doi:10.1002/sd.447

Bell, S., & Morse, S. (2003). *Measuring Sustainability Learning from doing*. London: Earthscan.

BSR/GlobeScan. (2012). *2012 BSR/GlobeScan State of Sustainable Business Poll*. Retrieved from http://www.globescan.com/commentary-and-analysis/press-releases/press-releases-2012/244-new-poll-of-business-leaders-highlights-sustainability-priorities-for-global-companies.html

Carboni, J., González, M., & Hodgkinson, J. (2013). The GPM® Guide to Sustainability. In *Project Management*. GPM Global.

Edum-Fotwe, F. T., & Price, A. D. F. (2009). A Social Ontology for Appraising Sustainability of Construction Projects and Developments. *International Journal of Project Management, 27*(4), 313–322. doi:10.1016/j.ijproman.2008.04.003

Eid, M. (2009). *Sustainable Development & Project Management*. Cologne, Germany: Lambert Academic Publishing.

Elkington, J. (1997). *Cannibals with Forks: The Triple Bottom Line of 21st Century Business*. Oxford, UK: Capstone Publishing Ltc.

Eskerod, P., & Huemann, M. (2013). Sustainable development and project stakeholder management: What standards say. *International Journal of Managing Projects in Business, 6*(1), 36–50. doi:10.1108/17538371311291017

Fernández-Sánchez, G., & Rodríguez-López, F. (2010). A methodology to identify sustainability indicators in construction project management—Application to infrastructure projects in Spain. *Ecological Indicators, 10*(6), 1193–1201. doi:10.1016/j.ecolind.2010.04.009

Freeman, R.E., Harrison, J.S., & Wicks, A.C. (2007). *Managing for Stakeholders: Survival, Reputation, and Success*. Yale University Press.

Freeman, R. E., Harrison, J. S., Wicks, A. C., Parmar, B. L., & De Colle, S. (2010). *Stakeholder Theory: The State of the Art*. Cambridge, UK: Cambridge University Press. doi:10.1017/CBO9780511815768

Friedman, M. (1970). The Social Responsibility of Business is to Increase its Profits. *The New York Times Magazine*.

Gareis, R. (1990). *Handbook of Management by Projects*. Academic Press.

Gareis, R., Huemann, M., & Martinuzzi, R.-A. (2013). Project Management & Sustainable Development Principles. Project Management Institute.

Global, G. P. M. (2014). *Sustainability Management Plan 1.2*. GPM Global.

Hardi, P., & Zdan, T. (Eds.). (1997). Assessing sustainable development: Principles in Practice. International Institute for Sustainable Development. Retrieved from http://www.iisd.org/pdf/bellagio.pdf

Hobday, M. (2000). The project-based organisation: An ideal form for managing complex products and systems?. *Research Policy, 29*(7-8), 871–893. doi:10.1016/S0048-7333(00)00110-4

Huemann, M. (2015). *Human Resource Management in the project-oriented Organization*. Farnham: Gower Publishing.

International Project Management Association. (2015). *Individual Competence Baseline version 4*. Nijkerk, The Netherlands: International Project Management Association.

Julian, S. D., Ofori-Dankwa, J. C., & Justis, R. T. (2008). Understanding strategic responses to interest group pressures. *Strategic Management Journal, 29*(9), 963–984. doi:10.1002/smj.698

Keeble, J. J., Topiol, S., & Berkeley, S. (2003). Using Indicators to Measure Sustainability Performance at a Corporate and Project Level. *Journal of Business Ethics*, *44*(2-3), 149–158. doi:10.1023/A:1023343614973

Keegan, A., & Turner, J. R. (2002). The management of innovation in project-based firms. *Long Range Planning*, *35*(4), 367–388. doi:10.1016/S0024-6301(02)00069-9

Knoepfel, H. (Ed.). (2010). *Survival and Sustainability as Challenges for Projects*. Zurich: International Project Management Association.

Labelle, F., & Leyrie, C. (2013, May). Stakepartner Management in Projects. *The Journal of Modern Project Management*.

Labuschagne, C., & Brent, A. C. (2006). Social indicators for sustainable project and technology life cycle management in the process industry. *International Journal of Life Cycle Assessment*, *11*(1), 3–15. doi:10.1065/lca2006.01.233

Libaert, T. (1998). Faire accepter un projet: principes et méthodes. *Communication et langages*, *117*, 76-90.

Lundin, R. A., & Söderholm, A. (1995). A theory of the temporary organization. *Scandinavian Journal of Management*, *11*(4), 437–455. doi:10.1016/0956-5221(95)00036-U

Ma, U. (2011). *No Waste; Managing Sustainability in Construction*. Farnham: Gower Publishing.

Maltzman, R., & Shirley, D. (2011). *Green Project Management*. Boca Raton, FL: CRC press.

Marcelino-Sádaba, S., Pérez-Ezcurdia, A., & González-Jaen, L. F. (2015). Using Project Management as a way to sustainability. From a comprehensive review to a framework definition. *Journal of Cleaner Production*, *99*, 1–16. doi:10.1016/j.jclepro.2015.03.020

Martens, M. L., & Carvalho, M. M. (2016). Key factors of sustainability in project management context: A survey exploring the project managers' perspective. *International Journal of Project Management*.

Martens, M. L., & Carvalho, M. M. (2017). Sustainability and Success Variables in the Project Management Context: An Expert Panel. *Project Management Journal*, *47*(6), 24–43.

Mettler, T., & Rohner, P. (2009). Situational Maturity Models as Instrumental Artifacts for Organizational Design. *Proceedings of the DESRIST'09*. doi:10.1145/1555619.1555649

Oehlmann, I. (2011). *The Sustainable Footprint Methodology*. Cologne, Germany: Lambert Academic Publishing.

Pasian, B., & Silvius, A. J. G. (2016). *A Review of Project Management Research in IRNOP and PMI Conferences from 2009 to 2014 to Identify Emerging Perspectives*. European Academy of Management (EURAM).

Shenhar, A., & Patanakul, P. (2014). Implementing Strategy through Projects. In J. R. Turner (Ed.), Handbook of Project Management. Gower Publishing.

Silvius, A. J. G. (2015a). Considering Sustainability in Project Management Processes. In K. D. Thomas (Ed.), *Handbook of Research on Sustainable Development and Economics*. IGI Global Publishing. doi:10.4018/978-1-4666-8433-1.ch014

Silvius, A. J. G. (2015b). *Model Project SMP*. Retrieved from http://www.slideshare.net/GilbertSilvius/model-project-smp

Silvius, A.J.G. (2016). Sustainability as a competence of Project Managers. *PM World Journal*, *5*(9), 1-13.

Silvius, A. J. G., Kampinga, M., Paniagua, S., & Mooi, H. (2017). Considering Sustainability in Project Management Decision Making; An investigation using Q-methodology. *International Journal of Project Management*.

Silvius, A.J.G., & Schipper, R. (2014). Sustainability in Project Management: A literature review and impact analysis. *Social Business*, *4*(1).

Silvius, A. J. G., & Schipper, R. (2015). Developing a Maturity Model for Assessing Sustainable Project Management. *Journal of Modern Project Management*, *3*(1), 16–27.

Silvius, A. J. G., & Schipper, R. (2016). Exploring the relationship between sustainability and project success - Conceptual model and expected relationships. *International Journal of Information Systems and Project Management*, *4*(3), 5–22.

Silvius, A. J. G., Schipper, R., Planko, J., van den Brink, J., & Köhler, A. (2012). *Sustainability in Project Management*. Farnham: Gower Publishing.

Silvius, A. J. G., & Tharp, J. (Eds.). (2013). *Sustainability Integration for Effective Project Management*. IGI Global Publishing. doi:10.4018/978-1-4666-4177-8

Sydow, J., Lindkvist, L., & DeFillippi, R. (2004). Editorial: Project organizations, embeddedness and repositories of knowledge. *Organization Studies*, *25*(9), 1475–1489. doi:10.1177/0170840604048162

Tam, G. C. K. (2017). *Managerial Strategies an Green Solutions for Project Sustainability*. IGI Global Publishing. doi:10.4018/978-1-5225-2371-0

Taylor, T. (2008). *A sustainability checklist for managers of projects*. Retrieved from http://www.pmforum.org/library/papers/2008/PDFs/Taylor-1-08.pdf

Tiron-Tudor, A., & Ioana-Maria, D. (2013). Project Success by Integrating Sustainability in Project Management. In A. J. G. Silvius & J. Tharp (Eds.), *Sustainability Integration for Effective Project Management*. IGI Global Publishing.

Turner, J. R., Anbari, F., & Bredillet, C. (2013). Perspectives on research in project management: The nine schools. *Global Business Perspectives*, *1*(1), 3–28. doi:10.1007/s40196-012-0001-4

Turner, J. R., & Müller, R. (2003). On the nature of the project as a temporary organization. *International Journal of Project Management*, *21*(3), 1–8. doi:10.1016/S0263-7863(02)00020-0

Turner, R., & Keegan, A. (1999). The versatile project-based organization: Governance and operational control. *European Management Journal*, *17*(3), 296–309. doi:10.1016/S0263-2373(99)00009-2

Turner, R. J. (2014). Projects and Their Management. In J. R. Turner (Ed.), Handbook of Project Management. Gower Publishing.

van Tulder, R., van Tilburg, R., Francken, M., & da Rosa, A. (2014). *Managing the Transition to a Sustainable Enterprise: Lessons from Frontrunner Companies*. Earthscan by Routledge.

Visser, W. T. (2002). Sustainability reporting in South Africa. *Corporate Environmental Strategy*, *9*(1), 79–85. doi:10.1016/S1066-7938(01)00157-9

Willard, B. (2005). *The next sustainability wave: building boardroom buy-in*. New Society Publishers.

World Commission on Environment and Development. (1987). *Our Common Future*. Oxford, UK: Oxford University Press.

Conclusion

Green production has been a popular research topic in recent years, not only for its great academic potential, but also for increasing environmental stress all over the world. Given the environmental and economic benefits of green production, this practice is considered an imperative strategic decision for enterprises, especially manufacturers. From the perspective of governments, some green-production-related laws and regulations have been enacted in countries all over the world to promote effective implementation of green production. Also, environmental awareness of consumers leads companies to invest in green production and provide green information to the market to gain competitive edge and expand market share. In recent years, several international automobile manufacturers have recently developed strict environmental protection measures on their production and operation. Furthermore, some apparel enterprise, for instance, H&M, Marks & Spencer and Levis, have taken many approaches to minimize environmental pollution in its production process by adopting new technologies.

This book provides relevant theoretical frameworks and the latest empirical research findings, and obtains some interesting and useful results in theory and practice on green production. The research conclusions in this book provide some guidelines on operations management for enterprises and governments. Implications of green production for the society and economy, particularly at the company and supply chain levels, is the topic covered in this book. This book consists of three sections: (1) Green Production and Application; (2) Green Environment and Management; (3) Sustainability and Management.

Section 1, "Green Production and Application," which consists of Chapters 1-5, focuses on the basic concept and application of green production. Section 1 starts with the concept of green production, then expands the concept from the company level to the supply chain level, and proposes green supply chain governance mechanism from the perspective of government, supplier, manufacturer, focal company, customer and NGO. The proposed green supply chain governance framework provides theoretical basis for the following contents of Section 1. Based on the green supply chain governance framework, some specific implementation steps of green production and green supply chain governance are studied in Chapter 2 – Chapter 5, including supply chain greenness assessment, green supplier selection, the application of cleaner production in the Brazilian sugar-energy sector, and sustainable manufacturing practices of India Small and Medium Enterprises (SMEs).

Then, the topic of the second section (referenced in Chapters 6-10) is Green Environment and Management. As the basis of Section 2, Chapter 6 proposes and introduces the concept of Environmental Management Systems (EMS), which includes economic, social and environmental considerations in all aspects of sustainability. Based on the concept of Environmental Management Systems (EMS), CSR-consumer binomial is analyzed to explore the role of consumer in Environmental Management Systems. And then, to improve the performance of Corporate Social Responsibility (CSR), some governance mechanisms of Corporate Social Irresponsibility (CSI) are explored. Focusing on the features of SMEs, sustainable business strategic planning is studied. Lastly, a case study of MengNiu Dairy is carried out to explore the influence of social trust on green innovation.

Based on Sections 1 and 2, Section 3 (referenced in Chapters 11-14) focuses on Sustainability and Management, which provide specific measures and operation support for various enterprises, organizations and industries based on their respective characteristics. Issues addressed in Section 3 include: green entrepreneurship in transitional economies, evaluation of renewable energy source in automobile industry, risk implications of heavy metal contamination in agricultural soil and crop productivity, and strategic postures for sustainability in the project-based organization. All the above research provides enterprises and policy makers with managerial suggestions on the implementation of green production.

In the fierce market competition, the 21st century enterprise faces new risks and challenges, and greenness becomes one of the unique competitive advantages of enterprises. High greenness of production and ultimate product can improve brand image, and promote sustainable development of companies, e.g., through the more efficient/green use of materials and energy, higher employee motivation, access to new market segments (such as green consumers), etc. From the perspective of enterprise practice, there is still much more topics to be studied regarding implementation of green production. For example, more important practical factors can be considered, including capital flow constraint, commercial and international trade, energy consumption, transportation, land use, etc. Furthermore, in the era of 'big data', the huge amount of data, structured and unstructured, may bring new opportunities and operation modes for green production of enterprises. The application of big data analysis helps enterprises collect and analyze data, and provides more intelligent selections and accurate decisions on green production.

Compilation of References

Aarseth, W., Ahola, T., Aaltonen, K., Økland, A., & Andersen, B. (2016). (in press). Project sustainability strategies: A systematic literature review. *International Journal of Project Management*.

Abaza, H., Bisset, R., & Sadler, B. (2004). *Environmental Impact Assessment and Strategic Environmental Assessment: Towards an Integrated Approach*. Geneva: UNEP.

Abdul Rashid, S. H., Evans, S., & Longhurst, P. (2008). A comparison of four sustainable manufacturing strategies. *International Journal of Sustainable Engineering*, *1*(3), 214–229. doi:10.1080/19397030802513836

Abraham, K., Sridevi, R., Suresh, B., & Damodharam, T. (2013). Effect of heavy metals (Cd, Pb, Cu) on seed germination of *Arachis hypogeae*. L. *Asian Journal of Plant Science and Research*, *3*, 10–12.

Abreu, M. C. S. D., Castro, F. D., Soares, F. D. A., & Silva Filho, J. C. L. (2012). A comparative understanding of corporate social responsibility of textile firms in Brazil and China. *Journal of Cleaner Production*, *20*(1), 119–126. doi:10.1016/j.jclepro.2011.08.010

Acharya, S., & Sharma, D. K. (2014). Study on the effects of heavy metals on seed germination and plant growth on *Jatropha curcas. International Journal of Agricultural Science Research*, *3*, 31–34.

Agamuthu, P., & Victor, D. (2011). Policy trends of extended producer responsibility in Malaysia. *Waste Management & Research*, *29*(9), 945–953. doi:10.1177/0734242X11413332 PMID:21771873

Agrawal, V. V., Ferguson, M., Toktay, L. B., & Thomas, V. M. (2012). Is leasing greener than selling? *Management Science*, *58*(3), 523–533. doi:10.1287/mnsc.1110.1428

Agrawal, V. V., & Ülkü, S. (2012). The role of modular upgradability as a green design strategy. *Manufacturing & Service Operations Management*, *15*(4), 640–648. doi:10.1287/msom.1120.0396

Ahmad, W. N. K. W., Rezaei, J., Tavasszy, L. A., & de Brito, M. P. (2016). Commitment to and preparedness for sustainable supply chain management in the oil and gas industry. *Journal of Environmental Management*, *180*, 202–213. doi:10.1016/j.jenvman.2016.04.056 PMID:27233046

Alcott, B. (2005). Jevons paradox. *Ecological Economics*, *54*(1), 9–21. doi:10.1016/j.ecolecon.2005.03.020

Aldrich, H. E., & Fiol, C. M. (1994). Fools rush in? The institutional context of industry creation. *Academy of Management Review*, *19*(4), 645–670.

Aliakbaria, A., & Seifbarghy, M. (2011). A supplier selection model for social responsible supply chain. *Journal of Optimization in Industrial Engineering*, *8*, 41–53.

Ali, B., Gill, R. A., Yang, S., Gill, M. B., Farooq, M. A., Liu, D., & Zhou, W. et al. (2015). Regulation of cadmium-induced proteomic and metabolic changes by 5-aminolevulinic acid in leaves of *Brassica napus* L. *PLoS ONE*, *10*(4), 1–23. doi:10.1371/journal.pone.0123328 PMID:25909456

Alkhader, A. M. F. (2015). The impact of phosphorus fertilizers on heavy metals content of soils and vegetables grown on selected farms in Jordan. *Agrotechnology*, *5*(01), 1–15. doi:10.4172/2168-9881.1000137

Allen, J. C., & Malin, S. (2008). Green entrepreneurship: A method for managing natural resources? *Society & Natural Resources*, *21*(9), 828–844. doi:10.1080/08941920701612917

Alloway, B. J. (2009). Soil factors associated with zinc deficiency in crops and humans. *Environmental Geochemistry and Health*, *31*(5), 537–548. doi:10.1007/s10653-009-9255-4 PMID:19291414

Alonso-Almeida, M. D. M., Bremser, K., & Llach, J. (2015). Proactive and reactive strategies deployed by restaurants in times of crisis: Effects on capabilities, organization and competitive advantage. *International Journal of Contemporary Hospitality Management*, *27*(7), 1641–1661. doi:10.1108/IJCHM-03-2014-0117

Al-Saleh, Y., & Mahroum, S. (2015). A critical review of the interplay between policy instruments and business models: Greening the built environment a case in point. *Journal of Cleaner Production*, *109*, 260–270. doi:10.1016/j.jclepro.2014.08.042

Alting, L. (Ed.). (1993). *Life-cycle design of products: a new opportunity for manufacturing enterprises, concurrent engineering: Automation, tools and techniques*. New York, NY: Wley.

Alting, L., & Jorgensen, J. (1993). Life cycle concept as a basic for sustainable industrial production. *CIRP Annals-Manufacturing Technology.*, *42*(1), 163–167. doi:10.1016/S0007-8506(07)62417-2

Alvarado-Herrera, A., Bigne, E., & Aldas-Manzano, J. (2015). A Scale for Measuring Consumer Perceptions of Corporate Social Responsibility Following the Sustainable Development Paradigm. *Journal of Business Ethics*, *140*(2), 1–20.

Alves Filho, M. (2007). Colhedora de cana preserva emprego e reduz desperdício. *Jornal da Unicamp*. Retrieved from http://www.unicamp.br/unicamp/unicamp_hoje/ju/julho2007/ju365pag05.html

AlWaer, H., Sibley, M., & Lewis, J. (2008). Different Stakeholder Perceptions of Sustainability Assessment. *Architectural Science Review*, *51*(1), 48–59. doi:10.3763/asre.2008.5107

Anastas, P. T., & Zimmerman, J. B. (2006). The twelve principles of green engineering as a foundation for sustainability. *Sustainability Science and Engineering: Defining Principles*, 11-31.

Andersen, M., & Skjoett-Larsen, T. (2009). Corporate social responsibility in global supply chains. *Supply Chain Management*, *14*(2), 75–86. doi:10.1108/13598540910941948

Anderson, B. S., Kreiser, P. M., Kuratko, D. F., Hornsby, J. S., & Eshima, Y. (2015). Reconceptualizing entrepreneurial orientation. *Strategic Management Journal*, *36*(10), 1579–1596. doi:10.1002/smj.2298

Arkani, S., & Theobald, R. (2007). Building civil society: Is there a role for the corporate sector? *Journal of Corporate Citizenship*, 37–50.

Armstrong, J. S. (1977). Social Irresponsibility in Management. *Journal of Business Research*, *5*(3), 185–213.

Armstrong, J. S., & Green, K. C. (2013). Effects of corporate social responsibility and irresponsibility policies. *Journal of Business Research*, *66*(10), 1922–1927.

Arsal, I., Woosnam, K. M., Baldwin, E. D., & Backman, S. J. (2010). Residents as travel destination information providers: An online community perspective. *Journal of Travel Research*, *49*(4), 400–413. doi:10.1177/0047287509346856

Arunakumara, K. K. I. U., Walpola, B. C., & Yoon, M.-H. (2013). Alleviation of Phyto-toxicity of Copper on Agricultural Plants. *Journal of the Korean Society for Applied Biological Chemistry*, *56*(5), 505–517. doi:10.1007/s13765-013-3126-5

Asati, A., Pichhode, M., & Kumar, N. (2016). Effect of heavy metals on plants: An overview. *International Journal of Application or Innovation in Engineering & Management*, *5*, 56–66.

Ashraf, R., & Ali, T. A. (2007). Effect of heavy metals on soil microbial community and mung beans seed germination. *Pakistan Journal of Botany*, *39*, 629–636.

Atasu, A., Özdemir, Ö., & Van Wassenhove, L. N. (2013). Stakeholder Perspectives on E-Waste Take-Back Legislation. *Production and Operations Management*, *22*(2), 382–396. doi:10.1111/j.1937-5956.2012.01364.x

Atasu, A., Sarvary, M., & Van Wassenhove, L. N. (2008). Remanufacturing as a marketing strategy. *Management Science*, *54*(10), 1731–1746. doi:10.1287/mnsc.1080.0893

Atasu, A., & Souza, G. C. (2013). How does product recovery affect quality choice? *Production and Operations Management*, *22*(4), 991–1010. doi:10.1111/j.1937-5956.2011.01290.x

Atasu, A., & Subramanian, R. (2012). Extended Producer Responsibility for E-Waste: Individual or Collective Producer Responsibility? *Production and Operations Management*, *21*(6), 1042–1059. doi:10.1111/j.1937-5956.2012.01327.x

Atasu, A., Van Wassenhove, L. N., & Sarvary, M. (2009). Efficient Take-Back Legislation. *Production and Operations Management*, *18*(3), 243–258. doi:10.1111/j.1937-5956.2009.01004.x

Atul, B. B., Govindan, K., & Satish, V. B. (2013). Analytical hierarchy process-based framework for VMI adoption. *International Journal of Production Research*, *51*(4), 963–978. doi:10.1080/00207543.2011.650795

Azzone, G., & Noci, G. (1998). Identifying effective PMSs for the deployment of green manufacturing strategies. *International Journal of Operations & Production Management*, *18*(4), 308–335. doi:10.1108/01443579810199711

Bae, J., Benoit, D. L., & Watson, A. K. (2016). Effect of heavy metals on seed germination and seedling growth of common ragweed and roadside ground cover legumes. *Environmental Pollution*, *213*, 112–118. doi:10.1016/j.envpol.2015.11.041 PMID:26882164

Barbieri, J. C. (2007). *Gestão Ambiental Empresarial: conceitos, modelos instrumentos*. São Paulo, Brasil: Saraiva.

Baron, R. M., & Kenny, D. A. (1986). The moderator–mediator variable distinction in social psychological research: Conceptual, strategic, and statistical considerations. *Journal of Personality and Social Psychology*, *51*(6), 1173–1182. doi:10.1037/0022-3514.51.6.1173 PMID:3806354

Barro, R. J. (1991). A cross-country study of growth, saving, and government. In *National saving and economic performance* (pp. 271–304). University of Chicago Press.

Basil, D. Z., & Weber, D. (2006). Values motivation and concern for appearances: The effect of personality traits on responses to corporate social responsibility. *International Journal of Nonprofit and Voluntary Sector Marketing*, *11*(11), 61–72. doi:10.1002/nvsm.38

Baskaran, V., Nachiappan, S., & Rahman, S. (2011). Supplier assessment based on corporate social responsibility criteria in Indian automotive and textile industry sectors. *International Journal of Sustainable Engineering*, *4*(4), 359–369. doi:10.1080/19397038.2011.579360

Baumgartner, R. J., & Ebner, D. (2010). Corporate Sustainability Strategies: Sustainability Profiles and Maturity Levels. *Sustainable Development*, *18*(2), 76–89. doi:10.1002/sd.447

Baycu, G., Rognes, S. E., Hakan, O., Gören-Saglam, N., Csatári, I., & Szabo, S. (2017). Abiotic stress effects on the antioxidative response profile of *Albizia julibrissin* Durazz. (Fabaceae). *Brazilian Journal of Botany*, *40*(1), 21–32. doi:10.1007/s40415-016-0318-3

Beamon, B. M. (1999). Designing the green supply chain. *Logistics Information Management*, *12*(4), 332-342.

Becker-Blease, J. R., & Sohl, J. E. (2015). New venture legitimacy: The conditions for angel investors. *Small Business Economics*, *45*(4), 735–749. doi:10.1007/s11187-015-9668-7

Becker, H. A., & Vanclay, F. (2003). *The international handbook of social impact assessment*. Cheltenham, UK: Edward Elgar. doi:10.4337/9781843768616

Bell, F. G., Bullock, S. E. T., Halbich, T. F. J., & Lindsay, P. (2001). Environmental impacts associated with an abandoned mine in the Witbank Coalfield, South Africa. *International Journal of Coal Geology*, *45*(2-3), 195–216. doi:10.1016/S0166-5162(00)00033-1

Bell, S., & Morse, S. (2003). *Measuring Sustainability Learning from doing*. London: Earthscan.

Bentler, P. M., & Chou, C. P. (1987). Practical issues in structural modeling. *Sociological Methods & Research*, *16*(1), 78–117. doi:10.1177/0049124187016001004

Berens, G., Riel, C. B. M. V., & Bruggen, G. H. V. (2005). Corporate Associations and Consumer Product Responses: The Moderating Role of Corporate Brand Dominance. *Journal of Marketing*, *69*(3), 35–48. doi:10.1509/jmkg.69.3.35.66357

Beske, P., Koplin, J., & Seuring, S. (2008). The use of environmental and social standards by German first-tier suppliers of the Volkswagen AG. *Corporate Social Responsibility and Environmental Management*, *15*(2), 63–75. doi:10.1002/csr.136

Bhattacharya, C. B., & Sen, S. (2003). *Consumer-Company Identification: A Framework for Understanding Consume*. Academic Press.

Bhattacharyya, P., Chakrabarti, K., Chakraborty, A., Tripathy, S., & Powell, M. A. (2008). Fractionation and bioavailability of Pb in municipal solid waste compost and Pb uptake by rice straw and grain under submerged condition in amended soil. *Geosciences Journal*, *12*(1), 41–45. doi:10.1007/s12303-008-0006-9

Bhutta, K. S., & Huq, F. (2002). Supplier selection problem: A comparison of the total cost of ownership and analytic hierarchy process approaches. *Supply Chain Management*, *7*(3), 126–135. doi:10.1108/13598540210436586

Biehal, G. J., & Sheinin, D. A. (2007). The Influence of Corporate Messages on the Product Portfolio. *Journal of Marketing*, *71*(2), 12–25. doi:10.1509/jmkg.71.2.12

Big River Newspaper. (2006). Retrieved from http://news.sohu.com/20060425/n242984539.shtml

Bi, S. (2011). *Research on Green Marketing Strategy of Enterprises in the Background of Ecological Civilization*. Ocean University of China.

Blake, P., & El Mansour, B. (2012). *Exploring Corporate Social Responsibility Leadership*. Retrieved from http://www.ufhrd.co.uk/wordpress/wp-content/uploads/2012/11/UFHRD2012Sust2.pdf

Boothroyd, G., & Dewhurst, P. (1994). *Adecade of DFMA Research, Conference Proceedings, The 1994 International Forum on Design for Manufacture and Assembly*. Newport.

Boulding, W., & Christen, M. (2001). First-mover disadvantage. *Harvard Business Review*, *79*(9), 20–21.

Boynd, H. William, & Massy. (2002). Marketing Management. New York: Harcourt Brace Jovanvich.

Branco, M. C., & Rodrigues, L. L. (2006). Corporate social responsibility and resource-based perspectives. *Journal of Business Ethics*, *69*(2), 111–132. doi:10.1007/s10551-006-9071-z

Braun, S. (2006). Core labor standards and FDI: Friends or foes? the case of child labor. *Review of World Economics*, *142*(4), 765–791. doi:10.1007/s10290-006-0092-6

Bronn, P. S., & Vrioni, A. B. (2001). Corporate social responsibility and cause-related marketing: An overview. *International Journal of Advertising*, *20*(2), 207–223. doi:10.1080/02650487.2001.11104887

Brooks, S., Wang, X., & Sarker, S. (2012). Unpacking green IS: a review of the existing literature and directions for the future. In *Green Business Process Management* (pp. 15–37). Springer Berlin Heidelberg. doi:10.1007/978-3-642-27488-6_2

Brown, T. J., & Dacin, P. A. (1997). The company and the product: Corporate associations and consumer product responses. *Journal of Marketing*, *61*(1), 68–84. doi:10.2307/1252190

Brunetti, A., Kisunko, G., & Weder, B. (1997). *Institutions in transition: reliability of rules and economic performance in former socialist countries* (Vol. 9708). World Bank Publications.

Brunetti, A., & Weder, B. (1998). Investment and institutional uncertainty: A comparative study of different uncertainty measures. *Weltwirtschaftliches Archiv*, *134*(3), 513–533. doi:10.1007/BF02707928

Bruton, G., Khavul, S., Siegel, D., & Wright, M. (2015). New financial alternatives in seeding entrepreneurship: Microfinance, crowdfunding, and peer-to-peer innovations. *Entrepreneurship Theory and Practice*, *39*(1), 9–26. doi:10.1111/etap.12143

BSR/GlobeScan. (2012). *2012 BSR/GlobeScan State of Sustainable Business Poll.* Retrieved from http://www.globescan.com/commentary-and-analysis/press-releases/press-releases-2012/244-new-poll-of-business-leaders-highlights-sustainability-priorities-for-global-companies.html

Buhmann, K. (2006). Corporate social responsibility in China: Current issues and their relevance for implementation of law. *Copenhagen Journal of Asian Studies*, *22*(1), 62–91.

Buyukozkan, G., & Çifçi, G. (2012). A novel hybrid MCDM approach based on fuzzy DEMATEL, fuzzy ANP and fuzzy TOPSIS to evaluate green suppliers. *Expert Systems with Applications*, *39*(3), 3000–3011. doi:10.1016/j.eswa.2011.08.162

Cao, M., & Zhang, Q. (2011). Supply chain collaboration: Impact on collaborative advantage and firm performance. *Journal of Operations Management*, *29*(3), 163–180. doi:10.1016/j.jom.2010.12.008

Carasco, E. F., & Singh, J. B. (2008). Human rights in global business ethics codes. *Business and Society Review*, *113*(3), 347–374. doi:10.1111/j.1467-8594.2008.00324.x

Carboni, J., González, M., & Hodgkinson, J. (2013). The GPM® Guide to Sustainability. In *Project Management*. GPM Global.

Carrasco, M. I., & Buendía, M. I. (2013). Corporate social responsibility: A crossroad between changing values, innovation and internationalisation. *European Journal of International Management*, *7*(3), 295–314.

Casillejo Tinedo, A., & Stensson, F. (2011). *Returnable plastic packaging flow in the automotive industry-an evaluation of the washing from a green logistics' perspective*. Academic Press.

Central Statistics Office. (2013). *Energy Statistics 2013*. Retrieved from http://mospiold.nic.in/Mospi_New/upload/Energy_Statistics_2013.pdf

CETESB - Companhia Ambiental do Estado de São Paulo. (n.d.). *A Produção Mais Limpa (P+L) no Setor Sucroalcooleiro – informações gerais.* Retrieved from http://www.agencia.cnptia.embrapa.br/Repositorio/Adubacao_organica_producao_mais_limpaID-37HFh1RpEg.pdf

Chan, F. T. S., Kumar, N., Tiwari, M. K., Lau, H. C. W., & Choy, K. L. (2008). Global supplier selection: A fuzzy-AHP approach. *International Journal of Production Research*, *46*(14), 3825–3857. doi:10.1080/00207540600787200

Chatterjee, J., & Chatterjee, C. (2000). Phytotoxicity of cobalt, chromium and copper in cauliflower. *Environmental Pollution*, *109*(1), 69–74. doi:10.1016/S0269-7491(99)00238-9 PMID:15092914

Chen, X. (2010). *Exploration of Chinese Charity Marketing Model.* Academic Press.

Chen, Z. (2010). Charity marketing: Concept, Theory and Literature Review. *Market Modernization, 2010*(20), 34-35.

Chen, C. H., Bang, N., & Klaus, P. (2015). Exploring electronic word-of-mouth (eWOM) in the consumer purchase decision-making process: The case of online holidays - evidence from United Kingdom (UK) consumers. *Journal of Travel & Tourism Marketing*, *32*(8), 953–970. doi:10.1080/10548408.2014.956165

Cheraghi, S. H., Dadashzadeh, M., & Subramanian, M. (2011). Critical success factors for supplier selection: An update. *Journal of Applied Business Research*, *20*(2), 91–108. doi:10.19030/jabr.v20i2.2209

Che, Z. H. (2010). A genetic algorithm-based model for solving multi-period supplier selection problem with assembly sequence. *International Journal of Production Research*, *48*(15), 4355–4377. doi:10.1080/00207540903049399

Chibuike, G. U., & Obiora, S. C. (2014). Heavy Metal Polluted Soils: Effect on Plants and Bioremediation Methods. *Applied and Environmental Soil Science*, *2014*, 1–12. doi:10.1155/2014/752708

Choi, T. M., Li, Y., & Xu, L. (2013). Channel leadership, performance and coordination in closed loop supply chains. *International Journal of Production Economics*, *146*(1), 371–380. doi:10.1016/j.ijpe.2013.08.002

Christofaki, M. I. (2011). *Effect of heavy metal stress in plant metabolism of solanaceous plant species with emphasis on nitrogen assimilation* (PhD thesis). Cranfield University.

Chuan, M. C., Shu, G. Y., & Liun, J. C. (1996). Solubility of heavy metals in a contaminated soil: Effects of redox potential and pH. *Water, Air, and Soil Pollution*, *90*(3-4), 543–556. doi:10.1007/BF00282668

Clayton, M. C. (2016). Know Your Customers "Jobs to Be Done". *Harvard Business Review*, *94*(9), 54–62.

Cohen, B., & Winn, M. I. (2007). Market imperfections, opportunity and sustainable entrepreneurship. *Journal of Business Venturing*, *22*(1), 29–49. doi:10.1016/j.jbusvent.2004.12.001

CONAB. (n.d.). Retrieved from http://www.conab.gov.br/conabweb/download/safra/1graos_08.09.pdf

Cope, J. (2005). Toward a dynamic learning perspective of entrepreneurship. *Entrepreneurship Theory and Practice, 29*(4), 373-397.

Costa, G., & Morel, J. L. (1994). Water relations, gas exchange and amino acid content in Cd-treated lettuce. *Plant Physiology and Biochemistry*, *32*, 561–570.

Costanza, R., & Daly, H. E. (1992). Natural Capital and Sustainable Development. *Conservation Biology*, *6*(1), 37–46. doi:10.1046/j.1523-1739.1992.610037.x

Cottrell, T., & Sick, G. (2002). Real options and follower strategies: The loss of real option value to first-mover advantage. *The Engineering Economist*, *47*(3), 232–263. doi:10.1080/00137910208965035

Covin, J. G., & Slevin, D. P. (1989). Strategic management of small firms in hostile and benign environments. *Strategic Management Journal*, *10*(1), 75–87. doi:10.1002/smj.4250100107

Cox, J. F., Blackstone, J. H., & Spencer, M. S. (Eds.). (1995). *American Production and Inventory Control Society*. Falls Church, VA: APICS Dictionary.

Curtis, H., & Walker, J. (2001). *Design council European survey of manufacturing companies' attitudes towards design for sustainability*. London: Design Council.

Czinkota, M., Kaufmann, H. R., & Basile, G. (2014). The relationship between legitimacy, reputation, sustainability and branding for companies and their supply chains. *Industrial Marketing Management*, *43*(1), 91–101. doi:10.1016/j.indmarman.2013.10.005

Darnall, N., Henriques, I., & Sadorsky, P. (2008). Do environmental management systems improve business performance in an international setting?. *Journal of International Management*, *14*(4), 364–376. doi:10.1016/j.intman.2007.09.006

Das, D., & Dutta, P. (2015). Design and analysis of a closed-loop supply chain in presence of promotional offer. *International Journal of Production Research*, *53*(1), 141–165. doi:10.1080/00207543.2014.942007

Dawkins, C. E. (2012). Labored relations: Corporate citizenship, labor unions, and freedom of association. *Business Ethics Quarterly*, *22*(3), 473–500. doi:10.5840/beq201222331

Dawson, J. F., & Richter, A. W. (2006). Probing three-way interactions in moderated multiple regression: Development and application of a slope difference test. *The Journal of Applied Psychology*, *91*(4), 917–926. doi:10.1037/0021-9010.91.4.917 PMID:16834514

De Giovanni, P. (2014). Environmental collaboration in a closed-loop supply chain with a reverse revenue sharing contract. *Annals of Operations Research*, *220*(1), 135–157. doi:10.1007/s10479-011-0912-5

De Sequeira, J. E. (2009). El principio de responsabilidad de Hans Jonas. *Bioethikos*, *3*(2), 171–193.

Dean, T. J., & McMullen, J. S. (2007). Toward a theory of sustainable entrepreneurship: Reducing environmental degradation through entrepreneurial action. *Journal of Business Venturing*, *22*(1), 50–76. doi:10.1016/j.jbusvent.2005.09.003

del Mar García de losSalmones, M. (2005). Influence of Corporate Social Responsibility on Loyalty and Valuation of Services. *Journal of Business Ethics*, *61*(4), 369–385. doi:10.1007/s10551-005-5841-2

Delmar, F., & Shane, S. (2004). Legitimating first: Organizing activities and the survival of new ventures. *Journal of Business Venturing*, *19*(3), 385–410. doi:10.1016/S0883-9026(03)00037-5

Dembe, A. E. (2009). Ethical issues relating to the health effects of long working hours. *Journal of Business Ethics*, *84*(2), 195–208. doi:10.1007/s10551-008-9700-9

Demiral, T., & Türkan, I. (2005). Comparative lipid peroxidation, antioxidant defense systems and proline content in roots of two rice cultivars differing in salt tolerance. *Environmental and Experimental Botany*, *53*(3), 247–257. doi:10.1016/j.envexpbot.2004.03.017

Demirevska-kepova, K., Simova-Stoilova, L., Stoyanova, Z., Holzer, R., & Feller, U. (2004). Biochemical changes in barely plants after excessive supply of copper and manganese. *Environmental and Experimental Botany*, *52*(3), 253–266. doi:10.1016/j.envexpbot.2004.02.004

Deng, Q., & Ji, S. (2015). Organizational Green IT Adoption: Concept and Evidence. *Sustainability*, *7*(12), 16737–16755. doi:10.3390/su71215843

Dessy, S. E., & Vencatachellum, D. (2003). Explaining cross-country differences in policy response to child labor. *The Canadian Journal of Economics. Revue Canadienne dEconomique*, *36*(1), 1–20. doi:10.1111/1540-5982.00001

Dey, S., Mazumder, P. B., & Paul, S. B. (2014). Effect of copper on growth and chlorophyll content in tea plants (*Camellia sinensis)* (l.) O. Kuntze. *International Journal of Research in Applied Natural and Social Sciences*, *2*, 223–230.

Dickson, G. W. (1996). An analysis of vendor selection systems and decision. *Journal of Purchasing*, *2*(15), 1377–1382.

Dietz, K. J., Baier, M., & Krämer, U. (1999). *Free radicals and reactive oxygen species as mediators of heavy metal toxicity in plants: from molecules to ecosystems*. Berlin: Springer- Verlag. doi:10.1007/978-3-662-07745-0_4

Digalwar, K., A., R. Tagalpallewar, A., & K. Sunnapwar, V. (. (. (2013). Green manufacturing performance measures: An empirical investigation from Indian manufacturing industries. *Measuring Business Excellence*, *17*(4), 59–75. doi:10.1108/MBE-09-2012-0046

Ding, G. (2010). Research on Creative Connotation. *Journal of Chongqing University of Technology: Social Science*, *24*(11), 77–80.

Doganova, L., & Karnøe, P. (2015). Clean and profitable: entangling valuations in environmental entrepreneurship. *Moments of Valuation: Exploring Sites of Dissonance*, 229-48.

Donaire, D. (1999). *Gestão ambiental na empresa*. São Paulo, Brasil: Atlas.

Dou, Y., Zhu, Q., & Sarkis, J. (2015). Integrating Strategic Carbon Management into Formal Evaluation of Environmental Supplier Development Programs. *Business Strategy and the Environment*, *24*(8), 873–891. doi:10.1002/bse.1851

Dowling, J., & Pfeffer, J. (1975). Organizational legitimacy: Social values and organizational behavior. *Pacific Sociological Review*, *18*(1), 122–136. doi:10.2307/1388226

Duarte, S., Cabrita, R., & Machado, V. C. (2011, January). Exploring lean and green supply chain performance using balanced scorecard perspective. *Proceedings of the 2011 international conference on industrial engineering and operations management*, 520-525.

Du, S., Bhattacharya, C. B., & Sen, S. (2011). Corporate Social Responsibility and Competitive Advantage: Overcoming the Trust Barrier. *Management Science*, *57*(9), 1528–1545. doi:10.1287/mnsc.1110.1403

Ebbs, S. D., & Kochian, L. V. (1997). Toxicity of zinc and copper to Brassica species: Implications for phytoremediation. *Journal of Environmental Quality*, *26*(3), 776–781. doi:10.2134/jeq1997.00472425002600030026x

Edum-Fotwe, F. T., & Price, A. D. F. (2009). A Social Ontology for Appraising Sustainability of Construction Projects and Developments. *International Journal of Project Management*, *27*(4), 313–322. doi:10.1016/j.ijproman.2008.04.003

Ehsanifar, A. (2013). Applying fuzzy DEMATEL method to analyze supplier selection criteria (case study: WagonPars Company). *International Research Journal of Finance & Economics*, *115*, 76–86.

Eid, M. (2009). *Sustainable Development & Project Management*. Cologne, Germany: Lambert Academic Publishing.

Eleren, A., & Yilmaz, C. (2011). Selection of suppliers by fuzzy TOPSIS model: Sample study from Turkey. *International Journal of Business & Social Science*, *2*(22), 189–200.

El-Khatib, A. A. (2003). The response of some common Egyptian plants ozone and their use a biomonitors. *Environmental Pollution*, *124*(3), 419–428. doi:10.1016/S0269-7491(03)00045-9 PMID:12758022

Elkington, J. (1997). *Cannibals with Forks: The Triple Bottom Line of 21st Century Business*. Oxford, UK: Capstone Publishing Ltc.

EPA. (2007). *Learn About Environmental Management Systems*. Retrieved April 3, 2017, from https: //www.epa. gov/ems /learn-about-environmental-management-systems

Eshima, Y., & Anderson, B. S. (2016). Firm growth, adaptive capability, and entrepreneurial orientation. *Strategic Management Journal*.

Eskerod, P., & Huemann, M. (2013). Sustainable development and project stakeholder management: What standards say. *International Journal of Managing Projects in Business*, *6*(1), 36–50. doi:10.1108/17538371311291017

Euromonitor International. (2016). *Toys and Games: Global Trends, Developments and Prospects*. Available at: http://www.euromonitor.com/toys-and-games-global-trends-developments-and-prospects/report

Fahimnia, B., Sarkis, J., & Davarzani, H. (2015). Green supply chain management: A review and bibliometric analysis. *International Journal of Production Economics*, *162*, 101–114. doi:10.1016/j.ijpe.2015.01.003

Fatma, M., Rahman, Z., & Khan, I. (2016). Measuring consumer perception of CSR in tourism industry: Scale development and validation. *Journal of Hospitality and Tourism Management*, *27*, 39–48. doi:10.1016/j.jhtm.2016.03.002

Felício, J. A., Ieva Meidutė, I., & Kyvik, Ø. (2016). Global mindset, cultural context, and the internationalization of SMEs. *Journal of Business Research*, *69*(11), 4924–4932. doi:10.1016/j.jbusres.2016.04.054

Feng, Y., Tian, Y., & Zhu, Q. (2014). A combined input–output/decision making trial and evaluation laboratory method for evaluating effect of the remanufacturing sector development. *Journal of Cleaner Production*, *114*, 103–113. doi:10.1016/j.jclepro.2014.10.048

Ferdows, K., Lewis, M. A., & Machuca, J. A. (2004). Rapid-fire fulfillment. *Harvard Business Review*, *82*(11), 104–117.

Fernández-Sánchez, G., & Rodríguez-López, F. (2010). A methodology to identify sustainability indicators in construction project management—Application to infrastructure projects in Spain. *Ecological Indicators*, *10*(6), 1193–1201. doi:10.1016/j.ecolind.2010.04.009

Fernandez, Z., & Nieto, M. J. (2006). Impact of ownership on the international involvement of SMEs. *Journal of International Business Studies*, *37*(3), 340–351. doi:10.1057/palgrave.jibs.8400196

Ferreira, A. I., & Ribeiro, I. (2017). Are you willing to pay the price? The impact of corporate social (ir)responsibility on consumer behavior towards national and foreign brands. *Journal of Consumer Behaviour*, 16.

Fishbein, M., & Azjen, I. (1975). *Beliefs, Attitude, Intention and Behavior: An introduction to theory and research*. Addison- Wesley.

Fisher, G., Kotha, S., & Lahiri, A. (2016). Changing with the times: An integrated view of identity, legitimacy, and new venture life cycles. *Academy of Management Review*, *41*(3), 383–409. doi:10.5465/amr.2013.0496

Foolmaun, R. K., & Ramjeawon, T. (2008). Life Cycle Assessment (LCA) of PET bottles and comparative LCA of three disposal options in Mauritius. *International Journal of Environment and Waste Management*, *2*(1), 125–138. doi:10.1504/IJEWM.2008.016997

Fornell, C., & Larcker, D. F. (1981). Evaluating structural equation models with unobservable variables and measurement error. *JMR, Journal of Marketing Research*, *18*(1), 39–50. doi:10.2307/3151312

Franceschi, D. & Kahn, J.R., 2003. Beyond strong sustainability. *International Journal of Sustainable Development & World Ecology*, *10*(3), 211–220. Available at:.10.1080/13504500309469799

Fraser, S., & Greene, F. J. (2006). The effects of experience on entrepreneurial optimism and uncertainty. *Economica*, *73*(290), 169–192.

Freeman, R.E., Harrison, J.S., & Wicks, A.C. (2007). *Managing for Stakeholders: Survival, Reputation, and Success.* Yale University Press.

Freeman, R. E., Harrison, J. S., Wicks, A. C., Parmar, B. L., & De Colle, S. (2010). *Stakeholder Theory: The State of the Art.* Cambridge, UK: Cambridge University Press. doi:10.1017/CBO9780511815768

Friedman, M. (1970). The Social Responsibility of Business is to Increase its Profits. *The New York Times Magazine.*

Frondel, M., Horbach, J., & Rennings, K. (2008). What triggers environmental management and innovation? Empirical evidence for Germany. *Ecological Economics, 66*(1), 153–160. doi:10.1016/j.ecolecon.2007.08.016

Fülöp, G., & Pelcz, I. G. (2010). Concept of environmentally-conscious strategic management. *European Integration Studies, 8*(1), 143-172.

Gajewska, E., Sklodowska, M., Slaba, M., & Mazur, J. (2006). Effect of nickel on antioxidative enzyme activities, proline and chlorophyll contents in wheat shoots. *Biologia Plantarum, 50*(4), 653–659. doi:10.1007/s10535-006-0102-5

Gao, P., Hensley, R., & Zielke, A. (2014). A road map to the future for the auto industry. *The McKinsey Quarterly, 10,* 1–11.

Gao, Y. (2011). *Research on Marketing Channel Innovation of MengNiu Dairy Company.* Jilin University.

Gap Inc. (2017). *Addressing the Impacts of Our Supply Chain.* Retrieved February 13, 2017, from: http://www.gapinc-sustainability.com/environment/protecting-our-shared-environment/addressing-impacts-our-supply-chain

Gareis, R., Huemann, M., & Martinuzzi, R.-A. (2013). Project Management & Sustainable Development Principles. Project Management Institute.

Gareis, R. (1990). *Handbook of Management by Projects.* Academic Press.

George, G., Robley Wood Jr, D., & Khan, R. (2001). Networking strategy of boards: Implications for small and medium-sized enterprises. *Entrepreneurship & Regional Development, 13*(3), 269–285. doi:10.1080/08985620110058115

Gharakhani, D. (2012). The evaluation of supplier selection criteria by fuzzy DEMATEL method. *Journal of Applied Sciences Research, 2*(4), 3215–3224.

Gibbs, D., & ONeill, K. (2014). Rethinking sociotechnical transitions and green entrepreneurship: The potential for transformative change in the green building sector. *Environment & Planning A, 46*(5), 1088–1107. doi:10.1068/a46259

Giddings, B., Hopwood, B., & OBrien, G. (2002). Environment, economy and society: Fitting them together into sustainable development. *Sustainable Development, 10*(4), 187–196. doi:10.1002/sd.199

GIIRS. (2017). *Emerging Market Assessment Resource Guide: Implementing an Environmental Management System.* Retrieved April 17, 2017, from http: //www. bcorporation. net/sites/ default/files /documents /bestpractices /em_implementing_ environmental_management_system.pdf

Global, G. P. M. (2014). *Sustainability Management Plan 1.2.* GPM Global.

Goering, G. E. (2012). Corporate social responsibility and marketing channel coordination. *Research in Economics, 66*(2), 142–148.

Golden, L. (2009). A brief history of long work time and the contemporary sources of overwork. *Journal of Business Ethics, 84*(2), 217–227. doi:10.1007/s10551-008-9698-z

Golden, L., & Figart, D. (2000). Doing something about long hours. *Challenge, 43*(6), 15–37. doi:10.1080/05775132.2000.11472177

Goldsmith, R. E., & Horowitz, D. (2006). Measuring Motivations for Online Opinion Seeking. *Journal of Interactive Advertising*, *6*(2), 2–14. doi:10.1080/15252019.2006.10722114

González-Benito, J., & González-Benito, Ó. (2005). Environmental proactivity and business performance: An empirical analysis. *Omega*, *33*(1), 1–15. doi:10.1016/j.omega.2004.03.002

González, M., & Guzmán, G. I. (2006). *Tras los pasos de la insustentabilidad. Agricultura y medio ambiente en perspectiva histórica*. Barcelona: Icaria Editorial.

Govindan, K., Khodaverdi, R., & Jafarian, A. (2013). A fuzzy multi criteria approach for measuring sustainability performance of a supplier based on triple bottom line approach. *Journal of Cleaner Production*, *47*(47), 345–354. doi:10.1016/j.jclepro.2012.04.014

Govindan, K., Khodaverdi, R., & Vafadarnikjoo, A. (2015). Intuitionistic fuzzy based DEMATEL method for developing green practices and performances in a green supply chain. *Expert Systems with Applications*, *42*(20), 7207–7220. doi:10.1016/j.eswa.2015.04.030

Govindan, K., Rajendran, S., Sarkis, J., & Murugesan, P. (2015). Multi criteria decision making approaches for green supplier evaluation and selection: A literature review. *Journal of Cleaner Production*, *98*, 66–83. doi:10.1016/j.jclepro.2013.06.046

Govindan, K., Shankar, M., & Kannan, D. (2016). (in press). Supplier selection based on corporate social responsibility practices. *International Journal of Production Economics*. doi:10.1016/j.ijpe.2016.09.003

Gretzel, U., Yoo, K. H., & Purifoy, M. (2007). *Online Travel Review Study: Role & Impact of.* Online Travel Reviews.

Guala, S. D., Vega, F. A., & Covelo, E. F. (2010). The dynamics of heavy metals in plant–soil interactions. *Ecological Modelling*, *221*(8), 1148–1152. doi:10.1016/j.ecolmodel.2010.01.003

Guhl, A. (2004). *Coffee and landscape change in the colombian countryside 1970-2002*. Miami, FL: University of Florida.

Guide, V. D. R. Jr, & Srivastava, R. (1997). Repairable inventory theory: Models and applications. *European Journal of Operational Research*, *102*(1), 1–20. doi:10.1016/S0377-2217(97)00155-0

Gupta, M. C. (1995). Environmental management and its impact on the operations function. *International Journal of Operations & Production Management*, *15*(8), 34–51. doi:10.1108/01443579510094071

Gutowski, T., Dahmus, J., & Thiriez, A. (2006, May). Electrical energy requirements for manufacturing processes. *13th CIRP International Conference on Life Cycle Engineering*, 31.

H&M Group. (2017). *Sustainability look good, do good, feel good*. Retrieved March 13, 2017, from: http://about.hm.com/en/sustainability.html

Hall, J. K., Daneke, G. A., & Lenox, M. J. (2010). Sustainable development and entrepreneurship: Past contributions and future directions. *Journal of Business Venturing*, *25*(5), 439–448. doi:10.1016/j.jbusvent.2010.01.002

Hall, J. L. (2002). Cellular mechanisms for heavy metal detoxification and tolerance. *Journal of Experimental Botany*, *53*(366), 1–11. doi:10.1093/jxb/53.366.1 PMID:11741035

Hamel, G. (2007). *The future of management*. Boston: Harvard Business School Press.

Haq, A. N., & Govindan, K. (2006). Design of an integrated supplier selection and multi-echelon distribution inventory model in a built-to-order supply chain environment. *International Journal of Production Research*, *44*(10), 1963–1985. doi:10.1080/00207540500381427

Hardi, P., & Zdan, T. (Eds.). (1997). Assessing sustainable development: Principles in Practice. International Institute for Sustainable Development. Retrieved from http://www.iisd.org/pdf/bellagio.pdf

Hare, P. D., Cress, W. A., & Van Staden, J. (1999). Proline synthesis and degradation: A model system for elucidating stress-related signal transduction. *Journal of Experimental Botany*, *50*, 413–434.

Hartman, C. L., & Stafford, E. R. (1997). Green alliances: Building new business with environmental groups. *Long Range Planning*, *30*(2), 184–149. doi:10.1016/S0024-6301(96)00111-2

Hassini, E., Surti, C., & Searcy, C. (2012). A literature review and a case study of sustainable supply chains with a focus on metrics. *International Journal of Production Economics*, *140*(1), 69–82. doi:10.1016/j.ijpe.2012.01.042

Hediger, W. (1999). Reconciling "weak" and "strong" sustainability. *International Journal of Social Economics*, *26*(7/8/9), 1120–1144.

Hediger, W. (2010). Welfare and capital-theoretic foundations of corporate social responsibility and corporate sustainability. *Journal of Socio-Economics*, *39*(4), 518–526.

Hegedus, A., Erdei, S., & Horvath, G. (2001). Comparative studies of H_2O_2 detoxifying enzymes in green and greening barley seedlings under cadmium stress. *Plant Science*, *160*(6), 1085–1093. doi:10.1016/S0168-9452(01)00330-2 PMID:11337065

Hemdi, A. R., Saman, M. Z., & Sharif, S. (2009). Proposed decision making framework for sustainable product design evaluation. *Asia Pacific Industrial Engineering & Management Systems Conference,* 1601-1609.

Hennig-Thurau, T., Gwinner, K. P., Walsh, G., & Gremler, D. D. (2004). Electronic word-of-mouth via consumer-opinion platforms: What motivates consumers to articulate themselves on the Internet? *Journal of Interactive Marketing*, *18*(1), 38–52. doi:10.1002/dir.10073

Hervani, A. A., Helms, M. M., & Sarkis, J. (2005). Performance measurement for green supply chain management. *Benchmarking: An International Journal*, *12*(4), 330–353. doi:10.1108/14635770510609015

Hewitt, E. J. (1983). Metals and micronutrients: uptake and utilization by plants. London: Academic Press.

Hobbs, S., & Cornwell, D. (1986). Child labor: An underdeveloped topic in psychology. *International Journal of Psychology*, *21*(1-4), 225–234. doi:10.1080/00207598608247587

Hobday, M. (2000). The project-based organisation: An ideal form for managing complex products and systems?. *Research Policy*, *29*(7-8), 871–893. doi:10.1016/S0048-7333(00)00110-4

Honeyman, K., & Goodman, J. (1991). Womens work, gender conflict, and labor markets in Europe, 15001900. *The Economic History Review*, *44*(4), 608–628. doi:10.2307/2597804

Hossain, M. A., Hoque, M. A., Burritt, D. J., & Fujita, M. (2014). *Proline Protects Plants Against Abiotic Oxidative Stress: Biochemical and Molecular Mechanisms. In Oxidative damage to plants* (pp. 477–52). Academic Press.

Hsueh, C. F., & Chang, M. S. (2008). Equilibrium analysis and corporate social responsibility for supply chain integration. *European Journal of Operational Research*, *190*(1), 116–129.

Hua, H. (2000). Discussion on 4P Theory and 4C Theory. *Journal of Beijing Institute of Finance and Commerce*, *2000*(4), 36-37.

Huang, M., & Wang, B. (2014). Evaluating green performance of building products based on gray relational analysis and analytic hierarchy process. *Environmental Progress & Sustainable Energy*, *33*(4), 1389–1395.

Huang, S. (2014). Research on Pricing and Social Responsibility Decision Models in Dual-channel Supply Chain. *Commercial Research*, *56*(5), 153–161.

Huang, T., Wu, F., Yu, J., & Zhang, B. (2015). Political risk and dividend policy: Evidence from international political crises. *Journal of International Business Studies*, *46*(5), 574–595. doi:10.1057/jibs.2015.2

Huang, W. (2008). *Marketing Strategy Analysis of MengNiu Dairy*. Xiamen University.

Huemann, M. (2015). *Human Resource Management in the project-oriented Organization*. Farnham: Gower Publishing.

Hunkeler, D., & Rebitzer, G. (2005). The Future of Life Cycle Assessment. *International Journal of Life Cycle Assessment.*, *10*(5), 305–308. doi:10.1065/lca2005.09.001

Hu, X. (2004). Corporate Sustainable Development and Responsibility. Journal of Chongqing University of Posts and Telecomm unications. *Social Sciences*, *16*(2), 123–125.

Hu, Z., Fang, F., Ben, D., Pu, G., & Wang, C. (2004). Net Energy, CO2 Emission and Life-cycle Cost Assessment of Cassava-based Ethanol as an Alternative Automotive Fuel in China. *Applied Energy*, *78*(3), 247–256. doi:10.1016/j.apenergy.2003.09.003

INDETEX. (2017). *Sustainability*. Retrieved March 13, 2017: https://www.inditex.com/es/sustainability

International Project Management Association. (2015). *Individual Competence Baseline version 4*. Nijkerk, The Netherlands: International Project Management Association.

Isaak, R. (1997). Globalisation and green entrepreneurship. *Greener Management International*, 80-90.

ISO 14001. (2017). *ISO 14001 Requirements*. Retrieved April 16, 2017 from http: //www.iso14001.com.au /iso-14001-requirements.html

Jabour, A. B. L. S., & Jabour, C. J. C. (2013). *Gestão ambiental nas organizações; fundamentos e tendências*. São Paulo, Brasil: Atlas.

Jadia, C. D., & Fulekar, M. H. (2008). Phytoremediation: The application of vermicompost to remove zinc, cadmium, copper, nickel and lead by sunflower plant. *Environmental Engineering and Management Journal*, *7*, 547–558.

Jamieson, S. (2004). Feminist theory, globalization and comparative labor law: Women workers in Australia and Ireland. *International Journal of Human Resource Management*, *15*(3), 459–465. doi:10.1080/0958519042000181197

Jan, S., & Parray, J. A. (2016). *Heavy metal uptake in plants approaches to heavy metal tolerance in plants* (pp. 1–18). Singapore: Springer. doi:10.1007/978-981-10-1693-6

Jian Liu, J., Fong-Yuen Ding, F. Y., Vinod Lall, V., Jian, L., Fong-Yuen, D., & Vinod, L. (2000). Using data envelopment analysis to compare suppliers for supplier selection and performance improvement. *Supply Chain Management: An International JournalSupply Chain Management*, *5*(3), 143–150. doi:10.1108/13598540010338893

Jiang, L. (2014). A Review of Foreign Literature of Corporate Social Irresponsibility. *Foreign Economics & Management*, *36*(2), 13–23.

Jiang, Z., Zhang, H., & Sutherland, J. W. (2011). Development of multi-criteria decision making model for remanufacturing technology portfolio selection. *Journal of Cleaner Production*, *19*(17-18), 1939–1945. doi:10.1016/j.jclepro.2011.07.010

Jia, P., Diabat, A., & Mathiyazhagan, K. (2015). Analyzing the SSCM practices in the mining and mineral industry by ISM approach. *Resources Policy*, *46*, 76–85. doi:10.1016/j.resourpol.2014.04.004

Jimena, J. (2010). Tax is just one more corporate social issue. *Canadian Mining Journal*, *131*(3), 9–9.

Jin, L. Y. (2006). The effects of consumers' company identification on their evaluation of products and behavioral responses. *Nankai Business Review*, *9*(3), 16–21.

John, R., Ahmad, P., Gadgil, K., & Sharma, S. (2009). Heavy metal toxicity: Effect on plant growth, biochemical parameters and metal accumulation by *Brassica juncea* L. *International Journal of Plant Production*, *3*, 65–76.

Jordao, C. P., Nascentes, C. C., Cecon, P. R., Fontes, R. L. F., & Pereira, J. L. (2006). Heavy metal availability in soil amended with composted urban solid wastes. *Environmental Monitoring and Assessment*, *112*(1-3), 309–326. doi:10.1007/s10661-006-1072-y PMID:16404548

Julian, S. D., Ofori-Dankwa, J. C., & Justis, R. T. (2008). Understanding strategic responses to interest group pressures. *Strategic Management Journal*, *29*(9), 963–984. doi:10.1002/smj.698

Kainuma, Y., & Tawara, N. (2006). A multiple attribute utility theory approach to lean and green supply chain management. *International Journal of Production Economics*, *101*(1), 99–108. doi:10.1016/j.ijpe.2005.05.010

Kapoor, D., Kaur, S., & Bhardwaj, R. (2014). Physiological and Biochemical Changes in *Brassica juncea* plants under Cd-Induced Stress. *BioMed Research International*, *2014*, 1–13. doi:10.1155/2014/726070 PMID:25133178

Kapoor, D., Rattan, A., Kaur, S., & Bhardwa, R. (2016). Influence of cadmium on antioxidative defence system, photosynthesis, level of osmolytes and ions uptake in *Brassica juncea*. *International Journal of Pharmacy and Pharmaceutical Sciences*, *8*, 1–13.

Kapur, A., Baldwin, C., Swanson, M., Wilberforce, N., McClenachan, G., & Rentschler, M. (2012). Comparative life cycle assessment of conventional and Green Seal-compliant industrial and institutional cleaning products. *International Journal of Life Cycle Assessment.*, *17*(4), 377–387. doi:10.1007/s11367-011-0373-8

Keeble, J. J., Topiol, S., & Berkeley, S. (2003). Using Indicators to Measure Sustainability Performance at a Corporate and Project Level. *Journal of Business Ethics*, *44*(2-3), 149–158. doi:10.1023/A:1023343614973

Keegan, A., & Turner, J. R. (2002). The management of innovation in project-based firms. *Long Range Planning*, *35*(4), 367–388. doi:10.1016/S0024-6301(02)00069-9

Kelly, B., Pástor, Ľ., & Veronesi, P. (2016). The price of political uncertainty: Theory and evidence from the option market. *The Journal of Finance*, *71*(5), 2417–2480. doi:10.1111/jofi.12406

Kemp, D., Bond, C. J., Franks, D. M., & Cote, C. (2010). Mining, water and human rights: Making the connection. *Journal of Cleaner Production*, *18*(15), 1553–1562. doi:10.1016/j.jclepro.2010.06.008

Kennish, L. (1992). Toxicity of heavy metals: Effect of Cr and Se on human health. *Journal of Indian Public Health Education*, *2*, 36–64.

Keunen, E., Peshev, D., Vangronsveld, J., Wim, V. D. E., & Cuypers, A. (2013). Plant sugars are crucial players in the oxidative challenge during abiotic stress: Extending the traditional concept. *Plant, Cell & Environment*, *36*(7), 1242–1255. doi:10.1111/pce.12061 PMID:23305614

Khaledabadi, H. J., & Magnusson, T. (2008). Corporate social responsibility and knowledge management implications in sustainable vehicle innovation and development. *Communications of the Ibima*, *6*(2), 8–14.

Khan, S., Cao, Q., Zheng, Y. M., Huang, Y. Z., & Zhu, Y. G. (2008). Health risks of heavy metals in contaminated soils and food crops irrigated with wastewater in Beijing, China. *Environmental Pollution*, *152*(3), 686–692. doi:10.1016/j.envpol.2007.06.056 PMID:17720286

Khoshgoftarmanesh, A. H., Hosseini, F., & Afyuni, M. (2011). Nickel supplementation effect on the growth, urease activity and urea and nitrate concentrations in lettuce supplied with different nitrogen sources. *Scientia Horticulturae, 130*(2), 381–385. doi:10.1016/j.scienta.2011.07.009

Kim, J., Lindgreen, A., & Lee, K. (2009). Current status of CSR in the realm of supply management: The case of the Korean electronics industry. *Supply Chain Management, 14*(2), 335–340.

Kim, S. (2010). Transferring Effects of CSR Strategy on Consumer Responses: The Synergistic Model of Corporate Communication Strategy. *Journal of Public Relations Research, 23*(2), 218–241. doi:10.1080/1062726X.2011.555647

Kim, W. C., & Mauborgne, R. (2005). *Blue ocean strategy: How to create uncontested market space and make the competition irrelevant.* Boston: Harvard Business School Press.

Kleindorfer, P. R., Singhal, K., & Wassenhove, L. N. (2005). Sustainable operations management. *Production and Operations Management, 14*(4), 482–492. doi:10.1111/j.1937-5956.2005.tb00235.x

Klewitz, J., & Hansen, E. (2014). Sustainability-oriented innovation of SMEs: A systematic review. *Journal of Cleaner Production, 65*, 57-75.

Knemeyer, A. M., Ponzurick, T. G., & Logar, C. M. (2002). A qualitative examination of factors affecting reverse logistics systems for end-of-life computers. *International Journal of Physical Distribution & Logistics Management, 32*(6), 455–479. doi:10.1108/09600030210437979

Knoepfel, H. (Ed.). (2010). *Survival and Sustainability as Challenges for Projects.* Zurich: International Project Management Association.

Kobrin, S. J. (2009). Private political authority and public responsibility: Transnational politics, transnational firms, and human rights. *Business Ethics Quarterly, 19*(3), 349–374. doi:10.5840/beq200919321

Kongar, E. (2005). Evaluation criteria for environmentally benign forward and reverse supply chain management. *International Journal of Operations and Quantitative Management, 11*(2), 147.

Kostova, T., & Zaheer, S. (1999). Organizational legitimacy under conditions of complexity: The case of the multinational enterprise. *Academy of Management Review, 24*(1), 64–81.

Kotler, P., & Armstrong, G. (2003). *Fundamentos del Marketing.* Pearson Education.

Kreiser, P. M. (2011). Entrepreneurial orientation and organizational learning: The impact of network range and network closure. *Entrepreneurship Theory and Practice, 35*(5), 1025–1050. doi:10.1111/j.1540-6520.2011.00449.x

Kuckertz, A., & Wagner, M. (2010). The influence of sustainability orientation on entrepreneurial intentions—Investigating the role of business experience. *Journal of Business Venturing, 25*(5), 524–539. doi:10.1016/j.jbusvent.2009.09.001

Kumar, V. (2012). *101 Design Methods: A Structured Approach for Driving Innovation in Your Organization.* Hoboken, NJ: John Wiley & Sons, Inc.

Kumar, D. T., Palaniappan, M., Kannan, D., & Shankar, K. M. (2014). Analyzing the CSR issues behind the supplier selection process using ISM approach. *Resources, Conservation and Recycling, 92*, 268–278. doi:10.1016/j.resconrec.2014.02.005

Kumar, P. (2015). Application of DEMATEL and MMDE for analyzing key influencing factors relevant to selection of supply chain coordination schemes. *International Journal of Information Systems and Supply Chain Management, 8*(2), 51–72.

Kuznetsov, V. V., & Shevyakova, N. I. (1997). Stress responses of tobacco cells to high temperature and salinity: Proline accumulation and phosphorylation of polypeptides. *Physiologia Plantarum*, *100*(2), 320–326. doi:10.1111/j.1399-3054.1997.tb04789.x

Labelle, F., & Leyrie, C. (2013, May). Stakepartner Management in Projects. *The Journal of Modern Project Management*.

Laberge, Y. (2007). Ecomanagerialism. In P. Robbins (Ed.), *Encyclopedia of Environment and Society* (pp. 593–594). Sage Publications.

Labuschagne, C., & Brent, A. C. (2006). Social Indicators for Sustainable Project and Technology Life Cycle Management in the Process Industry. *International Journal of Life Cycle Assessment*, *11*(1), 3–15. doi:10.1065/lca2006.01.233

Lafferty, B. A., & Goldsmith, R. E. (1999). Corporate Credibilitys Role in Consumers Attitudes and Purchase Intentions When a High versus a Low Credibility Endorser Is Used in the Ad. *Journal of Business Research*, *44*(2), 109–116. doi:10.1016/S0148-2963(98)00002-2

Lakhal, S. Y., H'Mida, S., Béland, M., & Kassogue, C. O. (2004). *Green Supply Chain Parameters for Companies in the Lumber Industry*. Supply Chain Discussion Paper.

Lankoski, L. (2000). *Determinants of environmental profit: An analysis of the firm-level relationship between environmental performance and economic performance*. Helsinki University of Technology.

Lee, C. W., Choi, J. M., & Pak, C. H. (1996). Micronutrient toxicity in seed geranium (*Pelargonium×hortorum* Baley). *Journal of the American Society for Horticultural Science*, *121*, 77–82.

Lee, S. K., Seo, K. L., & Sharma, A. (2013). Corporate social responsibility and firm performance in the airline industry: The moderating role of oil prices. *Tourism Management*, *38*(13), 20–30. doi:10.1016/j.tourman.2013.02.002

Lenntech. (2004). *Water Treatment and Air Purification*. Retrieved from www.excelwater.com/thp/filters/Water-Purification.htm

Lewis, S., Donkin, M. E., & Depledge, M. H. (2001). Hsp70 expression in *Enteromorpha intestinalis* (Chlorophyta) exposed to environmental stressors. *Aquatic Toxicology (Amsterdam, Netherlands)*, *51*(3), 277–291. doi:10.1016/S0166-445X(00)00119-3 PMID:11090890

Li, Y., Peng, M. W., & Macaulay, C. D. (2013). *Market–political ambidexterity during institutional transitions*. Academic Press.

Liang, Y., Ni, D., & Tang, X. (2013). Corporate Social Responsibility Based Model on Dual-channel Competition in Supply Chains. *Chinese Journal of Management Science*, *21*(S2), 453–460.

Libaert, T. (1998). Faire accepter un projet: principes et méthodes. *Communication et langages*, *117*, 76-90.

Lieberman, M. B., & Montgomery, D. B. (1988). First-mover advantages. *Strategic Management Journal*, *9*(S1), 41–58. doi:10.1002/smj.4250090706

Li, H. F., Gray, C., Mico, C., Zhao, F. J., & McGrath, S. P. (2009). Phytotoxicity and bioavailability of cobalt to plants in a range of soils. *Chemosphere*, *75*(7), 979–986. doi:10.1016/j.chemosphere.2008.12.068 PMID:19217640

Li, H., & Atuahene-Gima, K. (2001). Product innovation strategy and the performance of new technology ventures in China. *Academy of Management Journal*, *44*(6), 1123–1134. doi:10.2307/3069392

Lii, Y. S., & Lee, M. (2012). Doing Right Leads to Doing Well: When the Type of CSR and Reputation Interact to Affect Consumer Evaluations of the Firm. *Journal of Business Ethics*, *105*(1), 69–81. doi:10.1007/s10551-011-0948-0

Li, J. Y. (2013). *Study on Green Marketing Strategy of Hua Peng Food*. Central South University of Forestry and Technology.

Lin, C. S., Chang, R. Y., & Dang, V. T. (2015). An integrated model to explain how corporate social responsibility affects corporate financial performance. *Sustainability*, *7*(7), 8292–8311. doi:10.3390/su7078292

Lin-Hi, N., & Müller, K. (2013). The CSR bottom line: Preventing corporate social irresponsibility. *Journal of Business Research*, *66*(10), 1928–1936.

Liu, L., Sun, H., Chen, J., Zhang, Y., Li, D., & Li, C. (2014). Effects of cadmium (Cd) on seedling growth traits and photosynthesis parameters in cotton (*Gossypium hirsutum* L.). *Plant Omics Journal*, *7*, 284–290. doi:10.1111/tpj.12387

Li, W. A., Li, Y. J., & Shi, D. (2016). Theoretical research on supply chain governance: Concept, intention, and normative analysis framework. [in Chinese]. *Nankai Business Review*, *19*(1), 4–15.

Li, X. (2015). The Research and Prospect of Green Innovation Research. *Research and Development Management*, *27*(2), 1–11.

Li, X., Wang, C., & Xu, X. (2014). Rethinking new ventures cognitive legitimacy: An experimental study. *Systems Research and Behavioral Science*, *31*(3), 437–446. doi:10.1002/sres.2281

Li, Y. (2011). Research on the performance measurement of green supply chain management in China. *Journal of Sustainable Development*, *4*(3), 101. doi:10.5539/jsd.v4n3p101

Li, Y., Zhao, X., Shi, D., & Li, X. (2014). Governance of sustainable supply chains in the fast fashion industry. *European Management Journal*, *32*(5), 823–836. doi:10.1016/j.emj.2014.03.001

Lopez, L. E., & Roberts, E. B. (2002). First-mover advantages in regimes of weak appropriability: The case of financial services innovations. *Journal of Business Research*, *55*(12), 997–1005. doi:10.1016/S0148-2963(01)00200-4

Lounsbury, M., & Glynn, M. A. (2001). Cultural entrepreneurship: Stories, legitimacy, and the acquisition of resources. *Strategic Management Journal*, *22*(6-7), 545–564. doi:10.1002/smj.188

Lovins, A. (2003). *Natural capitalism and the new North Carolina economy*. Rocky Mountain Institute. Available from: http://www.rmi.org/images/ other/Energy/E03-11_NapCapNewNCEcon.pdf

Lowell Center for Sustainable Production. (1998). *Sustainable production: A working definition*. Informal meeting of the committee members. Author.

Lugon-Moulin, N., Zhang, M., Gadani, F., Rossi, L., Koller, D., Krauss, M., & Wagner, G. J. (2004). Critical review of the science and options for reducing cadmium in tobacco (*Nicotiana tabacum* L.) and other plants. *Advances in Agronomy*, *83*, 111–180. doi:10.1016/S0065-2113(04)83003-7

Lumpkin, G. T., & Dess, G. G. (1996). Clarifying the entrepreneurial orientation construct and linking it to performance. *Academy of Management Review*, *21*(1), 135–172.

Lundin, R. A., & Söderholm, A. (1995). A theory of the temporary organization. *Scandinavian Journal of Management*, *11*(4), 437–455. doi:10.1016/0956-5221(95)00036-U

Luthra S., Mangla S.K., Xu L., & Diabat A. (2016). Using AHP to evaluate barriers in adopting sustainable consumption and production initiatives in a supply chain. *International Journal of Production Economics*, *181*(Part B), 342-349.

Luthra, S., Garg, D., & Haleem, A. (2014). Green supply chain management: Implementation and performance–a literature review and some issues. *Journal of Advances in Management Research*, *11*(1), 20–46. doi:10.1108/JAMR-07-2012-0027

Luthra, S., Garg, D., & Haleem, A. (2016). The impacts of critical success factors for implementing green supply chain management towards sustainability: An empirical investigation of Indian automobile industry. *Journal of Cleaner Production, 121*, 142–158. doi:10.1016/j.jclepro.2016.01.095

Ma & Yao. (2014). Discussion on Green Marketing. *Management Supervisor, 2014*(5X), 275-275.

MacKinnon, D. W. (1962). The Personality Correlates of Creativity: A Study of American Architects. In G. S. Neilsen (Ed), *Proceeding of the XIV International Congress of Applied Psychology*, (vol. 2, pp. 11-39). Copenhagen: Munksgarrd.

Mahmood, T., Islam, K. R., & Muhammad, S. (2007). Toxic effects of heavy metals on early growth and tolerance of cereal crops. *Pakistan Journal of Botany, 39*, 451–462.

Makadok, R. (1998). Can first-mover and early-mover advantages be sustained in an industry with low barriers to entry/imitation? *Strategic Management Journal, 19*(7), 683–696. doi:10.1002/(SICI)1097-0266(199807)19:7<683::AID-SMJ965>3.0.CO;2-T

Makoi, J. H. J. R. (2016). Status of Fe, Cu, Zn and Mn and their relationship with some physical chemical properties in the soils of Sonjo and Lupiro alluvial/flood plains in Kilombero and Ulanga Districts, Morogoro, Tanzania. *Academia Journal of Biotechnology, 4*, 76–84.

Ma, L., Wang, Y. H., & Ma, M. S. (2016). The Tripartite Evolutionary Game Analysis of Collusive behavior Regulatory in EPR System. [in Chinese]. *Science and Management, 36*(3), 24–29.

Malar, S., Sahi, S. V., Favas, P. J. C., & Venkatachalam, P. (2014). Lead heavy metal toxicity induced changes on growth and antioxidative enzymes level in water hyacinths. *Botanical Studies (Taipei, Taiwan), 55*, 1–11. PMID:28597420

Maloni, M. J., & Brown, M. E. (2006). Corporate social responsibility in the supply chain: An application in the food industry. *Journal of Business Ethics, 68*(1), 35–52. doi:10.1007/s10551-006-9038-0

Maltzman, R., & Shirley, D. (2011). *Green Project Management*. Boca Raton, FL: CRC press.

Marcelino-Sádaba, S., Pérez-Ezcurdia, A., & González-Jaen, L. F. (2015). Using Project Management as a way to sustainability. From a comprehensive review to a framework definition. *Journal of Cleaner Production, 99*, 1–16. doi:10.1016/j.jclepro.2015.03.020

Marin, L., Ruiz, S., & Rubio, A. (2009). The Role of Identity Salience in the Effects of Corporate Social Responsibility on Consumer Behavior. *Journal of Business Ethics, 84*(1), 65–78. doi:10.1007/s10551-008-9673-8

Martens, M. L., & Carvalho, M. M. (2016). Key factors of sustainability in project management context: A survey exploring the project managers' perspective. *International Journal of Project Management*.

Martens, M. L., & Carvalho, M. M. (2017). Sustainability and Success Variables in the Project Management Context: An Expert Panel. *Project Management Journal, 47*(6), 24–43.

Martin, D., & Schouten, J. (2014). *Sustainable Marketing*. Shanghai Renmin Press.

Martineau, C., & Pastoriza, D. (2016). International involvement of established SMEs: A systematic review of antecedents, outcomes and moderators. *International Business Review, 25*(2), 458–470. doi:10.1016/j.ibusrev.2015.07.005

Martínez, A. M. Á. (2007). Un modelo de empresa innovadora y flexible: El caso Zara. *Revista de Investigaciones Políticas y Sociológicas, 6*(1), 69–80.

Masscheleyn, P., Pardue, J., Delaune, R., & Patrick, J. (1991). Effect of redox potential and pH on As speciation and solubility in a contaminated soil. *Environmental Science & Technology, 25*(8), 1414–1419. doi:10.1021/es00020a008

Mathiyazhagan, K., Diabat, A., Al-Refaie, A., & Xu, L. (2015). Application of analytical hierarchy process to evaluate pressures toimplement green supply chain management. *Journal of Cleaner Production, 107*, 229–236. doi:10.1016/j.jclepro.2015.04.110

Ma, U. (2011). *No Waste; Managing Sustainability in Construction.* Farnham: Gower Publishing.

Maurer, J. G. (1971). *Readings in organization theory: Open-system approaches.* Random House.

Mauro, P. (1995). Corruption and growth. *The Quarterly Journal of Economics, 110*(3), 681–712. doi:10.2307/2946696

Mavragani, A., Nikolaou, I. E., & Tsagarakis, K. P. (2016). Open Economy, Institutional Quality, and Environmental Performance: A Macroeconomic Approach. *Sustainability, 8*(7), 601. doi:10.3390/su8070601

Maxwell, D., & Van der Vorst, R. (2003). Developing sustainable products and services. *Journal of Cleaner Production, 11*(8), 883–895. doi:10.1016/S0959-6526(02)00164-6

McLaughlin, M. (2002). Heavy metals. In R. Lal (Ed.), *Encyclopedia of soil science* (pp. 650–653). New York: Marcel Dekker.

Mendel, R. R. (2011). Cell biology of molybdenum in plants. *Plant Cell Reports, 30*(10), 1787–1797. doi:10.1007/s00299-011-1100-4 PMID:21660547

Mettler, T., & Rohner, P. (2009). Situational Maturity Models as Instrumental Artifacts for Organizational Design. *Proceedings of the DESRIST'09*. doi:10.1145/1555619.1555649

Meyer, J. W., & Scott, W. R. (1983). Centralization and the legitimacy problems of local government. *Organizational Environments: Ritual and Rationality, 199*, 215.

Miemczyk, J., Johnsen, T. E., & Macquet, M. (2012). Sustainable purchasing and supply management: A structured literature review of definitions and measures at the dyad, chain and network levels. *Supply Chain Management: An International Journal, 17*(5), 478–496. doi:10.1108/13598541211258564

Mildenberger, U., & Khare, A. (2000). Planning for an environment-friendly car. *Technovation, 20*(4), 205–214. doi:10.1016/S0166-4972(99)00111-X

Millar, H. H., & Russell, S. N. (2011). The adoption of sustainable manufacturing practices in the Caribbean. *Business Strategy and the Environment, 20*(8), 512–526. doi:10.1002/bse.707

Miller, D. (2011). A reflection on EO research and some suggestions for the future. *Entrepreneurship Theory and Practice, 35*(5), 873–894. doi:10.1111/j.1540-6520.2011.00457.x

Miller, D., & Friesen, P. H. (1983). Strategy-making and environment: The third link. *Strategic Management Journal, 4*(3), 221–235. doi:10.1002/smj.4250040304

Miller, J. E., Hassett, J. J., & Koeppe, D. E. (1976). Uptake of cadmium by soybean as influenced by soil cation exchange capacity, pH and available phosphorus. *Journal of Environmental Quality, 5*(2), 157–160. doi:10.2134/jeq1976.00472425000500020009x

Ministerio de Comercio Industria y Turismo. (2012). *Informe Sostenibilidad 2012, Sector Sistema Moda.* Retrieved March 20, 2017 from: http://static.globalreporting.org/report-pdfs/2013/ef86ec29136d077c71c790ab7324d762.pdf

Mishina, Y., Dykes, B. J., Block, E. S., & Pollock, T. G. (2010). Why good firms do bad things: The effects of high aspirations, high expectations, and prominence on the incidence of corporate illegality. *Academy of Management Journal, 53*(4), 701–722. doi:10.5465/AMJ.2010.52814578

Mishra, A., & Choudhuri, M. A. (1999). Monitoring or phyto-toxicity of lead and mercury from germination and early seedling growth induces two rice cultivars. *Water, Air, and Soil Pollution, 114*(3/4), 339–346. doi:10.1023/A:1005135629433

Mithofer, A., Schulze, B., & Boland, W. (2004). Biotic and heavy metal stress response in plants: Evidence for common signals. *FEBS Letters, 566*(1-3), 1–5. doi:10.1016/j.febslet.2004.04.011 PMID:15147858

Modak, N. M., Panda, S., Sana, S. S., & Basu, M. (2014). Corporate social responsibility, coordination and profit distribution in a dual-channel supply chain. *Pacific Science Review, 16*(4), 235–249.

Mohr, L. A., & Webb, D. J. (2005). The effects of corporate social responsibility and price on consumer responses. *The Journal of Consumer Affairs, 39*(1), 121–147. doi:10.1111/j.1745-6606.2005.00006.x

Mora-Riapira, E. H., Vera-Colinab, M., & Melgarejo-Molina, Z. A. (2015). Planificación estratégica y niveles de competitividad de las Mipymes del sector comercio en Bogotá. *Estudios Gerenciales, 31*(134), 79–87. doi:10.1016/j.estger.2014.08.001

Morel, J. L. (1997). Bioavailability of trace elements to terrestrial plants. In J. Tarradellas, G. Bitton, & D. Rossel (Eds.), *Soil ecotoxicology* (pp. 141–176). CRS Press.

Morris, M. H., Kuratko, D. F., & Covin, J. G. (2010). *Corporate entrepreneurship & innovation*. Cengage Learning.

Moura, L. A. A. (2008). *Qualidade e gestão ambiental: Sustentabilidade e implantação da ISO 14.001* (5th ed.). São Paulo, Brasil: Juarez de Oliveira.

Muduli, K., Govindan, K., Barve, A., Kannan, D., & Geng, Y. (2013). Role of behavioural factors in green supply chain management implementation in Indian mining industries. *Resources, Conservation and Recycling, 76*, 50–60. doi:10.1016/j.resconrec.2013.03.006

Munda, G. (1997). Environmental economics, ecological economics, and the concept of sustainable development. *Environmental Values, 6*(2), 213–233. doi:10.3197/096327197776679158

Murphy, G. B., Trailer, J. W., & Hill, R. C. (1996). Measuring performance in entrepreneurship research. *Journal of Business Research, 36*(1), 15–23. doi:10.1016/0148-2963(95)00159-X

Murphy, P. E., & Schlegelmilch, B. B. (2013). Corporate social responsibility and corporate social irresponsibility: Introduction to a special topic section. *Journal of Business Research, 66*(10), 1807–1813.

Navis, C., & Glynn, M. A. (2011). Legitimate distinctiveness and the entrepreneurial identity: Influence on investor judgments of new venture plausibility. *Academy of Management Review, 36*(3), 479–499.

Ndubisi, N. O., & Nair, S. R. (2009). Green entrepreneurship (GE) and green value added (GVA): A conceptual framework. *International Journal of Entrepreneurship, 13*, 21.

Nelson, O. N., & Sumesh, R. N. (2009). Green entrepreneurship and green value added. *J. Int. J. Entrep, 13*, 21–31.

Neumayer, E. (2003). *Weak versus strong sustainability: exploring the limits of two opposing paradigms*. Northampton, UK: Edward Elgar.

Neumayer, E., & Soysa, I. D. (2007). Globalisation, womens economic rights and forced labor. *World Economy, 30*(10), 1510–1535. doi:10.1111/j.1467-9701.2007.01060.x

Newman, S. T., Nassehi, A., Imani-Asrai, R., & Dhokia, V. (2012). Energy efficient process planning for CNC machining. *CIRP Journal of Manufacturing Science and Technology, 5*(2), 127–136. doi:10.1016/j.cirpj.2012.03.007

Nguyen, X.C. (2015). Effect of Heavy Metals on plant growth and ability to use fertilizing substances to reduce heavy metal accumulation by *Brassica juncea* L. Czern. *Global Journal of Science Frontier Research: D Agriculture and Veterinary, 15*, 35-40.

Ni, D., Li, K. W., & Tang, X. (2010). Social responsibility allocation in two-echelon supply chains:Insights from wholesale price constracts. *European Journal of Operational Research, 207*(3), 1269–1279.

Ni, D., Li, X., & Tang, X. (2015). CSR Operations in Supply Chains: Mutual Incentive, CSR Allocation and Cooperation. *Chinese Journal of Management Science, 23*(9), 97–105.

Nidumolu, R., Prahalad, C. K., & Rangaswami, M. R. (2009). Why sustainability is now the key driver of innovation. *Harvard Business Review, 87*(9), 56–64.

Nies, D. H. (1999). Microbial heavy-metal resistance. *Applied Microbiology and Biotechnology, 51*(6), 730–750. doi:10.1007/s002530051457 PMID:10422221

Nilsen, H. R. (2010). The joint discourse reflexive sustainable development—From weak towards strong sustainable development. *Ecological Economics, 69*(3), 495–501. doi:10.1016/j.ecolecon.2009.11.011

Noctor, G., & Foyer, C. H. (1998). Ascorbate and glutathione: Keeping active oxygen under control. *Annual Review in Plant Physiology and Molecular Biology, 49*(1), 249–279. doi:10.1146/annurev.arplant.49.1.249 PMID:15012235

Nqa. (2017). *EMAS: EU Eco-Management and Audit Scheme*. Retrieved April 16, 2017 from https: //www.nqa. com/ en-gb/ certification/ standards/emas

Ntabe, E. N., LeBel, L., Munson, A. D., & Santa-Eulalia, L. A. (2015). A systematic literature review of the supply chain operations reference (SCOR) model application with special attention to environmental issues. *International Journal of Production Economics, 169*, 310–332. doi:10.1016/j.ijpe.2015.08.008

Öberseder, M., Schlegelmilch, B. B., Murphy, P. E., & Gruber, V. (2014). Consumers Perceptions of Corporate Social Responsibility: Scale Development and Validation. *Journal of Business Ethics, 124*(1), 101–115. doi:10.1007/s10551-013-1787-y

Odeyale, S. O. (2014). Performance appraisal for green/environmental friendliness of a supply chain department. *Journal of Industrial Engineering and Management, 7*(5), 1316. doi:10.3926/jiem.1057

OECD. (1998). *Waste minimisation profiles of OECD member countries*. Organisation for Economic Co-operation and Development.

Oehlmann, I. (2011). *The Sustainable Footprint Methodology*. Cologne, Germany: Lambert Academic Publishing.

Ogura, K. (2009). Long Working hours in Japan: An international comparison and research topics. *Journal of Economics, 36*(2), 23–45.

Oi, J. C. (1995). The role of the local state in Chinas transitional economy. *The China Quarterly, 144*, 1132–1149. doi:10.1017/S0305741000004768

Olasunkanmi, S., Jelili, A., Olubunmi, E., & Odeyale, L. (2014). Evaluation and selection of an effective green supply chain management strategy: A case study. *International Journal of Research Studies in Management, 3*(1), 27–39.

Olsthoorn, X., Tyteca, D., Wehrmeyer, W., & Wagner, M. (2001). Environmental indicators for business: A review of the literature and standardisation methods. *Journal of Cleaner Production, 9*(5), 453–463. doi:10.1016/S0959-6526(01)00005-1

Olugu, E. U., Wong, K. Y., & Shaharoun, A. M. (2011). Development of key performance measures for the automobile green supply chain. *Resources, Conservation and Recycling, 55*(6), 567–579. doi:10.1016/j.resconrec.2010.06.003

Oncel, I., Keles, Y., & Ustun, A. S. (2000). Interactive effects of temperature and heavy metal stress on the growth and some biochemical compounds in wheat seedlings. *Environmental Pollution, 107*(3), 315–320. doi:10.1016/S0269-7491(99)00177-3 PMID:15092977

Ordoobadi, S. (2009a). Application of Taguchi loss functions for supplier selection. *Supply Chain Management, 14*(1), 22–30. doi:10.1108/13598540910927278

Ordoobadi, S. (2009b). Development of a supplier selection model using fuzzy logic. *Supply Chain Management, 14*(4), 314–327. doi:10.1108/13598540910970144

O'Reilly, J. (2010). *The Green Rule: Measuring Progress*. Retrieved from http://www.inboundlogistics.com

Orlitzky, M., Siegel, D. S., & Waldman, D. A. (2011). Strategic corporate social responsibility and environmental sustainability. *Business & Society, 50*(1), 6–27. doi:10.1177/0007650310394323

Osterwalder, A., & Pigneur, Y. (2010). *Business model generation: a handbook for visionaries, game changers, and challengers*. John Wiley & Sons.

Oves, M., Khan, S., Qari, H., Felemban, N., & Almeelbi, T. (2016). Heavy metals: Biological importance and detoxification strategies. *Journal of Bioremediation and Biodegradation, 7*, 1–15.

Palmås, K., Andersson Schwarz, J., & Larsson, S. (2014). The liability of politicalness: Legitimacy and legality in piracy-proximate entrepreneurship. *International Journal of Entrepreneurship and Small Business, 22*(4), 408–425. doi:10.1504/IJESB.2014.064269

Pal, P., & Kumar, B. (2008). 16T: toward a dynamic vendor evaluation model in integrated SCM processes. *Supply Chain Management, 13*(6), 391–397. doi:10.1108/13598540810905642

Panda, S. (2014). Coordination of a socially responsible supply chain using revenue sharing contract. *Transportation Research Part E, Logistics and Transportation Review, 67*, 92–104.

Panda, S., Modak, N. M., Basu, M., & Goyal, S. K. (2015). Channel coordination and profit distribution in a social responsible three-layer supply chain. *International Journal of Production Economics, 168*, 224–233.

Pandey, N., & Sharma, C. P. (2002). Effect of heavy metals Co2+, Ni2+, and Cd2+ on growth and metabolism of cabbage. *Plant Science, 163*(4), 753–758. doi:10.1016/S0168-9452(02)00210-8

Papazoglou, E. G., Karantounias, G. A., Vemmos, S. N., & Bouranis, D. L. (2005). Photosynthesis and growth responses of giant reed (*Arundo donax* L.) to the heavy metals Cd and Ni. *Environment International, 31*(2), 243–249. doi:10.1016/j.envint.2004.09.022 PMID:15661290

Parekh, D., Puranik, R. M., & Srivastava, H. S. (1990). Inhibition of chlorophyll biosynthesis by cadmium in greening maize leaf segments. *Biochemie und Physiologie der Pflanzen, 186*(4), 239–242. doi:10.1016/S0015-3796(11)80078-8

Parisi, C., & Maraghini, M. P. (2010). Operationalising sustainability: How small and medium sized enterprises translate social and environmental issues into practice. In Business Performance Measurement and Management, (pp. 131-147). Springer Berlin Heidelberg.

Parsons, T. (1964). *Structure and process in modern societies*. Academic Press.

Pasian, B., & Silvius, A. J. G. (2016). *A Review of Project Management Research in IRNOP and PMI Conferences from 2009 to 2014 to Identify Emerging Perspectives*. European Academy of Management (EURAM).

Passariello, B., Giuliano, V., Quaresima, S., Barbaro, M., Caroli, S., Forte, G., & Iavicoli, I. et al. (2002). Evaluation of the environmental contamination at an abandoned mining site. *Microchemical Journal, 73*(1-2), 245–250. doi:10.1016/S0026-265X(02)00069-3

Pastrana, N., & Srirameshb, K. (2014). Corporate Social Responsibility: Perceptions and practices among SMEs in Colombia. *Public Relations Review, 40*(1), 14–24. doi:10.1016/j.pubrev.2013.10.002

Pataki, G. E., & Crotty, E. M. (2017). *Understanding and Implementing an Environmental Management System: A step by step guide*. New York State Department of Environmental Conversation Pollution Prevention Unit. Retrieved April 3, 2017, from http: //www.dec. ny.gov/ docs/permits_ej _operations_pdf/p2emsstep1.pdf

Patrick, H., Ronald, D., & Patrick, J. (1990). Transformations of Se, As affected by sediment oxidation–reduction potential and pH. *Environmental Science & Technology, 24*(1), 91–96. doi:10.1021/es00071a010

Paul, J., Parthasarathyc, S., & Gupta, P. (2017). Exporting challenges of SMEs: A review and future research agenda. *Journal of World Business, 52*(3), 327–342. doi:10.1016/j.jwb.2017.01.003

Payne, A. (2009). Corporate social responsibility and sustainable development. *Journal of Public Affairs, 6*(3-4), 286–297. doi:10.1002/pa.230

Pearson, R. (2007). Beyond women workers: Gendering CSR. *Third World Quarterly, 28*(28), 731–749. doi:10.1080/01436590701336622

Pedersen, E. R., & Andersen, M. (2006). Safeguarding corporate social responsibility (CSR) in global supply chains: How codes of conduct are managed in buyer-supplier relationships. *Journal of Public Affairs, 6*(3-4), 228–240. doi:10.1002/pa.232

Peña-Montoya, C. C. (2016). *Gestión sostenible de residuos sólidos industriales mediante la integración de estrategias asociadas con la lógistica de reversa* (Doctoral Thesis). Universidad del Valle, Colombia.

Peng, M. W., & Heath, P. S. (1996). The growth of the firm in planned economies in transition: Institutions, organizations, and strategic choice. *Academy of Management Review, 21*(2), 492–528.

Pérez, M.A., & Rojas, J.H. (2010). *Desarrollo sostenible. Principios, aplicaciones y lineamientos de la política en Colombia*. Programa editorial Universidad del Valle.

Politowicz, K. (2013). *Design Lifetimes: A Manifesto for Strategic Change*. London: University of the Arts London.

Posselt, G., Kellens, K., Thiede, S., Herrmann, C., Dewulf, W., & Duflou, J. R. (2013). Combining Machine Tool Builder and Operator Perspective towards Energy and Resource Efficiency in Manufacturing. In *Re-engineering Manufacturing for Sustainability* (pp. 209–214). Springer Singapore. doi:10.1007/978-981-4451-48-2_34

Poveda, C. A., & Young, R. (2015). Potential benefits of developing and implementing environmental and sustainability rating systems: Making the case for the need of diversification. *International Journal of Sustainable Built Environment, 4*(1), 1–11. doi:10.1016/j.ijsbe.2014.12.003

Preuss, L., & Brown, D. (2012). Business policies on human rights: An analysis of their content and Prieto-Carrón M. (2008). Women workers, industrialization, global supply chains and corporate codes of conduct. *Journal of Business Ethics, 83*(1), 5–17.

Puffer, S. M., McCarthy, D. J., & Boisot, M. (2010). Entrepreneurship in Russia and China: The impact of formal institutional voids. *Entrepreneurship Theory and Practice, 34*(3), 441-467.

Puncheva-Michelotti, P., Michelotti, M., & Gahan, P. (2010). The relationship between individuals recognition of human rights and responses to socially responsible companies: Evidence from Russia and Bulgaria. *Journal of Business Ethics, 93*(4), 583–605. doi:10.1007/s10551-009-0242-6

Puri, P., Compston, P., & Pantano, V. (2009). Life cycle assessment of Australian automotive door skins. *International Journal of Life Cycle Assessment, 14*(5), 420–428. doi:10.1007/s11367-009-0103-7

Qi, Y. (2010). Practice Low Carbon: MengNiu Dairy Build Green Industry Chain. *Advertiser Market Observer, 2010*(8), 50-50.

Rahman, H., Sabreen, S., Alam, S., & Kawai, S. (2005). Effects of nickel on growth and composition of metal micronutrients in barley plants grown in nutrient solution. *Journal of Plant Nutrition, 28*(3), 393–404. doi:10.1081/PLN-200049149

Rao, K. V. M., Raghavendra, A. S., & Reddy, K. J. (Eds.). (2006). *Physiology and Molecular Biology of Stress Tolerance in Plants*. Springer.

Rao, P. (2005). The greening of suppliers—in the South East Asian context. *Journal of Cleaner Production, 13*(9), 935–945. doi:10.1016/j.jclepro.2004.04.010

Raskin, I., & Ensley, B. D. (2000). *Phytoremediation of toxic metals: Using plants to clean up the environment*. New York: John Wiley & Sons.

Rathnayake, I. V., Megharaj, M., Krishnamurti, G. S., Bolan, N. S., & Naidu, R. (2013). Heavy metal toxicity to bacteria- are the existing growth media accurate enough to determine heavy metal toxicity. *Chemosphere, 90*(3), 1195–1200. doi:10.1016/j.chemosphere.2012.09.036 PMID:23040649

Ray, R. (2004). Child labor: A survey of selected Asian countries. *Asian-Pacific Economic Literature, 18*(2), 1–18. doi:10.1111/j.1467-8411.2004.00148.x

Rendell, E. G., & McGinty, K. A. (2004). *Environmental Management Systems: A Guidebook for Improving Energy and Environmental Performance in Local Government*. Five Winds International. Retrieved April 16, 2017, from http: //www. lgean.org/documents /EMSGuidebookPA.pdf

Riechmann, J., & Tickner, J. (2002). Un principio para reorientar las relaciones de la humanidad con la Biosfera. In J. Riechmann & J. Tickner (Eds.), El Principio de Precaución en medio ambiente y salud pública: de las definiciones a la práctica. Barcelona: Icaria.

Rockström, J. (2009). A safe operating space for humanity. *Nature, 461*, 472–475. 10.1038/461472a

Rockström, J., Lannerstad, M., & Falkenmark, M. (2007). Assessing the water challenge of a new green revolution in developing countries. *Proceedings of the National Academy of Sciences of the United States of America, 104*(15), 6253–6260. doi:10.1073/pnas.0605739104 PMID:17404216

Rondinelli, D. A., & Berry, M. A. (2000). Environmental citizenship in multinational corporations: Social responsibility and sustainable development. *European Management Journal, 18*(1), 70–84. doi:10.1016/S0263-2373(99)00070-5

Ronquim, C. C. (2010). *Queimada na colheita de cana-de-açúcar: impactos ambientais, sociais e econômicos*. Campinas: Embrapa Monitoramento por Satélite. Retrieved from https://www.cnpm.embrapa.br/projetos/carbcana/download/ser_doc_77_queima_cana.pdf

Roodhooft, F., & Konings, J. (1997). Vendor selection and evaluation an activity based costing approach. *European Journal of Operational Research, 96*(1), 97–102. doi:10.1016/0377-2217(95)00383-5

Ros, R., Cooke, D. T., Martínez-Cortina, C., & Picazo, I. (1992). Nickel and cadmium-related changes in growth, plasma membrane lipid composition, ATPase hydrolytic activity and proton-pumping of rice (*Oryza sativa* L. cv. Bahia) shoots. *Journal of Experimental Botany*, *43*(11), 1475–1481. doi:10.1093/jxb/43.11.1475

Rottig, D., & Reus, T. H. (2009, August). Institutional distance, organizational legitimacy, and the performance of foreign acquisitions in the United States. In Academy of Management Proceedings (Vol. 2009, No. 1, pp. 1-6). Academy of Management.

Russo, M. V., & Fouts, P. A. (1997). A resource-based perspective on corporate environmental performance and profitability. *Academy of Management Journal*, *40*(3), 534–559. doi:10.2307/257052

Rutkowska, B., Szulc, W., Bomze, K., Gozdowski, D., & Spychaj-Fabisiak, E. (2015). Soil factors affecting solubility and mobility of zinc in contaminated soils. *International Journal of Environmental Science and Technology*, *12*(5), 1687–1694. doi:10.1007/s13762-014-0546-7

Ryszko, A. (2016). Proactive Environmental Strategy, Technological Eco-Innovation and Firm Performance—Case of Poland. *Sustainability*, *8*(2), 156. doi:10.3390/su8020156

Salam, M. A. (2007). Social responsibility in purchasing: The case of Thailand. *International Journal of Procurement Management*, *1*(1), 97–116. doi:10.1504/IJPM.2007.015357

Salisbury, F. B., & Ross, C. W. (1992). *Plant Physiology* (4th ed.). Wadsworth Publisher.

Sammalisto, K., & Brorson, T. (2008). Training and communication in the implementation of environmental management systems (ISO 14001): A case study at the University of Gävle, Sweden. *Journal of Cleaner Production*, *16*(3), 299–309. doi:10.1016/j.jclepro.2006.07.029

Saridogan, M. (2012). The impact of green supply chain management on transportation cost reduction in Turkey. *International Review of Management and Marketing*, *2*(2), 112.

Sarita, S., Manisha, G., & Prakash, C. (1996). Bioaccumulation and biochemical effects of mercury in the plant *Bacopa monnieri* L. *Environmental Toxicology and Water Quality*, *11*(2), 105–112. doi:10.1002/(SICI)1098-2256(1996)11:2<105::AID-TOX5>3.0.CO;2-D

Sarkis, J. (1998). Evaluating environmentally conscious business practices. *European Journal of Operational Research*, *107*(1), 159–174. doi:10.1016/S0377-2217(97)00160-4

Sarkis, J. (1999). A methodological framework for evaluating environmentally conscious manufacturing programs. *Computers & Industrial Engineering*, *36*(4), 793–810. doi:10.1016/S0360-8352(99)00166-7

Sarkis, J. (2002). A strategic decision framework for green supply chain management. *Journal of Cleaner Production*, *11*(4), 397–409. doi:10.1016/S0959-6526(02)00062-8

Sarkis, J., & Weinrach, J. (2001). Using data envelopment analysis to evaluate environmentally conscious waste treatment technology. *Journal of Cleaner Production*, *9*(5), 417–427. doi:10.1016/S0959-6526(00)00084-6

Satpathy, I. (2005). *Environment Management*. New Delhi: Excel Books.

Schaltegger, S. (2002). A Framework for Ecopreneurship. *Greener Management International*, *38*(38), 45–58. doi:10.9774/GLEAF.3062.2002.su.00006

Schepers, J., Voordeckers, W., Steijvers, T., & Laveren, E. (2014). The entrepreneurial orientation–performance relationship in private family firms: The moderating role of socioemotional wealth. *Small Business Economics*, *43*(1), 39–55. doi:10.1007/s11187-013-9533-5

Scherer, A. G., Palazzo, G., & Seidl, D. (2013). Managing legitimacy in complex and heterogeneous environments: Sustainable development in a globalized world. *Journal of Management Studies*, *50*(2), 259–284. doi:10.1111/joms.12014

Schützendübel, A., & Polle, A. (2002). Plant responses to abiotic stresses: Heavy metal-induced oxidative stress and protection by mycorrhization. *Journal of Experimental Botany*, *53*, 1351–1365. PMID:11997381

Schwartz, C., Gerard, E., Perronnet, K., & Morel, J. L. (2001). Measurement of *in situ* phytoextraction of zinc by spontaneous metallophytes growing on a former smelter site. *The Science of the Total Environment*, *279*(1-3), 215–221. doi:10.1016/S0048-9697(01)00784-7 PMID:11712599

Schwartz, K., Tapper, R., & Font, X. (2008). A sustainable supply chain management framework for tour operators. *Journal of Sustainable Tourism*, *16*(3), 298–314. doi:10.1080/09669580802154108

Scott, L. (2010). *About Jenga: The Remarkable Business of Creating a Game That Became a Household Name*. Austin, TX: Greenleaf Book Group LLC.

Scott, W. R. (1995). *Institutions and organizations* (Vol. 2). Thousand Oaks, CA: Sage.

Seiffert, M. E. B. (2011). *Gestão ambiental: Instrumento, esferas de ação e educação ambiental*. São Paulo, Brasil: Atlas.

Sen, S., & Bhattacharya, C. B. (2001). Does Doing Good Always Lead to Doing Better? Consumer Reactions to Corporate Social Responsibility. *JMR, Journal of Marketing Research*, *38*(2), 225–243. doi:10.1509/jmkr.38.2.225.18838

Seuring, S. (2013). A review of modeling approaches for sustainable supply chain management. *Decision Support Systems*, *54*(4), 1513–1520. doi:10.1016/j.dss.2012.05.053

Shahid, M., Khalid, S., Abbas, G., & Dumat, C. (2015). Heavy metal stress and crop productivity. In K. R. Hakeem (Ed.), *Crop Production and Global Environmental Issues* (pp. 1–25). Springer International Publishing. doi:10.1007/978-3-319-23162-4_1

Shakoor, M. B., Ali, S., Hameed, A., Farid, M., Hussain, S., Yasmeen, T., & Abbasi, G. H. et al. (2014). Citric acid improves lead (Pb) phytoextraction in *Brassica napus* L. By mitigating Pb-induced morphological and biochemical damages. *Ecotoxicology and Environmental Safety*, *109*, 38–47. doi:10.1016/j.ecoenv.2014.07.033 PMID:25164201

Shane, S. (2000). Prior knowledge and the discovery of entrepreneurial opportunities. *Organization Science*, *11*(4), 448–469. doi:10.1287/orsc.11.4.448.14602

Shane, S., & Venkataraman, S. (2000). The promise of entrepreneurship as a field of research. *Academy of Management Review*, *25*(1), 217–226.

Sharma, P., Jha, A. B., Dubey, R. S., & Pessarakli, M. (2012). Reactive Oxygen Species, Oxidative Damage, and Antioxidative Defense Mechanism in Plants under Stressful Conditions. *Le Journal de Botanique*, *2012*, 1–26.

Sharma, S. S., & Dietz, K. J. (2006). The significance of amino acids and amino acid-derived molecules in plant responses and adaptation to heavy metal stress. *Journal of Experimental Botany*, *57*(4), 711–726. doi:10.1093/jxb/erj073 PMID:16473893

Sharma, S., & Vredenburg, H. (1998). Proactive corporate environmental strategy and the development of competitively valuable organizational capabilities. *Strategic Management Journal*, *19*(8), 729–753. doi:10.1002/(SICI)1097-0266(199808)19:8<729::AID-SMJ967>3.0.CO;2-4

Sheetal, K. R., Singh, S. D., Anand, A., & Prasad, S. (2016). Heavy metal accumulation and effects on growth, biomass and physiological processes in mustard. *Indian Journal of Plant Physiology*, *21*(2), 219–223. doi:10.1007/s40502-016-0221-8

Sheldon, C., & Yoxon, M. (2006). *Environmental management systems: a step-by-step guide to implementation and maintenance* (3rd ed.). Routledge.

Sheng, S., Zhou, K. Z., & Li, J. J. (2011). The effects of business and political ties on firm performance: Evidence from China. *Journal of Marketing*, *75*(1), 1–15. doi:10.1509/jmkg.75.1.1

Shenhar, A., & Patanakul, P. (2014). Implementing Strategy through Projects. In J. R. Turner (Ed.), Handbook of Project Management. Gower Publishing.

Shepherd, D. A., & Patzelt, H. (2011). The new field of sustainable entrepreneurship: Studying entrepreneurial action linking what is to be sustained with what is to be developed. *Entrepreneurship Theory and Practice*, *35*(1), 137–163. doi:10.1111/j.1540-6520.2010.00426.x

Shier, W. T. (1994). Metals as toxins in plants. *Journal of Toxicology. Toxin Reviews*, *13*(2), 205–216. doi:10.3109/15569549409089960

Shikida, P. F. A. (2013). Expansão canavieira no Centro-Oeste: Limites e potencialidades. Revista de Política Agrícola, 22(2), 122-137.

Shirokova, G., Bogatyreva, K., Beliaeva, T., & Puffer, S. (2016). Entrepreneurial orientation and firm performance in different environmental settings: Contingency and configurational approaches. *Journal of Small Business and Enterprise Development*, *23*(3), 703–727. doi:10.1108/JSBED-09-2015-0132

Sidhu, G. P. S. (2016). Heavy metal toxicity in soils: Sources, remediation technologies and challenges. *Advances in Plants & Agriculture Research*, *5*, 1–3.

Sidhu, G. P. S., Singh, H. P., Batish, D. R., & Kohli, R. K. (2016). Effect of lead on oxidative status, antioxidative response and metal accumulation in *Coronopus didymus*. *Plant Physiology and Biochemistry*, *105*, 290–296. doi:10.1016/j.plaphy.2016.05.019 PMID:27214085

Silajdžić, I., Kurtagić, S. M., & Vučijak, B. (2015). Green entrepreneurship in transition economies: A case study of Bosnia and Herzegovina. *Journal of Cleaner Production*, *88*, 376–384. doi:10.1016/j.jclepro.2014.07.004

Silva, V. L. M. M., Gomes, W. C., & Alsina, O. L. S. (2007). *Utilização do bagaço de cana-de-açúcar como biomassa adsorvente na adsorção de poluentes orgânicos. In Revista Eletrônica de Materiais e Processos* (Vol. 2, pp. 27–32). Retrieved from www2.ufcg.edu.br/revista-remap/index.php/REMAP/article/download/28/64

Silvius, A. J. G. (2015b). *Model Project SMP*. Retrieved from http://www.slideshare.net/GilbertSilvius/model-project-smp

Silvius, A.J.G. (2016). Sustainability as a competence of Project Managers. *PM World Journal*, *5*(9), 1-13.

Silvius, A.J.G., & Schipper, R. (2014). Sustainability in Project Management: A literature review and impact analysis. *Social Business*, *4*(1).

Silvius, A. J. G. (2015a). Considering Sustainability in Project Management Processes. In K. D. Thomas (Ed.), *Handbook of Research on Sustainable Development and Economics*. IGI Global Publishing. doi:10.4018/978-1-4666-8433-1.ch014

Silvius, A. J. G., Kampinga, M., Paniagua, S., & Mooi, H. (2017). Considering Sustainability in Project Management Decision Making; An investigation using Q-methodology. *International Journal of Project Management*.

Silvius, A. J. G., & Schipper, R. (2015). Developing a Maturity Model for Assessing Sustainable Project Management. *Journal of Modern Project Management*, *3*(1), 16–27.

Silvius, A. J. G., & Schipper, R. (2016). Exploring the relationship between sustainability and project success - Conceptual model and expected relationships. *International Journal of Information Systems and Project Management*, *4*(3), 5–22.

Silvius, A. J. G., Schipper, R., Planko, J., van den Brink, J., & Köhler, A. (2012). *Sustainability in Project Management*. Farnham: Gower Publishing.

Silvius, A. J. G., & Tharp, J. (Eds.). (2013). *Sustainability Integration for Effective Project Management*. IGI Global Publishing. doi:10.4018/978-1-4666-4177-8

Simón-Moya, V., Revuelto-Taboada, L., & Guerrero, R. F. (2014). Institutional and economic drivers of entrepreneurship: An international perspective. *Journal of Business Research*, *67*(5), 715–721. doi:10.1016/j.jbusres.2013.11.033

Singh, J. V., Tucker, D. J., & Meinhard, A. G. (1991). Institutional change and ecological dynamics. *The new institutionalism in organizational analysis*, 390-422.

Singh, A., Agrawal, S. B., & Rathore, D. (2005). Amelioration of Indian urban air pollution phytotoxicity in *Beta vulgaris* L. by modifying NPK nutrients. *Environmental Pollution*, *134*(3), 385–395. doi:10.1016/j.envpol.2004.09.017 PMID:15620584

Singh, R. P., & Agrawal, M. (2007). Effects of sewage sludge amendment on heavy metal accumulation and consequent responses of *Beta vulgaris* plants. *Chemosphere*, *67*(11), 2229–2240. doi:10.1016/j.chemosphere.2006.12.019 PMID:17289111

Singh, S., & Sinha, S. (2005). Accumulation of metals and its effects in *Brassica juncea* (L.) Czern. (cv. Rohini) grown on various amendments of tannery waste. *Ecotoxicology and Environmental Safety*, *62*(1), 118–127. doi:10.1016/j.ecoenv.2004.12.026 PMID:15978297

Sonya, S., Jordan, J., & Mazar, N. (2015). *Green Consumerism: Moral Motivations to a Sustainable Future*. Current Opinion in Psychology.

Soto-Acosta, P., Cismaru, D. M., Vătămănescu, E. M., & Ciochină, R. S. (2016). Sustainable Entrepreneurship in SMEs: A Business Performance Perspective. *Sustainability*, *8*(4), 342. doi:10.3390/su8040342

Sprynskyy, M., Kosobucki, P., Kowalkowski, T., & Buszewsk, B. (2007). Influence of clinoptilolite rock on chemical speciation of selected heavy metals in sewage sludge. *Journal of Hazardous Materials*, *149*(2), 310–316. doi:10.1016/j.jhazmat.2007.04.001 PMID:17513045

Sroufe, R. P., Melnyk, S. A., & Vastag, G. (1998). *Environmental Management Systems as a source of competitive advantage*. Michigan State University-Department of Marketing and Supply Chain Management.

Stamas, G. D. (1980). Percent working long hours shows first post-recession decline. *Monthly Labor Review*, *103*(5), 39–42.

Stam, W., & Elfring, T. (2008). Entrepreneurial orientation and new venture performance: The moderating role of intra- and extraindustry social capital. *Academy of Management Journal*, *51*(1), 97–111. doi:10.5465/AMJ.2008.30744031

Starr, J. A., & MacMillan, I. C. (1990). Resource cooptation via social contracting: Resource acquisition strategies for new ventures. *Strategic Management Journal*, 79–92.

Steffen, W., Richardson, K., Rockstrom, J., Cornell, S. E., Fetzer, I., Bennett, E. M., & Sorlin, S. et al. (2015). Planetary boundaries: Guiding human development on a changing planet. *Science*, *347*(6223), 1259855. doi:10.1126/science.1259855 PMID:25592418

Stevens, C. E., Xie, E., & Peng, M. W. (2015). Toward a legitimacy-based view of political risk: The case of Google and Yahoo in China. *Strategic Management Journal*.

Stevenson, F. J. (1982). *Humus chemistry, genesis, composition, reactions*. New York: John Wiley & Sons.

Stobart, A. K., Griffiths, W. T., Ameen-Bukhari, I., & Sherwood, R. P. (1985). The effect of Cd^{+2} on the biosynthesis of chlorophyll in leaves of barley. *Physiologia Plantarum, 63*(3), 293–298. doi:10.1111/j.1399-3054.1985.tb04268.x

Suchman, M. C. (1995). Managing legitimacy: Strategic and institutional approaches. *Academy of Management Review, 20*(3), 571–610.

Sulphey, M. M., & Safeer, M. M. (2015). *Introduction to Environment Management* (3rd ed.). Delhi: PHI Learning Pvt. Ltd.

Sun, H., & Li, H. (2012). Chinese Corporate Social Responsibility and the Way of Corporate Sustainable Development. *Reformation and Strategy, 28*(6), 158-160, 185.

Sundin, E. (2004). *Product and process design for successful remanufacturing*. Linköping, Sweden: Linköping University.

Sweetin, V. H., Knowles, L. L., Summey, J. H., & Mcqueen, K. S. (2013). Willingness-to-punish the corporate brand for corporate social irresponsibility. *Journal of Business Research, 66*(10), 1822–1830.

Sydow, J., Lindkvist, L., & DeFillippi, R. (2004). Editorial: Project organizations, embeddedness and repositories of knowledge. *Organization Studies, 25*(9), 1475–1489. doi:10.1177/0170840604048162

Tabachnick, B. G., Fidell, L. S., & Osterlind, S. J. (2001). *Using multivariate statistics*. Academic Press.

Tachizawa, T. (2011). *Gestão ambiental e responsabilidade social corporativa: estratégia de negócios focadas na realidade brasileira*. São Paulo, Brasil: Atlas.

Tam, G. C. K. (2017). *Managerial Strategies an Green Solutions for Project Sustainability*. IGI Global Publishing. doi:10.4018/978-1-5225-2371-0

Tan, J. (2001). Innovation and risk-taking in a transitional economy: A comparative study of Chinese managers and entrepreneurs. *Journal of Business Venturing, 16*(4), 359–376. doi:10.1016/S0883-9026(99)00056-7

Tanwer, A. K., Prajapati, D. R., & Singh, P. J. (2014). Green Supply Chain Management An Environmental Approach For Manufacturing Industry. *Proceedings of National Conference on Advancements and Futuristic Trends in Mechanical Engineering*. Retrieved from http://www.supply-chain.com/info/faq.html

Tao, W., & Wilson, C. (2016). The Impact of Corporate Communication on Company Evaluation: Examining the Message Effects of CSR. *Corporate Ability, and Hybrid Strategies, 10*(5), 426–444.

Taylor, T. (2008). *A sustainability checklist for managers of projects*. Retrieved from http://www.pmforum.org/library/papers/2008/PDFs/Taylor-1-08.pdf

Thakur, S., Singh, L., Wahid, Z. A., Siddiqui, M. F., Atnaw, S. M., & Din, M. F. M. (2016). Plant-driven removal of heavy metals from soil: Uptake, translocation, tolerance mechanism, challenges, and future perspectives. *Environmental Monitoring and Assessment, 188*(4), 206. doi:10.1007/s10661-016-5211-9 PMID:26940329

Theyel, G. (2006). Customer and supplier relations for environmental performance. In Greening the Supply Chain, (pp. 139-149). Springer London.

Thomas, L. G., & D'Aveni, R. A. (2004). *The Rise of Hypercompetition in the US Manufacturing Sector 1950-2002*. Tuck School of Business at Dartmouth.

Tian, Y., Govindan, K., & Zhu, Q. (2014). A system dynamics model based on evolutionary game theory for green supply chain management diffusion among Chinese manufacturers. *Journal of Cleaner Production, 80*(7), 96–105. doi:10.1016/j.jclepro.2014.05.076

Tinsley, S. (2002). EMS models for business strategy development. *Business Strategy and the Environment, 11*(6), 376–390. doi:10.1002/bse.340

Tiron-Tudor, A., & Ioana-Maria, D. (2013). Project Success by Integrating Sustainability in Project Management. In A. J. G. Silvius & J. Tharp (Eds.), *Sustainability Integration for Effective Project Management*. IGI Global Publishing.

Toledo, V. M. (2003). Hacia un modelo de conservacion bio-regional en las regiones tropicales de mexico: biodiversidad, sustentabilidad y pueblos indigenas. In *Hacia una Evaluación de las Áreas Naturales Protegidas del Trópico*. Xalapa, Veracruz, México: Universidad Veracruzana. Retrieved March 13, 2017 from: http://www.era-mx.org/biblio/politica/Toledo2003.pdf

Tracey, M., & Tan, C. L. (2001). Empirical analysis of supplier selection and involvement, customer satisfaction, and firm performance. *Supply Chain Management, 6*(4), 174–188. doi:10.1108/EUM0000000005709

Tramarico, C. L., Salomon, V. A. P., & Marins, F. A. S. (2016). Multi-criteria assessment of the benefits of a supply chain management training considering green issues. *Journal of Cleaner Production.*

Trudel, R., & Cotte, J. (2009). Does it pay to be good? *MIT Sloan Management Review, 50*(2), 61–68.

Tsai, W. H., & Hung, S. J. (2009). A fuzzy goal programming approach for green supply chain optimisation under activity-based costing and performance evaluation with a value-chain structure. *International Journal of Production Research, 47*(18), 4991–5017. doi:10.1080/00207540801932498

Turner, R. J. (2014). Projects and Their Management. In J. R. Turner (Ed.), Handbook of Project Management. Gower Publishing.

Turner, J. R., Anbari, F., & Bredillet, C. (2013). Perspectives on research in project management: The nine schools. *Global Business Perspectives, 1*(1), 3–28. doi:10.1007/s40196-012-0001-4

Turner, J. R., & Müller, R. (2003). On the nature of the project as a temporary organization. *International Journal of Project Management, 21*(3), 1–8. doi:10.1016/S0263-7863(02)00020-0

Turner, R., & Keegan, A. (1999). The versatile project-based organization: Governance and operational control. *European Management Journal, 17*(3), 296–309. doi:10.1016/S0263-2373(99)00009-2

Tuzkaya, G. (2013). An intuitionistic fuzzy Choquet integral operator based methodology for environmental criteria integrated supplier evaluation process. *International Journal of Environmental Science and Technology, 10*(3), 423–432. doi:10.1007/s13762-013-0180-9

Tuzkaya, G., Ozgen, A., Ozgen, D., & Tuzkaya, U. R. (2009). Environmental performance evaluation of suppliers: A hybrid fuzzy multi-criteria decision approach. *International Journal of Environmental Science and Technology, 6*(3), 477–490. doi:10.1007/BF03326087

UDOP. (n.d.). *Subprodutos da cana se tornaram fonte de receita*. Retrieved from http://www.udop.com.br

UNEP. (2017). *UNEP / FIDIC / ICLEI Urban Environmental Management: Environmental Management Training Resources Kit*. Retrieved April 17, 2017, from http: //www. unep.or.jp /ietc/Announcements /EMSkit_launch.asp

UNICA. (n.d.). *Bagaço de cana pode ganhar valor substituindo areia na construção civil*. Retrieved from http://www.unica.com.br/noticias/show.asp?nwsCode=%7B4E794FD3-7EC2-403C-A66D- 325016046000%7D

United Nations Environment Programme. (2012). *Towards a Life Cycle Sustainability Assessment*. New York, NY: United Nations.

United Nations General Assembly. (1987). Report of the World Commission on Environment and Development : Our Common Future. Oslo, Norway: Author.

Uygun, O., & Dede, A. (2016). Performance evaluation of green supply chain management using integrated fuzzy multi-criteria decision making techniques. *Computers & Industrial Engineering, 102*, 502–511. doi:10.1016/j.cie.2016.02.020

van Tulder, R., van Tilburg, R., Francken, M., & da Rosa, A. (2014). *Managing the Transition to a Sustainable Enterprise: Lessons from Frontrunner Companies*. Earthscan by Routledge.

Vandaie, R., & Zaheer, A. (2014). Surviving bear hugs: Firm capability, large partner alliances, and growth. *Strategic Management Journal, 35*(4), 566–577. doi:10.1002/smj.2115

Vidović, M., Morina, F., Milić-Komić, S., Vuleta, A., Zechmann, B., Prokić, Lj., & Jovanović, S. V. (2016). Characterisation of antioxidants in photosynthetic and non-photosynthetic leaf tissues of variegated *Pelargonium zonale* plants. *Plant Biology, 18*(4), 669–680. doi:10.1111/plb.12429 PMID:26712503

Vijayvagy, L. (2012). Decision framework for supplier selection through multi criteria evaluation models in supply chain. *International Journal of Management & Innovation, 4*(2), 16.

Violante, A., Cozzolino, V., Perelomov, L., Caporale, A. G., & Pigna, M. (2010). Mobility and bioavailability of heavy metals and metalloids in soil environments. *Journal of Soil Science Plant Nutrition, 10*(3), 268–292. doi:10.4067/S0718-95162010000100005

Visser, W. T. (2002). Sustainability reporting in South Africa. *Corporate Environmental Strategy, 9*(1), 79–85. doi:10.1016/S1066-7938(01)00157-9

Wagner, M. (2008). Empirical influence of environmental management on innovation: Evidence from Europe. *Ecological Economics, 66*(2), 392–402. doi:10.1016/j.ecolecon.2007.10.001

Wagner, M., & Schaltegger, S. (2004). The effect of corporate environmental strategy choice and environmental performance on competitiveness and economic performance: An empirical study of EU manufacturing. *European Management Journal, 22*(5), 557–572. doi:10.1016/j.emj.2004.09.013

Walder, A. G. (1995). Local governments as industrial firms: An organizational analysis of Chinas transitional economy. *American Journal of Sociology, 101*(2), 263–301. doi:10.1086/230725

Wales, W. J. (2016). Entrepreneurial orientation: A review and synthesis of promising research directions. *International Small Business Journal, 34*(1), 3–15. doi:10.1177/0266242615613840

Wan, S., & Liu, F. (2014). Empirical Analysis on the Relationship between Corporate Social Responsibility and Corporate Social Irresponsibility. *Finance and Accounting Monthly, 22*, 3-8.

Wang, L. (2014). Research on Interaction and Strategic Innovation of Green Logistics and Green Marketing. *Logistics Technology, 2014*(5), 25-26.

Wang, X., & Mao, W. (2015). Consumer Perceptions of Corporate Social Irresponsibility Initiatives: The Effects of Consumer Attribution and Expectation. *Journal of Beijing Institute of Technology (Social Sciences Edition), 17*(6), 74-80+112.

Wang, C. L., & Ahmed, P. K. (2007). Dynamic capabilities: A review and research agenda. *International Journal of Management Reviews, 9*(1), 31–51. doi:10.1111/j.1468-2370.2007.00201.x

Wang, H., & Han, B. (2016). Social Responsibility, Internal Control and Corporate Sustainable Development--Based on an Empirical Analysis of Chinese A-share Main Board Enterprises. *Journal of Beijing Technology and Business University, 31*(1), 75–84.

Wang, J., & Hou, T. (2009). Empirical Analysis of the Effects of Social Responsibility on Corporate Sustainable Development. *Science & Technology Progress and Policy*, *26*(18), 94–96.

Wang, S. H., Yang, Z. M., Yang, H., Lu, B., Li, S. Q., & Lu, Y. P. (2004). Copper-induced stress and antioxidative responses in roots of *Brassica juncea* L. *Botanical Bulletin of Academia Sinica*, *45*, 203–212.

Wang, W., & Deng, W. (2016). Comparison between the Reward-penalty Mechanism with the Tax-subsidy Mechanism for Reverse Supply Chains. *Chinese Journal of Management Science*, *24*(4), 102–110.

Wang, Z., Mathiyazhagan, K., Xu, L., & Diabat, A. (2015). A decision making trial and evaluation laboratory approach to analyze the barriers to Green Supply Chain Management adoption in a food packaging company. *Journal of Cleaner Production*, *117*, 19–28. doi:10.1016/j.jclepro.2015.09.142

Wan, S., & Liu, F. (2015). Research on the Causes and Economic Consequences of Corporate Social Irresponsibility. *Friends of Accounting*, *11*, 15–25.

Warsen, J., Laumer, M., & Momberg, W. (2011). Comparative Life Cycle Assessment of Remanufacturing and New Manufacturing of a Manual Transmission, Globalized Solutions for Sustainability in Manufacturing. *Proceedings of the 18th CIRP International Conference on Life Cycle Engineering.* Braunschweig: Technische Universitat Braunschweig.

Waylen, G. (2014). Informal institutions, institutional change, and gender equality. *Political Research Quarterly*, *67*(1), 212–223. doi:10.1177/1065912913510360

Weast, R. C. (1984). *CRC handbook of chemistry and physics* (64th ed.). Boca Raton, FL: CRC.

Weber, C. A., Current, J. R., & Benton, W. C. (1991). Vendor selection criteria and methods. *European Journal of Operational Research*, *50*(1), 2–18. doi:10.1016/0377-2217(91)90033-R

Wei, X. (2007). Charity marketing—Super Strategy of MengNiu Dairy. *Modern Communication, 2007*(1).

Weidema, B. P. (2006). The Integration of Economic and Social Aspects in Life Cycle Impact Assessment. *International Journal of Life Cycle Assessment*, *11*(1), 89–96. doi:10.1065/lca2006.04.016

Wei, J., Zhao, J., & Li, Y. (2014). Price and warranty period decisions for complementary products with horizontal firms cooperation/noncooperation strategies. *Journal of Cleaner Production*, *105*, 86–102. doi:10.1016/j.jclepro.2014.09.059

Weiß, P., & Bentlage, J. (2006). Environmental Management Systems and Certification. Baltic University Press.

Welford, R., & Frost, S. (2006). Corporate social responsibility in Asian supply chains. *Corporate Social Responsibility and Environmental Management*, *13*(3), 166–176. doi:10.1002/csr.121

Westhead, P., Ucbasaran, D., & Wright, M. (2005). Experience and cognition: Do novice, serial and portfolio entrepreneurs differ? *International Small Business Journal*, *23*(1), 72–98. doi:10.1177/0266242605049104

White, B. (1996). Globalization and the child labor problem. *Journal of International Development*, *8*(6), 829–839. doi:10.1002/(SICI)1099-1328(199611)8:6<829::AID-JID404>3.0.CO;2-Z

Wiklund, J., & Shepherd, D. (2003). Knowledge-based resources, entrepreneurial orientation, and the performance of small and medium-sized businesses. *Strategic Management Journal*, *24*(13), 1307–1314. doi:10.1002/smj.360

Willard, B. (2005). *The next sustainability wave: building boardroom buy-in.* New Society Publishers.

Williams, L. E., Pittman, J. K., & Hall, J. L. (2000). Emerging mechanisms for heavy metal transport in plants. *Biochimica et Biophysica Acta*, *1465*(1-2), 104–126. doi:10.1016/S0005-2736(00)00133-4 PMID:10748249

Winstanley, D., Clark, J., & Leeson, H. (2002). Approaches to child labor in the supply chain. *Business Ethics (Oxford, England), 11*(3), 210–223. doi:10.1111/1467-8608.00279

WIPO-World Intelectual Property Organization. (n.d.). *About the SMEs Section.* Retrieved March 22, 2017 from: http://www.wipo.int/sme/en/about_sme.html

Wójcik, M., & Tukiendorf, A. (2004). Phytochelatin synthesis and cadmium localization in wild type of *Arabidopsis thaliana. Plant Growth Regulation, 44*, 71–80. doi:10.1007/s10725-004-1592-9

Wojtaszek, P. (1997). Oxidative burst: An early plant response to pathogen infection. *The Biochemical Journal, 322*(3), 681–692. doi:10.1042/bj3220681 PMID:9148737

World Commission on Environment and Development (WCED). (1987). *Our common future.* Oxford, UK: Oxford University Press.

World Commission on Environment and Development. (1987). *Our Common Future.* Oxford, UK: Oxford University Press.

Wu, K. J., Liao, C. J., Tseng, M. L., & Chou, P. J. (2015). Understanding innovation for sustainable business management capabilities and competencies under uncertainty. *Sustainability, 7*(10), 13726–13760. doi:10.3390/su71013726

Wu, K. J., Tseng, M. L., & Vy, T. (2011). Evaluation the drivers of green supply chain management practices in uncertainty. *Procedia: Social and Behavioral Sciences, 25*, 384–397. doi:10.1016/j.sbspro.2012.02.049

Wu, M. W., & Shen, C. H. (2013). Corporate social responsibility in the banking industry: Motives and financial performance. *Journal of Banking & Finance, 37*(9), 3529–3547. doi:10.1016/j.jbankfin.2013.04.023

Wu, Y. (2005). *Marketing.* Wuhan University of Technology Press.

Xia, X., Govindan, K., & Zhu, Q. (2015). Analyzing internal barriers for automotive parts remanufacturers in China using grey-DEMATEL approach. *Journal of Cleaner Production, 87*(1), 811–825. doi:10.1016/j.jclepro.2014.09.044

Xie, H., Miao, L., & Kuo, P. J. (2011). Consumers responses to ambivalent online hotel reviews: The role of perceived source credibility and pre-decisional disposition. *International Journal of Hospitality Management, 30*(1), 178–183. doi:10.1016/j.ijhm.2010.04.008

Xie, P. H., & Zhou, Z. C. (2009). Empirical study on the relationship between corporate Social esponsibility and consumer purchase intention in China. *Nankai Business Review, 12*(1), 71–83.

Xin, J., & Liao, X. (2013). Sustainable Development and Social Responsibility of Enterprises. *Journal of Central South University of Forestry and Technology, 7*(3), 77–80.

Xu, L., & Peng, J. (2015). A Green Level Measure Model of Enterprise Supply Chain based on Fuzzy Information. *International Journal of u- and e- Service, Science and Technology, 8*(2).

Xu, L., Kumar, D. T., Shankar, K. M., Kannan, D., & Chen, G. (2013). Analyzing criteria and sub-criteria for the corporate social responsibility-based supplier selection process using AHP. *International Journal of Advanced Manufacturing Technology, 68*(1), 907–916. doi:10.1007/s00170-013-4952-7

Yakhou, M., & Dorweiler, V. P. (2004). Environmental accounting: An essential component of business strategy. *Business Strategy and the Environment, 13*(2), 65–77. doi:10.1002/bse.395

Yang, J. (2010). Discussion on Green Marketing. *Management & Technology of SME Every Ten Days, 2010*(6), 110-110.

Yang, C. S., Lu, C. S., Haider, J. J., & Marlow, P. B. (2013). The effect of green supply chain management on green performance and firm competitiveness in the context of container shipping in Taiwan. *Transportation Research Part E, Logistics and Transportation Review*, *55*, 55–73. doi:10.1016/j.tre.2013.03.005

Yang, H., Cui, L., & Zhu, Q. H. (2016). Full life cycle impact assessment of refrigerators considering their security processing proportion. [in Chinese]. *Systems Engineering*, (5): 136–143.

Yang, W., & Yang, S. (2016). An Empirical Study on the Relationship between Corporate Social Responsibility and Financial Performance under the Chinese Context——Based on Contrastive Analysis of Large, Small and Medium-size Listed Companies. *Chinese Journal of Management Science*, *24*(1), 143–150.

Yang, Y., Mei, Y., Zhang, C., Zhang, R., Liao, X., & Liu, Y. (2016). Heavy metal contamination in surface soils of the industrial district of Wuhan, China. *Human and Ecological Risk Assessment: An International Journal*, *22*(1), 126–140. doi:10.1080/10807039.2015.1056291

Yan, Z. (2006). *Research on Characteristics of Distribution Channels of MengNiu Dairy*. University of International Business and Economics.

Yaron, B., Calvet, R., & Prost, R. (1996). *Soil pollution processes and dynamics*. Berlin: Springer-Verlag. doi:10.1007/978-3-642-61147-6

Yin & Xin. (2015). A Comparative Study of 4P and 4C Marketing Theory. *Journal of Huaihai Institute of Technology (Humanities & Social Sciences Edition)*, *2015*(2), 88-90.

Yu, X. (1988). The definition, structure and hierarchy of creativity. *Contemporary Youth Research*, *1988*(8), 25-28.

Yucenur, G. N., Vayvay, O., & Demirel, N. C. (2011). Supplier selection problem in global supply chain by AHP and ANP approaches under fuzzy environment. *International Journal of Advanced Manufacturing Technology*, *56*(5-8), 823–833. doi:10.1007/s00170-011-3220-y

Zaier, H., Mudarra, A., Kutscher, D., de la Campa, M. R. F., Abdelly, C., & Sanz-Medel, A. (2010). Induced lead binding phytochelatins in *Brassica juncea* and *Sesuvium portulacastrum* investigated by orthogonal chromatography inductively coupled plasma-mass spectrometry and matrix assisted laser desorption ionization-time of flight-mass spectrometry. *Analytica Chimica Acta*, *671*(1-2), 48–54. doi:10.1016/j.aca.2010.04.054 PMID:20541642

Zengin, F. K., & Munzuroglu, O. (2005). Effects of some heavy metals on content of chlorophyll, proline and some anti-oxidant chemicals in bean (*Phaseolus vulgaris* l.) seedlings. *Acta Biologica Cracoviensia. Series; Botanica*, *47*, 157–164.

Zhang, R. (2011). International Green Marketing Strategy of China. *China Business & Trade*, *2011*(24), 20-21.

Zhang, J., & Kirkham, M. B. (1994). Drought stress induced changes in activities of superoxide dismutase, catalase and peroxide in wheat species. *Plant & Cell Physiology*, *35*(5), 785–791. doi:10.1093/oxfordjournals.pcp.a078658

Zhang, M., Tse, Y. K., Doherty, B., Li, S., & Akhtar, P. (2016). Sustainable supply chain management: Confirmation of a higher-order model. *Resources, Conservation and Recycling*. doi:10.1016/j.resconrec.2016.06.015

Zhang, R., Liu, B., & Wang, W. (2012). Pricing decisions in a dual channels system with different power structures. *Economic Modelling*, *29*(2), 523–533. doi:10.1016/j.econmod.2011.08.024

Zhang, W. H., & Tyerman, S. D. (1999). Inhibition of water channels by $HgCl_2$ in intact wheat root cells. *Plant Physiology*, *120*(3), 849–857. doi:10.1104/pp.120.3.849 PMID:10398721

Zhao, X. K., Li, Y. J., & Shi, D. (2015). Game analysis on supply chain governance mechanism based on extended producer responsibility. [in Chinese]. *Journal of Systems Engineering*, *30*(2), 231–239.

Zhao, X. K., Li, Y. J., Wang, W., & Abdullah, A. T. M. (2014). Framework for sustainability implementation of EPR: Supply chain governance perspective. *International Conference on Service Systems and Service Management. IEEE*,1-5. doi:10.1109/ICSSSM.2014.6943415

Zhao, Z., Zhang, Y., Lei, X., Xia, L., Jiang, X., Hussain, J., & Fawundu, E. (2016). Effects of irrigation and drainage modes on the residual characteristics of heavy metals in soil. *Soil Air Water, 44*(3), 291–298. doi:10.1002/clen.201400368

Zheng, P. (2015). Sustainable Development and Green Marketing. *China Economist, 2015*(7), 271-271.

Zhou, Z. S., Huang, S. Q., Guo, K., Mehta, S. K., Zhang, P. C., & Yang, Z. M. (2007). Metabolic adaptations to mercury-induced oxidative stress in in roots of *Medicago sativa* L. *Journal of Inorganic Biochemistry, 101*(1), 1–9. doi:10.1016/j.jinorgbio.2006.05.011 PMID:17084899

Zhu, Q.H., Feng, Y.T., &Tian, Y.H. (2014). Economic Impact Analysis of remanufacturing Industry in China Based on Input-output Method. *China Soft Science,* (6), 34-43. (in Chinese)

Zhu, Q., Geng, Y., Fujita, T., & Hashimoto, S. (2010). Green supply chain management in leading manufacturers: Case studies in Japanese large companies. *Management Research Review, 33*(4), 380–392. doi:10.1108/01409171011030471

Zhu, Q., Liu, J., & Lai, K. H. (2016). Corporate social responsibility practices and performance improvement among Chinese national state-owned enterprises. *International Journal of Production Economics, 171*, 417–426. doi:10.1016/j.ijpe.2015.08.005

Zhu, Q., & Sarkis, J. (2004). Relationships between operational practices and performance among early adopters of green supply chain management practices in Chinese manufacturing enterprises. *Journal of Operations Management, 22*(3), 265–289. doi:10.1016/j.jom.2004.01.005

Zhu, Q., Sarkis, J., & Geng, Y. (2005). Green supply chain management in China: Pressures, practices and performance. *International Journal of Operations & Production Management, 25*(5), 449–468. doi:10.1108/01443570510593148

Zhu, Q., Sarkis, J., & Lai, K. H. (2007a). Green supply chain management: Pressures, practices and performance within the Chinese automobile industry. *Journal of Cleaner Production, 15*(11), 1041–1052. doi:10.1016/j.jclepro.2006.05.021

Zhu, Q., Sarkis, J., & Lai, K. H. (2007b). Initiatives and outcomes of green supply chain management implementation by Chinese manufacturers. *Journal of Environmental Management, 85*(1), 179–189. doi:10.1016/j.jenvman.2006.09.003 PMID:17084502

Zhu, Q., & Zhang, Q. (2015). Evaluating practices and drivers of corporate social responsibility: The Chinese context. *Journal of Cleaner Production, 100*, 315–324. doi:10.1016/j.jclepro.2015.03.053

Zikmund, W. G., Babin, B. J., Carr, J. C., & Griffin, M. (2013). *Business research methods.* Cengage Learning.

Zimmerman, M. A., & Zeitz, G. J. (2002). Beyond survival: Achieving new venture growth by building legitimacy. *Academy of Management Review, 27*(3), 414–431.

About the Contributors

Sang-Bing Tsai is currently the Professor and the Lecture Professor at University of Electronic Science and Technology of China Zhongshan Institute & Civil Aviation University of China. He is both Technology Management and Business Management PhD. His major is technology management and computer science. He has published many research articles in SCI / SSCI journals, including Journal of Global Information Management, International Journal of Web Services Research, Advances in Mechanical Engineering, IEEE Transactions on Systems Man Cybernetics-Systems, Plos One, Environment and Planning B-Planning & Design, Renewable & Sustainable Energy Reviews, etc. His recent research interests in Information Management, Web Service Management, Production Management, Fuzzy System and Applied Mathematics. He had much academic experience and industry experience. He is on the SSCI/SCI/EI International Journal editorial team as an editor; Journal of Global Information Management (Associate Editor) (SSCI); Sustainability (Lead Guest Editor) (SSCI/SCI); Sustainability (Editor) (SSCI/SCI); SpringerPlus (Editor) (SCI); International Journal of Digital Crime and Forensics (Associate Editor) (EI); International Journal of Knowledge Management (International Advisory Board Editor) (EI); International Journal of Information and Communication Technology Education (International Advisory Board Editor) (EI).

* * *

Andres Aguilera-Castillo currently lectures in the School of Business at Universidad EAN. Among his research interests are the relation of technology in labour markets, sustainability in firms from emerging countries and the uses of simulations and games in education and corporate training.

Archana is presently the Ph.D. scholar of the Department of Plant Science, MJP Rohilkhand University, Bareilly. She has achieved gold medal during her M.Sc. program of the same University. She has 4 year research experience and is working on "Myco-remediation of heavy metals from soil and water systems of Bareilly. She has published 03 papers in referred journals. She has been awarded by the INSPIRE fellowship by Department of Science and Technology, New Delhi. She has also qualified the CSIR NET (Life Sciences).

Xuanming Bai is a Nankai University graduate student.

Jéssica Brunhara graduated in Administration by the Toledo University Center (2014). Extension course in Organizational Sustainability by the Universidade do Oeste Paulista de Presidente Prudente

(2015). I studied subjects of Methodology of Scientific Research and Special Topics in Environment and Regional Development, as special student of the Masters in Environment and Regional Development at the University of the West Paulista University - UNOESTE of Presidente Prudente (2016). Graduating post-graduate in Production and Logistics Management at UNOESTE de Presidente Prudente (2018). MSc in Agribusiness and Development from the Paulista State University Júlio de Mesquita Filho (2019). Has experience in the administrative area.

Li-Chung Chang is currently an associate professor at the Department of Management, University of Electronic Science and Technology of China (UESTC), Zhongshan Institute. He received his Ph.D. degree in Business Administration from National Chengchi University (Taiwan) MBA and a bachelor degree in Business Administration from National Sun Yat-Sen University (Taiwan). He was a director research fellow at Topology Technology, an associate research fellow at TIER (Taiwan Institute of Economic Research) and an analyst at TECO Group and OSE (Orient Semiconductor Electronics). His research interest in knowledge management, learning, innovation, and entrepreneurship.

Rosana Corrêa is an Academic Master in Business Administration from UNISAL (2009), Specialization in Communication and Digital Marketing from Unoeste (2016-2017), Specialization in Human Resources and Organizational Psychology from UniToledo (2014) and a degree in Business Administration from FAFIPE (2006). Administrative manager at Dynabase Comércio de Produtos Químicos Ltda (2003-2007). PAE-SEBRAE regional development agent from Penápolis in partnership with FUNEPE (2008-2011). Professor at the Faculty of Philosophy, Sciences and Letters of Penápolis, of the courses of Administration, Accounting Sciences, Information Systems and Technology in Agribusiness (2008-2011). Professor at the Toledo Teaching Society of Araçatuba (2012-Present), Administration, Biomedicine, Journalism, Nutrition, Mechanical Engineering, Electrical Engineering, Civil Engineering, Technology in Analysis and Systems Development and Human Resource Management. Professor of the administration course at IESP-FAAR (2013-Present). Main disciplines: entrepreneurship, marketing management, management tools (OS & M), agribusiness management, people management, environmental management and business plan.

Carlos Fajardo Toro received a PhD degree in Computer Science (Tools for developing intelligent software) and the degree of MSc in Management and Logistics and Supply Chain of the University of Vigo (Spain). He serves currently as a full time teacher at the School of Business Administration - EAN (Bogota, Colombia).

Yang Gao is the Doctor in School of Bussiness at the Dalian University of Technology in China. He has researched in the field of strategy management and entrepreneurship for 8 years. He has presided over multiple projects at the national level of China. As a research author, he has published many academic papers. He is also the reviewers of several international journals.

A. K. Jaitly is presently Professor and Head Department of Plant Science, MJP Rohilkhand University Bareilly. Prof. Jaitly has done his M.Sc. and Ph.D. in Botany from Lucknow University. He has 36 and 31 years of research and teaching experience respectively. He has more than 75 publications in reputed National and International journals. He has been awarded with Biotechnology National Associate, Biotechnology Overseas Award, UGC National Associateship and was selected as Fellow of Society for Plant Research. He attended many conferences in India and abroad. Prof. Jaitly has successfully guided more than 15 students for PhD.

K. Jayakrishna is currently working as Assistant professor, VIT University. He has completed his doctorate on Investigations on sustainable product development approach in automotive organizations from National Institute of Technology-Trichy, Masters in Production engineering from P.S.G College of Technology and Bachelors in Mechanical from Adhiyamman College of Engineering. He has published 19 International journals papers and 48 papers in International Conferences in leading SCI/ SCOPUS Indexed journals, he has also edited one book, co-authored three book chapters. He is also an editorial board member and reviewer of 6 international journals. His areas of research interests include lean manufacturing, sustainable manufacturing, fit manufacturing and rapid prototyping.

Vimal Kek is an Assistant Professor, Department of Mechanical Engineering, Sri Venkateswara College of Engineering, Sriperabudur, Tamil Nadu, India. He completed his Ph.D. and M.Tech from Production Engineering Department, National Institute of Technology, Tiruchirappalli, Tamil Nadu. He received his Bachelor's degree in Production Engineering (Sandwich Programme) from PSG College of Technology, Coimbatore, India. He has published 19 papers in International Journals and 30 papers in International Conferences. His areas of research interests include lean manufacturing, sustainable manufacturing, neural network and fuzzy logic.

Huseyin Selcuk Kilic is currently an instructor in the Department of Industrial Engineering in Marmara University. He received the BSc, MSc and PhD degrees in Industrial Engineering from Istanbul Technical University. He studied in-plant logistics design for his PhD dissertation. His main research areas are plant logistics, reverse logistics, lean production, decision making and ergonomics. He has research papers in journals that include Applied Mathematical Modelling, Computers and Industrial Engineering, International Journal of Advanced Manufacturing Technology, and Assembly Automation.

Wei Liu is the Doctor in Business School, the University of Sydney. He focuses on the study of automotive corporation, regional development and industrial evaluation.

Xiaobing Lv is a graduate student in logistics and supply chain management of Nankai University.

Hailing Ma is a student in the School of Science, Jinggangshan University. He focuses on the study of mathematical modeling.

Sergio Mazini is a Master in Production Engineering from UNESP, has specialized in analysis, development and management of systems projects by UNILINS and Graduate Technologist in Data Processing by FATEB. Course coordinator UniToledo of Production Engineering, University Professor of Business Administration courses Business, Information Systems and Production Engineering. Consultant and analyst in project development, implementation and management of integrated management systems, ORION Solutions Management. Has business consulting expertise, business management, production engineering, developing information systems, management innovation, knowledge management and enterprise portals.

Claudia Peña-Montoya is an Industrial Engineer, Universidad Autonoma de Occidente (Colombia), M.Sc in Management, Lancaster University, (UK), Logistics Especialist, Universidad del Valle (Colombia) and Ph.D in Engineer Universidad del Valle (Colombia). Researcher and fulltime lecturer at the Operations and Systmens Department, Engineering Faculty, Universidad Autonoma de Occidente, Cali, Colombia Topics of interest: Logistics, Supply chain issues, reverse logistics, sustainability, waste management.

Mauricio Quintero-Angel is an Agricultural Engineering, Ph.D. in Environmental Sciences, Universidad del Valle- Colombia. Over the last few years he has become interested in the socio-ecological interface under global change. Based on frameworks from ecological economics, agroecology and political ecology, he is interested in the analysis of nature-society interactions, particularly in the use and transformation of nature, associated environmental conflicts and the development and dissemination of sustainable technologies.

Ron Schipper (1971) is an experienced project and program manager, principal consultant at Van Aetsveld and an independent researcher. `Different projects for a better life` captures his drive, defining and realizing sustainability strategies through projects his daily practice. He is one of the well-known recognized experts in the field of sustainability in project management with over a dozen academic papers and several books published and various practical contributions towards organizations adopting this topic. He was awarded the 20114 Sustainability Award of the international Green Project Management organization. Ron is also an external examiner in the Master of Informatics and Master of Project Management in HU University of Applied Science. He holds a master degree in Computer Science.

Hongbo Shi is a professor in the School of Economics and Management, Harbin Institute of Technology. He focuses on the study of industrial economics, economic development and regional development.

Xiaoran Shi is an assistant professor in the School of Management at the Tianjin University of Technology, China. Her research interests include supply chain management, energy supply chain, big data technology. She received her Ph.D degree from the University of Miami, USA.

Gilbert Silvius (1963) is professor of project and programme management at LOI University of Applied Sciences, research associate at the University of Johannesburg and fellow at Turku University of Applied Sciences. He lectures at several universities across Europe and is considered one of the leading experts in the field of sustainability in project management. Gilbert has published over a 30 papers and several books on this topic and was awarded the 2013 Sustainability Award of the international Green Project Management organization. As a practitioner, Gilbert has over 20 years hands-on experience in organizational change and IT projects. He is also an internationally experienced consultant/trainer in project management and member of the international enable2change network of project management experts. Gilbert holds a PhD in information sciences from Utrecht University and masters' degrees in economics and business administration.

Mete Han Topgul, MSc, is currently working as a Data Manager in the Information Technology Department of PepsiCo Turkey. M. Topgul received his MSc. Degree in Engineering Management from Marmara University, in 2004. Throughout his industrial engineering education, his research interests are Supply Chain Management and Production Management.

Gülfem Tuzkaya, Ph.D., is currently working as an Associate Professor in the Department of Industrial Engineering, Marmara University. Dr. G. Tuzkaya received her MSc. Degree in Industrial Engineering from Yildiz Technical University, in 2002, MBA degree from Istanbul Technical University, in 2005, and Ph.D. degree in Industrial Engineering from Yildiz Technical University, in 2008. Her research interests are in the areas of reverse-forward logistics systems design, facility layout design and operations research applications for operations management problems.

Yu Xie is a master student under advisory of Dr. Lei Xu in School of Management, Tianjin University of Technology. His research interests include supply chain management, operations management as well as crowdfunding of big data.

Lei Xu, associate professor, is the chair of department of marketing and the assistant dean in School of Management, Tianjin University of Technology.

Li Yu is Economics Doctor and Specialized in corporate social responsibilities, entrepreneurial management and entrepreneurial education. Graduating from Beijing Normal University. She is a lecturer in Qilu University of Techonology and the director at Teaching and Researching of Founder Institute in School of entrepreneurship. Meanwhile, she is a post doctor in School of Management, Shandong University in China.

Index

T

U

W

www.ingramcontent.com/pod-product-compliance
Lightning Source LLC
LaVergne TN
LVHW062142300926
849161LV00010B/2598